EUROPEAN LANDSCAPE DYNAMICS

CORINE LAND COVER DATA

Edited by:

JAN	FERANEC
TOMAS	SOUKUP
GERARD	HAZEU
GABRIEL	JAFFRAIN

CRC Press
Taylor & Francis Group
Boca Raton London New York

CRC Press is an imprint of the
Taylor & Francis Group, an **informa** business

Cover design by Martina Drbalkova.

CRC Press
Taylor & Francis Group
6000 Broken Sound Parkway NW, Suite 300
Boca Raton, FL 33487-2742

First issued in paperback 2019

ISBN-13: 978-0-4822-4466-3 (hbk)
ISBN-13: 978-0-367-87021-8 (pbk)

Library of Congress Cataloging-in-Publication Data

Names: Feranec, J. (Jan), 1951- editor.
Title: European landscape dynamics : CORINE land cover data / edited by Jan
Feranec, Tomas Soukup, Gerard Hazeu, and Gabriel Jaffrain.
Description: Boca Raton, FL : CRC Press, 2016. | Includes bibliographical
references and index.
Identifiers: LCCN 2016009525 | ISBN 9781482244663 (hardcover : alk. paper)
Subjects: LCSH: Landscape assessment--Europe. | Land cover--Europe. | Land
use--Europe. | CORINE Biotopes Project
Classification: LCC GF91.E85 E87 2016 | DDC 304.2094--dc23
LC record available at https://lccn.loc.gov/2016009525

Visit the Taylor & Francis Web site at
http://www.taylorandfrancis.com

and the CRC Press Web site at
http://www.crcpress.com

EUROPEAN LANDSCAPE DYNAMICS

CORINE LAND COVER DATA

Ut Omnes Videant

Contents

Section I   Introduction

Section II   Methodology

Section III   European CLC Data Layers

Section IV   Case Studies: Solution of the European Environmental Problems Using the CORINE Land Cover Data

Section V   CLC Perspective

Foreword

"Across Europe and the world, accelerating rates of urbanisation, changing demographic and consumption patterns, technological changes, deepening market integration, and climate change place unprecedented demands on land. Yet the availability of land is finite. This imbalance is unsustainable," as stated in the latest report on the State and Outlook of Europe's Environment published by the European Environment Agency (EEA) in 2015.

Limiting "land take"—defined as the amount of agriculture, forest, and semi-natural land taken by urban and other artificial land development—is already an important policy objective at a national or subnational level. At the European Union (EU) level, the 2020 road map to a resource-efficient Europe proposes a number of milestones on land and soil. Furthermore, the target as set out in the 7th Environment Action Programme of the European Union is to achieve no net land take by 2050.

The regular CoORDination of INformation on the Environment (CORINE) Land Cover (CLC) inventories proved to be an essential source for monitoring land cover (LC) and land use (LU) change in Europe. Since 1986 national teams from 39 European countries that are members of the European Environmental Information and Observation Network (EIONET) have been working closely together with the EEA and in particular with the CLC technical unit to monitor changes within the European landscape.

After the feasibility study carried out in 1985 as part of the European experimental program for CORINE, Portugal was the first country to map its entire territory based on the standard European CORINE nomenclature and methodology for LC mapping. Other EU member states followed soon after, and CLC became one of the flagship activities of the EEA. This resulted ultimately in the systematic availability of long-term time series of high-quality comparable trans-boundary data of LC and LC changes across Europe for all EEA member and collaborating countries. A stable European nomenclature with the flexibility of refinement toward national or local needs contributed to the success of CLC as a reference dataset for monitoring the dynamics of the European landscape across borders.

Over time, the methodology for mapping LC and LC changes was adapted to benefit from the technological improvements of satellite-based observations, image processing, in situ data handling, and Geographic Information Systems (GIS). Consequently the visual interpretation of printouts of low-resolution Landsat MSS satellite imagery was gradually replaced by advanced computer-assisted photointerpretation of multitemporal high-resolution satellite data made accessible through a European spatial data infrastructure. In parallel, from 2006 onwards, the CLC inventories were synchronized with the LU and LC area frame surveys LUCAS, which provide indispensable in situ data on ground truth for verification and statistical validation of the CLC data.

As a result of these improvements, but also due to the consolidation of a real LC expert network across Europe, the data collection and processing time for a full coverage of the European territory covering 6 million km^2 has been reduced from 10 years in the 1990s to three years for the survey of 2012. The EU Copernicus program launched in 2014 is aiming to accelerate the process even further by reducing the time between satellite data acquisition and the final product to 1 year, and, more importantly, offering decision makers and industries up-to-date information for better policies and competitive business initiatives. In addition, complementary high-resolution data on LC characteristics related to imperviousness, forest types, wetness, and so forth are today collected on a regular basis through the Copernicus land monitoring service to provide more detailed information about the composition of each CLC unit. The new Sentinel 2 satellites are especially equipped with tailored spectral bands and broad geographical coverage for improving the frequency and the quality of the images used for land monitoring.

Today, CLC is recognized as a well-established, reliable information source to support environment policies and beyond.

In line with its open data policy, the EEA, in close partnership with EIONET, is making these data available to a very broad user community free of charge. As a result, the CLC data are year after year one of the most downloaded datasets from the EEA website. This results in a large uptake by the public as well as by the commercial sector, which is using the CLC data in combination with other biophysical and socioeconomic data for a wide range of applications. The range of applications is very broad covering environmental impact assessment, nature conservation, natural capital accounting, climate change adaptation, agriculture, transport, energy, education, and many others.

This endeavor of harmonized LC mapping and change monitoring from the Nordic boreal forest to the southern coastal Mediterranean regions and from the western Atlantic coastal region to the mountains in continental Europe co-funded by the EU and the national authorities has been possible only thanks to the dedication and passion of several hundred experts, scientists, and students who have been working together with great enthusiasm and perseverance. More than 100 public and private organizations were and still are involved in the development, production, maintenance, and dissemination of the data.

A special thank you goes to the editors and contributors of this publication who took the challenge and well succeeded to bring together a wealth of information in a journey from methodology, available data, case studies, and outlook that unquestionably will serve as a reference to the present and future generation of users of national and European CLC data across the world.

Chris Steenmans
Head of Programme, European Environment Agency

Preface

Four unique pan-European CoORdination of INformation on the Environment (CORINE) Land Cover datasets—CLC1990, CLC2000, CLC2006, and CLC2012—and three datasets concerning changes—$CLCC_{1990-2000}$, $CLCC_{2000-2006}$, and $CLCC_{2006-2012}$—offer options for the observation and study of the European landscape via land cover (LC) and its changes. The aforementioned datasets contain information about the development of the European landscape during almost 25 years (1990–2012), a sufficient interval for the analysis and assessment of not only changes for one period (e.g., 1990–2000) but also for the estimates of long-term trends (1990–2000–2006–2012) in the development of landscape. The editors of this book, in response to a suggestion from Taylor & Francis, decided to present a comprehensive methodology for the identification, analysis, and assessment of CLC and its changes, including the summarized results and examples of possible solutions to environmental problems of Europe, by applying the quoted data.

The number of European Environment Agency (EEA) Data Service downloads during the last few years (2011—37,572; 2012—40,910 and 2013—20,809) confirm great interest in the CLC products. The increasing number of journal articles with the frequency of terms *CORINE land cover* and *CORINE land cover change* indexed by SCOPUS, the largest abstract and citation database of peer-reviewed literature, also proves the use of the CLC and CLCC data in research and applied tasks (see Figures P.1 and P.2).

Statistics concerning the use of terms *CLC* and *CLC change* in the literature reveal that the data became an integral part of research and applied studies. It is difficult to obtain an overview about the generation, processing, and use of CLC data from such a large number of publications. The intention of this book is to provide a compound of the CLC data use in landscape monitoring; approaches to their acquisition, processing, and availability; and their use in the pan-European environmental problems.

In general, CLC and CLCC data, in spite of limited spatial resolution, offer information about the occurrence of CLC classes (location of classes and their changes) and about their size and the type of changes (which area of class or its part changed into another class). Thematically speaking, CLC and CLCC data represent an important information source for the assessment of indicators of the developing landscape, assessment of landscape fragmentation, and modeling of various processes taking place (e.g., runoff processes, a variety of elements related to the morphology of cities, etc.) on the pan-European and partially national and regional levels. The CLC and CLCC data also embody a valuable contribution to the Global Earth Observation System of Systems (GEOSS), which strives for the comprehension and anticipation of the dynamic processes ongoing on Earth.

The target readership of this book includes researchers, environmentalists, cartographers, territorial planners, managers, students of geosciences, and all those who are interested in the European landscape and its changes and development. The book may be of special value for remote sensing specialists involved with the identification, analysis, and assessment of the LC/LU (land use) and its changes using satellite images.

The book is divided into five sections. The first section contains two chapters that describe the monitoring systems recording the state of LC in Europe but also in broader contexts, especially the CLC project.

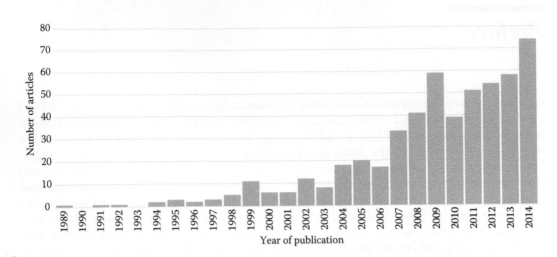

FIGURE P.1
Number of journal articles using the term *CORINE land cover* extracted from SCOPUS.

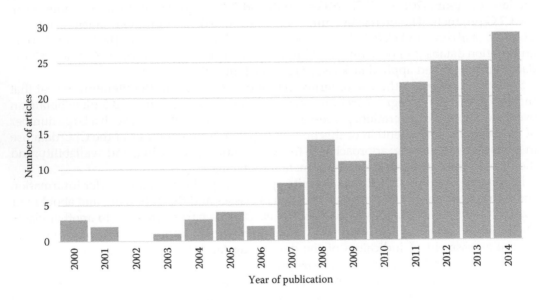

FIGURE P.2
Number of journal articles using the term *CORINE land cover change* extracted from SCOPUS.

Seven chapters of the second section are devoted to the CLC nomenclature methodology, characteristics of the used satellite images, their interpretation, precision assessment, CLC products, and the dissemination policy.

The following seven chapters of Section III mainly provide statistical characteristics and map representations of the frequency and areas of 44 CLC classes in Europe in the years 1990, 2000, 2006, and 2012 but also summarized statistical characteristics concerning the changes of $CLC_{1990-2000}$, $CLC_{2000-2006}$, and $CLC_{2006-2012}$ and their synoptic map representations by means of seven land cover flows (LCFs).

The fourth section demonstrates examples of solutions to various environmental problems in Europe using the CLC data. One out of nine chapters in this section explains implementation of CLC outside of Europe.

The fifth section, broken into three chapters, outlines the perspective of CLC in the near future. One chapter deals with approaches to derive the European CLC from more detailed national LC (bottom-up approach). In addition, examples are presented in which member states created a fourth or fifth CLC nomenclature level to add more detailed national LC. In the other chapters the relationship of CLC data to the Infrastructure for Spatial Information in Europe (INSPIRE) and a sketch of the future of land monitoring in Europe are presented.

The conclusions summarize the contribution of CLC and CLCC data to the cognition of the European landscape dynamics in 1990–2012.

Jan Feranec, Tomas Soukup, Gerard Hazeu, and Gabriel Jaffrain

Acknowledgments

The editors thank Yves Heymann, Chris Steenmans, Guy Croisille, and Michel Bossard, the pioneers of the CORINE land cover (CLC) projects, and the authors of the "red manual" of *CORINE land cover—Technical guide* (1994). Thanks are due to Eric Evrard and Jan Kolar and all participants of the seminar (February 24–27, 1992, Joint Research Centre, Ispra, Italy) devoted to this project who have explained to us for the first time the issue of the then recently launched big pan-European environmental activity. It motivated us and greatly influenced our careers.

Many thanks to the European Environment Agency for its vision and effort in support of CLC activities in Europe.

Special thanks to all co-authors of this monograph for their interesting and inspiring texts contributed in spite of their many work responsibilities.

The editors express their gratitude to all the collaborators and members of the national CLC teams who participated either in training seminars or in the completion of the individual CLC data layers and in the accompanying discussions.

They are also grateful to the anonymous reviewers of the initial proposal of this monograph to CRC Press, Taylor & Francis, who confirmed the significance of this project and supported the intention to publish it.

Thanks to reviewers Jan Kolar and Ivan Bicik for reading the manuscript and their valuable suggestions.

The editors are indebted to their employers for providing them the conditions enabling them to write and edit this monograph.

We owe thanks to Robert Pazur, who helped us with the technical aspects of the manuscript completion, and to Hana Contrerasova for translations and linguistic help.

Warm thanks go to our families for their moral support, tolerance, and encouragement.

Editors

Jan Feranec holds degrees in physical and regional geography. He is a scientific worker at the Institute of Geography of the Slovak Academy of Sciences (IG SAS) in Bratislava, Slovak Republic, where he held the post of director from July 2002 to June 2006. He was appointed associate professor at the Comenius University in Bratislava in 1996. He is today the supervisor of PhD studies for the Programme of Physical Geography, Geoecology, and Geoinformatics at IG SAS and the vice-chair of the current Commission on Sensor-Driven Mapping International Cartographic Association. His research interests include land cover/use, change mapping by satellite images, urban areas analysis by VHR images, remote sensing data interpretation and landscape assessment, and identification and analysis of the abandonment of arable lands and grasslands. He has been involved in many international projects, such as the Czechoslovakia-Canada Cooperation in Remote Sensing, Ecotons of the Rivers Danube and Morava, MERA, Phare Topic Link on Land Cover, BIOPRESS, and the Urban Atlas (2009–2011 and 2013–2016). From 2001 to 2009 he was a member of the central CORINE Land Cover Projects (CLC1990 and CLC2000) Technical Team.

Tomas Soukup is project manager and a senior consultant in remote sensing, GIS, and land monitoring for GISAT s.r.o. company in Prague, Czech Republic. His fields of study include geodesy, cartography, GIS, and remote sensing, and he has worked extensively over the last 20 years on the utilization of Earth observation (EO) data for various practical applications in the context of land monitoring, including urban studies, regional planning, environmental assessment, emergency response, and security support. Since the CORINE Land Cover (CLC) 1990 mapping campaign in the Czech Republic and Slovakia, he has collaborated with the European Environmental Agency (EEA) in the framework of European Topic Centers (PTL/LC, ETC/TE, ETC/LUSI, ETC/SIA, ETC/ULS), and he is a member of the central CLC Technical Team. In the last decade, Mr. Soukup has been active in several application projects developing, testing, and providing European GMES/Copernicus service capacities in land monitoring (e.g., SAGE, GUS, GSELand, Geoland2, EUROSION, BIOPRESS, UrbanAtlas+, HLANDATA, HELM, GRAAL, URBIS, MELODIES…). He also participates with the National Reference Centre Land Cover in EIONET, a partnership network of the EEA, and he is also a member of the EEA EIONET Action Group on Land monitoring in Europe (EAGLE), which supports implementation of Copernicus services.

Gerard Hazeu is a physical geographer working at the Alterra–Wageningen University and Research Centre, where he is a GIS and remote sensing researcher/project leader with specific interest in land cover/use change, agriculture, and biodiversity issues. Since 2001 he has been involved in several European land monitoring projects, such as FP5 BIOPRESS, FP7 Geoland2, FP7 HELM (Harmonised European Land Monitoring), and the Eurostat grant on pilot studies on the provision of harmonised land use/land cover statistics. Dr. Hazeu is responsible for the CORINE Land Cover updates in the Netherlands (2000, 2006, and 2012). He worked closely with EEA in the European Topic Centres (ETC/LUSI, ETC/SIA, and ETC/ULS). He is also responsible for the Dutch national land cover database (LGN), which every four years produces an actual representation of the national land cover. Furthermore, he represents the Netherlands regarding National Reference Centre

Land Cover in EIONET, which is a partnership network of the EEA, and he is involved in the INSPIRE TWG Land Use. Currently, he is involved in several EAGLE (EIONET Action Group on Land Monitoring in Europe) contracts drafting the framework for future land cover monitoring in Europe. Also, he supports the implementation of the Copernicus land continental and local component by providing service specifications and ensuring continuity of existing GIO land services.

Gabriel Jaffrain is educated in geography, environment and spatial planning by remote sensing. His studies at the University of Rennes included biology of organisms and ecology. He has been a project manager and a land cover/land use expert at IGN France International for more than 20 years and is now the referent for the environmental issues and land cover mapping topics. He is involved in numerous land occupation and environment projects in France and around the world (50 countries). As an experienced European expert for the CORINE Land Cover Projects since 1994, since 2001 he has been a member of the technical team of the European Topic Centre ETC/TE, ETC/LUSI, and ETC/SIA (Spatial Information and Analysis) within the EEA. Today he is involved in the production of land cover accounting in the context of SEEA Part2 on ecosystem capital accounting. He is also responsible for quality insurance/quality control on the mapping of tropical forests and forest change in the REDD+ international program in the Congo Basin. In Europe, he is involved in the Urban Atlas and Copernicus projects as a supervisor.

Contributors

Dania Abdul-Malak
University of Malaga
Malaga, Spain

Stephan Arnold
Federal Statistical Office
Wiesbaden, Germany

Antonio Arozarena
National Geographic Institute
Madrid, Spain

Gebhard Banko
Environmental Agency Austria (EAA)
Vienna, Austria

Claudia Baranzelli
European Commission, Joint Research
    Centre (JRC)
Institute for Environment and Sustainability
Sustainability Assessment Unit
Ispra, Italy

Ana Luisa Barbosa
European Commission, Joint Research
    Centre (JRC)
Institute for Environment and Sustainability
Sustainability Assessment Unit
Ispra, Italy

Ricardo Barranco
European Commission, Joint Research
    Centre (JRC)
Institute for Environment and Sustainability
Sustainability Assessment Unit
Ispra, Italy

Filipe Batista e Silva
European Commission, Joint Research
    Centre (JRC)
Institute for Environment and Sustainability
Sustainability Assessment Unit
Ispra, Italy

Katarzyna Biala
European Environment Agency
Copenhagen, Denmark

György Büttner
European Environment Agency
Copenhagen, Denmark

Markus Erhard
European Environment Agency
Copenhagen, Denmark

Jan Feranec
Institute of Geography
Slovak Academy of Sciences
Bratislava, Slovak Republic

Herbert Haubold
Environmental Agency Austria (EAA)
Vienna, Austria

Gerard Hazeu
Alterra–Wageningen UR
Wageningen, The Netherlands

Chris Jacobs-Crisioni
European Commission, Joint Research
    Centre (JRC)
Institute for Environment and Sustainability
Sustainability Assessment Unit
Ispra, Italy

Jochen A. G. Jaeger
Concordia University Montreal
Department of Geography, Planning
    and Environment
Montreal, Quebec, Canada

Gabriel Jaffrain
IGN France International
Paris, France

Marketa Jindrova
GISAT
Prague, Czech Republic

Katerina Jupova
GISAT
Prague, Czech Republic

Felix Kienast
Swiss Federal Research Institute WSL
Birmensdorf and Zurich, Switzerland

Stefan Kleeschulte
Space4environment Sàrl
Niederanven, Luxembourg

Miroslav Kopecky
GISAT
Prague, Czech Republic

Barbara Kosztra
Institute of Geodesy, Cartography and
    Remote Sensing
Budapest, Hungary

Carlo Lavalle
European Commission, Joint Research
    Centre (JRC)
Institute for Environment and Sustainability
Sustainability Assessment Unit
Ispra, Italy

Luis F. Madriñán
Cerrejón limited
Bogotá, D.C., Colombia

Joachim Maes
European Commission, Joint Research
    Centre (JRC)
Institute for Environment and Sustainability
Sustainability Assessment Unit
Ispra, Italy

Ivone Pereira Martins
European Environment Agency
Copenhagen, Denmark

Gergely Maucha
Institute of Geodesy, Cartography, and
    Remote Sensing
Budapest, Hungary

Branislav Olah
Technical University in Zvolen
Zvolen, Slovak Republic

Erika Orlitova
GISAT
Prague, Czech Republic

Carolina Perpiña Castillo
European Commission, Joint Research
    Centre (JRC)
Institute for Environment and Sustainability
Sustainability Assessment Unit
Ispra, Italy

Ana Maria Ribeiro de Sousa
European Environment Agency
Copenhagen, Denmark

Christophe Sannier
Systemes Inform Refer Spatiale (SIRS)
Villeneuve d'Ascq, France

Christian Schwick
Swiss Federal Research Institute WSL
Birmensdorf and Zurich, Switzerland

Geoff Smith
Specto Natura Ltd.
Impington, United Kingdom

Tomas Soukup
GISAT
Prague, Czech Republic

Chris Steenmans
European Environment Agency
Copenhagen, Denmark

Geir-Harald Strand
Norwegian Institute of Bioeconomy
    Research (NIBIO)
Ås, Norway

Nuria Valcárcel
National Geographic Institute
Madrid, Spain

Sara Vallecillo
European Commission, Joint Research
    Centre (JRC)
Institute for Environment and Sustainability
Sustainability Assessment Unit
Ispra, Italy

Ine Vandecasteele
European Commission, Joint Research
    Centre (JRC)
Institute for Environment and Sustainability
Sustainability Assessment Unit
Ispra, Italy

Jean-Louis Weber
European Environment Agency Scientific
    Committee
Copenhagen, Denmark

Grazia Zulian
European Commission, Joint Research
    Centre (JRC)
Institute for Environment and Sustainability
Sustainability Assessment Unit
Ispra, Italy

Abbreviations

AEI	Agro-Environment Indicator
ANPE	National Agency for Environmental Protection (Agence Nationale de Protection de l'Environnment)
BACI	before-after-control-impact
BBG	Bestand BodemGebruik
BD	Bird Directive
BDOT	Base de Donnée d'Occupation des Terres
BEV	Austrian Federal Office of Metrology and Surveying
CACED	CentrAmerican Commission of the Environment and Development
CAN	Andean Community of Nations
CAP	Common Agricultural Policy
CAPI	Computer Assisted Photo-Interpretation
CBC	Cross-Boundary-Connections procedure
CBD	Convention on Biological Diversity
CCM	Catchment Characterisation and Modelling
CE	Council of Europe
CEC	Commission of European Communities
CEOS	Committee on Earth Observation Satellites
CICPN	Commission Interministérielle des Comptes du Patrimoine Naturel
CIRAD	Centre de Coopération International en Recherche Agronomique pour le Développement
CIRAD-TERA	International Cooperation Centre in Agronomic Research for Development (land environment and people)
CLC	CORINE Land Cover
CLCC	CORINE Land Cover Change
CLUE	Conversion of Land Use and its Effects
CMEF	Common Monitoring and Evaluation Framework
CORINE	CoORdination of INformation on the Environment
CORMAGDALENA	Corporacion Autonoma Regional del Rio Grande de La Magdalena
DG AGRI	Directorate General for Agriculture
DG ENV	Directorate General for Environment
DG REGIO	Directorate General for Regional and Urban Policy
DLM-DE	Digital Land Cover Model Germany
DTL	Density of Transportation Lines
DUE	Data User Element
EAGLE	EIONET Action Group on Land Monitoring in Europe
EAP	Environmental Action Programme
EC	European Commission
ECA or SECA	(Simplified) Ecosystem Capital Accounts
ECNC	European Centre of Nature Conservation
ECU	Ecosystem Capability Unit

EDc	Education per capita
EEA	European Environment Agency
EEA-38	EEA member and cooperating countries that participated in the assessment
EEc	Environmental Expenditure per capita
EEN	Evaluation Expert Network
EIONET	European Environment Information and Observation Network
ENCA	Ecosystem Natural Capital Accounting
ENCA-QSP	Ecosystem Natural Capital Accounting: A Quick Start Package
ENVISAT	Environmental Satellite
EO	Earth Observation
ES map	Ecosystem map
ESA	European Space Agency
ESRI	Environmental Systems Research Institute
ESTIMAP	Ecosystem Services Mapping tool
ETC/LUSI	European Topic Centre for Land Use and Spatial Information
ETC/SIA	European Topic Centre for Spatial Information and Analysis
ETC/TE	European Topic Centre on Terrestrial Environment
ETC/ULS	European Topic Centre on Urban, Land and Soil systems
ETRS	European Terrestrial Reference System
EU	European Union
EU-27	Member countries of the EU before 2013
EU-15	Member countries of the EU before 2004
EU-N12	Member countries that joined the EU in 2004 and 2007
EU COST	European Cooperation in Science and Technology
EU ICZM	EU Integrated Coastal Zone Management
EUNIS	European Nature Information System
EUROSTAT	European Statistical Office
EVS	European Vegetation Survey
FADN	Farm Accountancy Data Network
FAO	Food and Agriculture Organization
FG	Fragmentation Geometry
FOEN	Swiss Federal Office for the Environment
FTY	Forest Type
GDP	Gross Domestic Product per capita
GE	Google Earth
GEOSS	Global Earth Observation System of Systems
GIO	GMES Initial Operations
GIS	Geographic Information System
GISCO	Geographic Information System at the COmmission
GLCF	Global Land Cover Facility
GLCN Programme	Global Land Cover Network
GLM	Generalized Linear Models
GMES	Global Monitoring for Environment and Security
GOFC-GOLD	Global Observation of Forest and Land Cover Dynamics
GRA	Permanent Grassland
GTOS	Global Terrestrial Observation System
GVA	Gross Value Added
HD	Habitat Directive

HELM	Harmonised European Land Monitoring
HNV	High Nature Value
HRL	High Resolution Layer
IA	Impact Assessment
IACS	Integrated Administration and Control System
IBA	Important Bird Area
ICZM	Integrated Coastal Zone Management
IDEAM	Instituto De Hidrologia, Meteorologia y Medio Ambiente
IENE	Infra Eco Network Europe
IFEN	Institut Francais de l'Environment
IGAC	Instituto Geografico Agustin Codazzi
IGBP	International Geosphere-Biosphere Programme
IGM	Geographical Institute of Mali
IGN FI	Institut Géographique National France International
IGOL	Integrated Global Observations of Land
IGOS	Integrated Global Observations Strategy
IHDP	International Human Dimensions Programme
INSEE	Institut National de la Statistique et des Etudes Economiques
INSPIRE	Infrastructure for Spatial Information in the European Community
IPCC	Intergovernmental Panel on Climate Change
IRS	Indian Remote Sensing Satellite
IRS AWiFS	Indian Remote Sensing Advanced Wide Field Sensor
ISO	International Organisation for Standardization
JRC	Joint Research Centre
LABES	Swiss Landscape Monitoring System
LAEA	Lambert Azimuthal Equal Area
Landsat	Land Remote-Sensing Satellite
Landsat MSS	Landsat Multispectral Scanner
Landsat TM	Landsat Thematic Mapper
LC	Land Cover
LCCS	Land Cover Classification System
LCEU	Land Cover Ecosystem Units
LCF	Land Cover Flow
LCM	Land Cover Map
LCML	Land Cover Meta Language
LEAC	Land and Ecosystem Accounting
LEAD	Leadership for Environment and Development
LGN	National LU database Netherlands
LISA	Land Information System Austria
LLC	LUCAS Land Cover
LLF	Landscape Linear Features
LLU	LUCAS Land Use
LPIS	Land Parcel Identification System
LU	Land Use
LUCAS	Land Use/Cover Area Frame Survey
LUISA	Land-Use-based Integrated Sustainability Assessment
LUMP	Land Use Modelling Platform
MAES	Mapping and Assessment of Ecosystems and their Services

MEDGEOBASE	Geographic database on land use planning along the Maroccan coast
MERIS	Medium Resolution Imaging Spectrometer
MMU	Minimum Mapping Unit
MMW	Minimum Mapping Width
MOLAND	Monitoring Land Use/Cover Dynamics
MONET	Swiss Monitoring System of Sustainable Development
MS	Member State
MSFD	Marine Strategy Framework Directive
NATURA 2000	is an EU-wide network of protected areas established pursuant to the Birds Directive and the Habitats Directive, collectively known as EU "nature legislation"
NDVI	Normalised Difference Vegetation Index
NGP	National Geomatics Plan
NPLM2	National Program for Land Management in Burkina Faso
NRC-EIONET	National Reference Centre-EIONET
NUTS	Nomenclature of Statistical Territorial Units
NVCS	United States National Vegetation Classification Standard
NWRMs	Natural Water Retention Measures
OECD	Organisation for Economic Co-operation and Development
PBAs	Prime Butterfly Areas
PD	Population Densities
PEBLDS	Pan-European Biological and Landscape Diversity Strategy
PEEN	Pan-European Ecological Network
PHARE	Poland and Hungary: Assistance for Restructuring their Economies
PLCC	Pure Land Cover Components
PNGIM	Programme National de Gestion de l'Information sur le Milieu
PSU	Primary Sampling Units
PTA	percent total agreement
PWB	Permanent Water Bodies
QA	Quality assurance
QC	Quality control
QGLUc	Quantity of goods loaded and unloaded per capita
RDP	Rural Development Programme
RERM	Roadmap to Resource Efficient Europe
SAIDE	decision-support system for the management of natural resources and the environment (Systeme d'Alde à la Décision pour la gestion des ressources naturelles et de l'Environnement)
SCOPUS	The largest abstract and citation database of peer-reviewed literature produced by the Elsevier Co.
SEBI	Streamlining European Biodiversity Indicators
SELU	socio-ecological landscape units
SFSO	Swiss Federal Statistical Office
SHERPA	Suivi Hydrologique et Environnemental pour l'Amérique Centrale
SIOSE	Sistema de Información de Ocupación del Suelo en España
SIRS	Systèmes d'Information à Référénce Spatiale

SOeS	Service des Observations et des statistiques Ministère de l'Ecologie et du développement durable et de l'énergie
SPOT	Satellite Pour l'Observation de la Terre
SSU	Secondary Sampling Units
SW	South-West
TCD	Tree Cover Density
TEEB	The Economics of Ecosystems and Biodiversity
TERM	Transport and Environment Reporting Mechanism
TEWN	Trans-European Wildlife Networks Project
TWG-LC	Thematic Working Group for Land Cover
UML	Unified Modelling Language
UN	United Nations
UNCEEA	United Nations Committee of Experts on Economic-Environmental Accounting
UNECE	UN Economic Commission for Europe
UNEP	United Nations Environmental Programme
USGS	United States Geological Survey
VHR	Very High Resolution
VPD	volume passenger density
WB	World Bank
WEI	Water Exploitation Index
WET	Wetlands
WFD	Water Framework Directive
WWF	World Wildlife Fund

Section I

Introduction

1

Overview of Land Cover and Land Use Monitoring Programs

Herbert Haubold and Jan Feranec

CONTENTS

1.1 Background

Land cover (LC) and land use (LU) monitoring is utilized at all political levels, from local to global. It proceeds from various motivations, is based on many approaches, and shows a large range of sophistication and longevity. The resulting diverse landscape analyses and assessments generate a range of issues regarding the compatibility of data, the access conditions, the topical focus, and so forth, requiring that the many actors involved join forces and collaborate. Their data must, at least to a degree, be compatible so they can be shared, exchanged, and used for long periods of time, all of which will create added value to data produced for one purpose and make it possible to use them for many more.

Provision of the data possessing the aforementioned attributes, especially those for large territories, is often problematic by conventional terrestrial approaches or would require enormous effort. Techniques of remote sensing, however, provide the data in digital form recorded via electromagnetic radiation that covers the whole Earth surface and makes them a powerful tool for the preparation of monitoring landscape systems. Today, the satellite and aerial remote sensing data have become the indivisible part of LC and LU monitoring, especially at global, pan-European, and national levels. The aim of this part of the book is to describe the present monitoring activities that record the status of LU and LC in the pan-European and wider contexts.

Monitoring is defined as a procedure that involves the systematic measurement of a targeted object in time (at least two times) to enable assessment of change and trends in quantity and/or quality of the targeted objects and finally an understanding of the processes that are behind these changes (Mücher, 2009: 9–10). Geist (2006) provides an overview of LC and LU monitoring programs. LU and LC change is a long-term process owing to societal and natural changes. For that reason the value of monitoring programs increases with longer time series of data, with a larger number of observations representing LC and LU at specific moments.

1.2 LU/LC Monitoring Programs

The history of landscape monitoring in Europe was documented by the Harmonised European Land Monitoring (HELM) project. Current and past initiatives were systematically listed and described (Blanes and Green, 2012). All of the HELM results, including this report, are summarized in Ben-Asher et al. (2013). The text that follows is partly based on HELM work but also contains updated and further information relevant to this book. Table 1.1 provides an overview of the referenced LC and LU monitoring initiatives.

1.2.1 CORINE

The project referred to as CLC (CoORdination of Information on the Environment [CORINE] Land Cover), whose data constitute a "backbone" of this book, was initiated in the European Union in 1985 as a centralized, remote sensing based LC mapping effort resulting in the CLC1990 LC dataset. CORINE was thus the first pan-European LC mapping process with a coherent nomenclature and has been ever since the de facto standard for a pan-European land monitoring system. CLC is based on Landsat, Indian Remote Sensing (IRS), and Satellite Pour l'Observation de la Terre (SPOT, Satellite for observation of Earth) data and also on ortho-photos and topographic maps.

Updates of the dataset were coordinated by the European Environment Agency (EEA), whereby the member states (MS) were increasingly involved in data production and also financing. In this way, CLC2000 was produced and also CLC2006, which was, for the first time, carried out under the umbrella of the then Global Monitoring for Environment and

TABLE 1.1

Pan-European and Global Land Monitoring Initiatives

Title of Monitoring Program	Beginning of Activity	Reference Years	Spatial Resolution
CORINE Land Cover (CLC)	1985	1990, 2000, 2006, 2012	25 ha
Copernicus High Resolution Layers (5 themes)	2011	2006/2009/2012 (varies per theme)	1 ha
Copernicus local component (urban atlas, riparian zones)	2009, 2014	2006, 2012, 2012	0.25–1 ha and 0.5 ha
MOLAND	1998	Varies per study area	1 ha
Forest cover maps for Europe	1990	1990, 2000, 2006	25 m
LUCAS	2001	2006, 2009, 2012	270,000 points
Copernicus global component (five topics)	2009		1 km (varies per product)
GlobCorine	2002	2005, 2009	300 m
GlobCover	2002	2005, 2009	300 m
Global Land Cover 2000 (GLC 2000)	1995	2000	30 arc seconds

Security (GMES) Core Service land. Currently, CLC2012 is produced under the umbrella of Copernicus, the GMES successor.

In spite of its novelty and innovative character three decades ago, CORINE no longer fulfills the expectations users have regarding a professional land monitoring system. The minimum mapping unit (MMU) of 25 ha is too coarse to make the maps useful at the national or regional level, as all features below this size are not captured at all. This includes, for example, 29% of all European municipalities. Classes are defined in an incoherent way; the mixed classes especially encompass quite inhomogeneous landscapes in one and the same polygon. In addition, LC and LU are not properly distinguished. Furthermore, no attributes or descriptions can be added to the polygons, so once they are classified, no further specifications can be added.

1.2.2 Other Pan-European Land Monitoring Initiatives

The Copernicus Land Monitoring Services, led by the EEA, also deliver five pan-European High-Resolution Layers (HRLs) initiated under the GMES Initial Operations (GIO) program that lasted from 2011 to 2013. The HRLs are currently being completed and do not yet have full European coverage. The themes are Imperviousness (artificial surfaces), Forests (Tree Cover Density and Forest Type), Permanent Grassland, Wetlands, and Permanent Water Bodies. The data sources are 20-m resolution satellite images primarily from IRS-P6, Resourcesat, and RapidEye, and maps with an MMU of 1 ha are produced in a semiautomated way.

The Urban Atlas, also produced as part of the EEA—coordinated Copernicus land services, was started as a joint initiative of the European Commission Directorate General for Regional and Urban Policy and the Directorate General for Enterprise and Industry. It provides LC maps with 17 urban and 10 rural and (semi-)natural classes, with MMUs of 0.25 and 1 ha, respectively. Data sources are primarily SPOT-5, Formosat-2, Kompsat-2, and ALOS satellite data. The first version was produced in 2006 and the second one in 2012. This effort is linked to the Urban Audit of EUROSTAT and now covers most municipalities with more than 50,000 inhabitants.

Monitoring Land Use/Cover Dynamics (MOLAND) is a research program of the Institute for Environment and Sustainability of the Joint Research Centre (JRC), including several companies to which work is contracted. MOLAND covers 40 representative urban-dominated study areas across Europe. Each urban study area consists of interlinked components. The first is database preparation, including the determination of past and present LU evolution in selected areas (urban, region, corridor, and so forth). Second, statistics and socioeconomic datasets are merged into the maps. Subsequently, indicators are developed to understand the territorial and sustainable development across Europe. Finally, scenarios are forecast regarding urban evolution to assess the sustainability of the development of European regions. Future LU development is thereby anticipated, including its impact on the environment as guidance for environmental policy.

The Forest cover maps for Europe and Forest type maps for Europe are also produced by the JRC. They distinguish forest and nonforest and specify broadleaf, coniferous, and mixed forests. Maps are produced through a fully automatic image processing method specifically developed for this purpose and represent the years 1990, 2000, and 2006. Data sources are Landsat imagery, and since 2006 also LISS-III and SPOT satellite data. CLC was used to improve the product further.

For completeness, the Land Use/Cover Area Frame Statistical Survey (LUCAS) should also be mentioned here, although it is not a remote sensing based program. It is an area frame survey of 270,000 points in a regular grid coordinated by European Statistical Office (EUROSTAT). It is not a remote sensing based program, as it deals with field observations. LUCAS provides data for monitoring of agroenvironmental indicators in Europe since 2004 through LC and LU mapping in the EU by means of statistical methods. The work started in 2001 within a framework of a cooperative effort between EUROSTAT and DG AGRI (General Directory of the EU for Agriculture) supported by the JRC.

1.2.3 Global Land Monitoring Initiatives

Although the topic of this book is pan-European land monitoring, the most important international initiatives are outlined in the text that follows. Partly they are supervised by European entities and in all cases they cover Europe as well.

GlobCorine was a project funded by the ESA Data User Element (DUE). Data are produced semiautomatically based on ENVISAT MERIS satellite imagery. GlobCorine built upon the experience and resources available through the GlobCover project. GlobCorine used 14 classes in a nomenclature similar to that of CLC and the reference years are 2005 and 2009. In spite of its name, GlobCorine covers only the pan-European and the entire Mediterranean area and the European part of Russia.

GlobCover was funded through the same project, reflects the same reference years, and is also based on ENVISAT MERIS data. It was produced by an international consortium and used regional LC products such as CLC for Europe as reference data. Several different products were delivered with a 300 by 300 m pixel size, including the Global and Regional GlobCover Land Cover maps. For the former, 22 classes were defined, whereas the latter has up to 51 second-level classes for higher thematic accuracy.

Global Land Cover (GLC) 2000 is produced by the JRC based on the FAO Land Cover Classification System (LCCS), which allows for a combination of locality-specific thematic inputs and a generalized global classification. GLC focuses on the boundaries between ecosystems such as forest, grassland, and cultivated systems. The data source for GLC 2000 is VEGA2000, a dataset of 14 months of preprocessed daily global data acquired by the VEGETATION instrument of the SPOT-4 satellite. GLC 2000 is not a monitoring activity, as it is not planned to repeat the assessment; however, it is among the core LC datasets and it was used for the Millennium Ecosystems Assessment. It also provides an essential framework for land monitoring initiatives.

A detailed review of the development of global LC mapping is provided by Mora et al. (2014).

1.2.4 Global Initiatives Related to Land Monitoring Harmonization on a Meta Level

The initiatives that follow are not specific land monitoring projects or programs but address the meta level of land monitoring, such as standardization.

The Global Land Cover Network (GLCN) Programme jointly founded by the Food and Agriculture Organization of the United Nations (FAO) and United Nations Environment Programme (UNEP) and the government of Italy in 2002, has the aim to harmonize LC monitoring on regional, national, and global levels, thereby focussing on developing countries.

The Integrated Global Observations of Land (IGOL) Programme is the terrestrial component of the Integrated Global Observations Strategy (IGOS), the aim of which is to monitor the effects of society on nature and vice versa. One of the tasks is to identify the scope of annual deforestation and LU change on Earth.

The Global Observation of Forest and Land Cover Dynamics (GOFC-GOLD) was originally developed as a pilot project by the Committee on Earth Observation Satellites (CEOS) as part of its Integrated Global Observing Strategy and is now a panel of the Global Terrestrial Observing System (GTOS). The program's main focus is the sustainable management of terrestrial resources and the terrestrial carbon budget and, for this purpose, it integrates satellite imagery and in situ data regarding the Earth's vegetation cover. Its current aim is a global LC mapping with a coarse resolution (250–1000 m) and a 5-year cycle combined with periodic mapping and monitoring of forested areas at fine resolution of 25 m.

The Global Land Cover Facility (GLCF), hosted at the Geography Department of the University of Maryland, focuses on determining LC and LC change around the world. The facility develops and distributes remotely sensed satellite data and products related to LC from the local to global scales.

The Global Land Project focuses on the research framework for the near future for land systems and is funded jointly by the International Geosphere–Biosphere Programme (IGBP) and the International Human Dimensions Programme (IHDP). Its main objective is to investigate the feedback of the Earth system to human activities on land, and the effects of global change on human–environment interaction.

1.3 Perspectives

Monitoring of LU and LC is, in spite of its importance and its many applications, insufficiently organized between the European countries and the European entities. Efforts are duplicated in that the same areas are monitored several times and data that are already produced at national or regional levels cannot be used as inputs for European monitoring programs owing to interoperability issues resulting from data models, spatial and temporal resolutions, repeat frequencies of data gathering, and national and institutional data policies.

Fortunately, land monitoring in Europe is currently undergoing significant changes. European land monitoring products are becoming more and more useful for the national and regional levels as well, especially owing to their higher resolutions and because their contents are to a larger degree coordinated with the member states' interests. Likewise, an increasing number of MS run their own professional land monitoring programs that are explicitly targeted, among other purposes, to contribute to CORINE and the possible successor program.

References

Ben-Asher, Z., Gilbert, H., Haubold, H., Smith, G., Strand, G.-H. 2013. HELM—Harmonised European Land Monitoring: Findings and recommendations of the HELM Project. Tel-Aviv: The HELM Project. Available at www.FP7HELM.eu/results.

Blanes, N., Green, T. 2012. Panorama of European land monitoring. HELM Project Report. Available at www.FP7HELM.eu/results.

Geist, H. (Ed.). 2006. *Our Earth's changing land: An encyclopedia of land-use and land-cover change*, Vol. 2. Westport, CT: Greenwood Press.

Mora, B., Tsendbazar, N.-E., Herold, M., Arino, O. 2014. Global land cover mapping: Current status and future trends. In I. Manakos and M. Braun (Eds.), *Land use and land cover mapping in Europe—Practices and trends* (pp. 11–30). Dordrecht: Springer.

Mücher, S. 2009. *Geo-spatial modelling and monitoring of European landscapes and habitats using remote sensing and field surveys*. Alterra Scientific Contributions, 31. Wageningen: Alterra.

2

Project CORINE Land Cover

Jan Feranec

CONTENTS

2.1 Introduction

Landscape is exposed to ever greater effects of the expanding settlements, transport infrastructure, intensification of farming, and extreme natural processes (such as floods and wind calamities). This trend has prompted launching of different programs with the aim of monitoring the aforementioned and similar effects. One of them is the Co-ORdination of INformation on the Environment (CORINE) program approved by the European Commission (EC) on June 27, 1985 with the objective of creating an information system about the state of the environment in the European Union (EU) and particularly

- To compile information on the state of the environment with regard to certain topics that have priority for all the member states (MS) of the EU
- To coordinate the compilation of data and the organization of information within the MS or at the international level
- To ensure consistency of such information and compatibility of data (Heymann et al., 1994)

The task of the European Environment Agency (EEA) after its founding in 1990 was the expert coordination of the CORINE program activities. The EEA established the European Environment Information and Observation Network (EIONET), which integrated all results of the CORINE program into a common European system.

At the meeting of environmental ministers of the European countries in 1991 in Dobříš, Czech Republic, it was decided that the implementation of the Biotopes, Corinair, and Land Cover (LC) projects under the CORINE program (in the majority of MS) would also be extended to 13 states of Central and Eastern Europe. The aim of the CLC projects was to derive* a single database for LC of Europe (see Chapter 4), that is, the physical status

---

* The terms *LC mapping* and *LC change mapping* are often incorrectly used to describe the main CLC objectives (e.g., Heymann et al., 1994; Nunes de Lima, 2005). Mapping is the design, compilation, and production of maps (Dent, 1996, p. 4) whereas the CLC is an inventory of LC classes stored in the database of the corresponding time horizon. These CLC data can then be presented as thematic maps at pan-European scales and so forth.

of landscape objects, principally their physiognomic and partially functional characteristics. Satellite images of the Earth surface became the primary source of such derivation. National teams produced the databases for their own countries, and these data were integrated at the European level.

2.2 From CLC1990 to CLC2012

The CLC1990 project was implemented under the auspices of the EC, and 27 countries participated in it between 1986 and 1998. Table 2.1 presents the technical specifications of this and the other three CLC projects:

- The satellite data used
- Time consistency
- Geometric accuracy of satellite images
- CLC minimum mapping unit (MMU is 25 ha; this means that objects with an area smaller than 25 ha cannot be present in the database. The width of a linear element is 100 m; it means that objects, mainly motorways and rivers, with a width less than 100 m cannot be present in the database.)
- Geometrical accuracy of CLC data
- Thematic accuracy
- Change detection
- Production time
- Documentation
- Access to the data
- Number of European countries involved

The CLC nomenclature consists of 44 classes. LC areas having a size smaller than the MMU are generalized. The generalization is based on similarity between a small object (size less than MMU) and the valid objects in the neighborhood (Bossard et al., 2000); for example, a small vineyard is joined to neighboring nonirrigated arable land rather than to a discontinuous urban fabric. This means of generalization is relatively easier for an experienced photointerpreter than the automated generalization (Büttner, 2014). It must be mentioned that the 25-ha MMU is compulsory in the European CLC datasets. However, in some countries, such as Finland and Sweden, the semiautomated method of interpretation (see Section 5.5 in Chapter 5) was applied to generate LC data with an MMU less than 25 ha. When integrated into the standard European CLC dataset (CLC2000) they were generalized into objects with the 25-ha MMU (CLC2000 Finland Final Report, 2005; Engberg, 2005). As Table 2.1 shows, the basic parameters of CLC did not change during generation of four data layers, supporting their compatibility. Table 2.2 contains an overview of countries participating in individual CLC projects. The usefulness of the CLC database led to an increasing number of countries joining the different CLC projects. Detailed characteristics of individual CLC data layers are provided in Chapters 10 through 16.

TABLE 2.1

Evolution of the CLC Projects

	CLC1990 Specifications	CLC2000 Specifications	CLC2006 Specifications	CLC2012 Specifications
Satellite data	Landsat-4/5 TM (in a few cases Landsat MSS) Single date	Landsat-7 ETM Single date	SPOT-4 and/or IRS LISS III two dates Dual date	IRS P6 LISS III and RapidEye Dual date
Time consistency	1986–1998	2000 ± 1 year	2006 ± 1 year	2011–2012
Geometric accuracy satellite images	≤50 m	≤25 m	≤25 m	≤25 m
CLC minimum detecting unit/ width	25 ha/100 m	25 ha/100 m	25 ha/100 m	25 ha/100 m
Geometric accuracy of CLC data	100 m	Better than 100 m	Better than 100 m	Better than 100 m
Thematic accuracy CLC data	≥85% (not validated)	≥85% (validated; see Büttner and Maucha, 2006)	≥85% (not checked)	≥85%
Change mapping CLCC	Not implemented	Boundary displacement min. 100 m; change area for existing polygons ≥5 ha; isolated changes ≥25 ha	Boundary displacement minimum 100 m; **all** changes ≥5 ha are to be mapped	Boundary displacement min. 100 m; **all** changes ≥5 ha are to be mapped
Thematic accuracy, CLCC	–	Not checked	≥85% (Büttner et al., 2011)	≥85%
Production time	10 years	4 years	3 years	2 years
Documentation	Incomplete metadata	Standard metadata	Standard metadata	Standard metadata
Access to the data (CLC, CLCC)	Unclear dissemination policy	Dissemination policy agreed from the start	Free access for all users	Free access for all users
Number of European countries involved	27	39	39	39

Source: With kind permission from Springer Science+Business Media: *Land use and land cover mapping in Europe: Practices & trends*, I. Manakos and M. Braun (Eds.), CORINE land cover and land cover change products, 2014, pp. 55–74, Büttner, G. Remote Sensing and Digital Image Processing, 18.

The EEA and the Joint Research Centre (JRC) of the European Commission launched the IMAGE2000 and CLC2000 (I&CLC2000) project. The I&CLC 2000 project aimed to provide a satellite image mosaic of Europe (IMAGE2000), an up-to-date LC database for the year 2000 (CLC2000), and information on general LC changes in Europe between 1990 and 2000 (Steenmans and Perdigão, 2001; Feranec et al., 2007). For CLC change (CLCC) mapping, it was necessary to reduce the MMU for changes down to 5 ha to produce the policy-relevant information at the European scale. This resulted in a much more detailed CLCC layer than is possible in the CLC status layers (MMU ratio is 25/5 = 5) (Büttner et al., 2002). Initiated in 2000 in MS, the project was extended in 2001 to the Accession countries and covered 39 countries (see Tables 2.1 and 2.2; Büttner et al., 2004).

Work on the third CLC2006 project was conducted under the guidance of the EEA, European Space Agency (ESA), and the EC; 39 countries participated in the period

TABLE 2.2

Participants of the CLC Projects

Country	CLC 1990	Change 1990–2000	CLC 2000	Change 2000–2006	CLC 2006	Change 2006–2012	CLC 2012
Albania[c]	No	No	Yes	Yes	Yes	Yes	Yes
Austria[a]	Yes	Yes	Yes	Yes	Yes	Yes	Yes
Belgium[a]	Yes	Yes	Yes	Yes	Yes	Yes	Yes
Bosnia/Hercegovina[c]	No	No	Yes	Yes	Yes	Yes	Yes
Bulgaria[a]	Yes	Yes	Yes	Yes	Yes	Yes	Yes
Croatia[a]	Yes	Yes	Yes	Yes	Yes	Yes	Yes
Cyprus[a]	No	No	Yes	Yes	Yes	Yes	Yes
Czech Republic[a]	Yes	Yes	Yes	Yes	Yes	Yes	Yes
Denmark[a]	Yes	Yes	Yes	Yes	Yes	Yes	Yes
Estonia[a]	Yes	Yes	Yes	Yes	Yes	Yes	Yes
Finland[a]	No	No	Yes	Yes	Yes	Yes	Yes
France[a]	Yes	Yes	Yes	Yes	Yes	Yes	Yes
Germany[a]	Yes	Yes	Yes	Yes	Yes	Yes	Yes
Greece[a]	Yes	Yes	Yes	Yes	Yes	Yes	Yes
Hungary[a]	Yes	Yes	Yes	Yes	Yes	Yes	Yes
Iceland[b]	No	No	Yes	Yes	Yes	Yes	Yes
Ireland[a]	Yes	Yes	Yes	Yes	Yes	Yes	Yes
Italy[a]	Yes	Yes	Yes	Yes	Yes	Yes	Yes
Kosovo[c]	No	No	Yes	Yes	Yes	Yes	Yes
Latvia[a]	Yes	Yes	Yes	Yes	Yes	Yes	Yes
Liechtenstein[b]	No	No	Yes	Yes	Yes	Yes	Yes
Lithuania[a]	Yes	Yes	Yes	Yes	Yes	Yes	Yes
Luxembourg[a]	Yes	Yes	Yes	Yes	Yes	Yes	Yes
Macedonia FYR[c]	No	No	Yes	Yes	Yes	Yes	Yes
Malta[a]	Yes	Yes	Yes	Yes	Yes	Yes	Yes
Montenegro[c]	Yes	Yes	Yes	Yes	Yes	Yes	Yes
Netherlands[a]	Yes	Yes	Yes	Yes	Yes	Yes	Yes
Norway[b]	No	No	Yes	Yes	Yes	Yes	Yes
Poland[a]	Yes	Yes	Yes	Yes	Yes	Yes	Yes
Portugal[a,e]	Yes	Yes	Yes	Yes	Yes	Yes	Yes
Romania[a]	Yes	Yes	Yes	Yes	Yes	Yes	Yes
Serbia[c]	Yes	Yes	Yes	Yes	Yes	Yes	Yes
Slovakia[a]	Yes	Yes	Yes	Yes	Yes	Yes	Yes
Slovenia[a]	Yes	Yes	Yes	Yes	Yes	Yes	Yes
Spain[a,f]	Yes	Yes	Yes	Yes	Yes	Yes	Yes
Sweden[a]	No	No	Yes	Yes	Yes	Yes	Yes
Switzerland[b]	No	No	Yes	Yes	Yes	Yes	Yes
Turkey[b]	Yes	Yes	Yes	Yes	Yes	Yes	Yes
United Kingdom[a,d]	No	Yes	Yes	Yes	Yes	Yes	Yes
Total	**27**	**28**	**39**	**39**	**39**	**39**	**39**

[a] EU member country (member of EEA as well) = 28.

[b] EEA member country = 5 (All collected CLC characteristics concerning Turkey are contained in corresponding Tables 10.2, 11.1, 12.1, 13.1, 14.1, and 15.1. They were not included into the overall characteristics of the European landscape.)

[c] EEA cooperating country = 6.

[d] United Kingdom of Great Britain and Northern Ireland.

[e] Including Azores and Madeira.

[f] Including Canary Islands.

2006–2008. Tables 2.1 and 2.2 provide an overview of all technical parameters of this project and list the participating states. The CLC2006 data layer was generated automatically (with optimal human interaction) by combining the revised CLC2000 and detected the $CLCC_{2000-2006}$ (Büttner et al., 2010). Another outcome of this project was the generation of the $CLCC_{2000-2006}$ layer. Chapters 12 and 15 describe characteristics of the CLC2006 products.

The most recent CLC2012 project is part of the Copernicus Land Monitoring Services, formerly the Global Monitoring for Environment and Security (GMES) Initial Operation (GIO) Land framework, under the auspices of the EEA, with participation of 39 countries. Tables 2.1 and 2.2 provide an overview of all technical parameters of this project and list the participating states. Products of the CLC2012 project, apart from the datasets of CLC2012 and $CLCC_{2006-2012}$, are also five high-resolution data layers (see Chapter 9): artificial surfaces (identification of degree of imperviousness), forest areas (identification of tree cover density and forest type), agricultural areas (identification of permanent grassland), wetlands (identification of the presence of surface water), and water bodies (identification of permanent water bodies) (Büttner and Kosztra, 2011).

2.3 Summary

The successful accomplishment of four CLC projects has offered opportunities to analyze and assess European landscape changes resulting from different socioeconomic and natural processes from the 1990s to 2012. Four CLC data layers simultaneously became an important part of the Copernicus Land Monitoring Services, which will further unfold, besides other measures, an efficient provision of geoinformation services principally in the European context. Through the Copernicus Land Monitoring Services the CLC data also became an integral part of the Global Earth Observation System of Systems (GEOSS), which aims to harmonize observations of Earth on the global level.

References

Bossard, M., Feranec, J., Otahel, J. 2000. *CORINE Land cover technical guide.* Addendum 2000, Technical Report 40. Copenhagen: European Environment Agency.

Büttner, G. 2014. CORINE land cover and land cover change products. In I. Manakos and M. Braun (Eds.), *Land use and land cover mapping in Europe: Practices and trends* (pp. 55–74). Remote Sensing and Digital Image Processing, 18. Dordrecht: Springer.

Büttner, G., Feranec, J., Jaffrain, G. 2002. The CORINE Land Cover update 2000. Technical guidelines. EEA Technical Report, 89. Copenhagen: European Environment Agency.

Büttner, G., Feranec, J., Jaffrain, G., Mari, L., Maucha, G., Soukup, T. 2004. The CORINE Land Cover 2000 project. In R. Reuter (Ed.), *EARSeL eProceedings* (Vol. 3, pp. 331–346). Paris: European Association of Remote Sensing Laboratories.

Büttner, G., Kosztra, B. 2011. CLC2012, Addendum to CLC2006 Technical Guidelines. European Environment Agency/European Topic Centre for Spatial Information and Analysis working document.

Büttner, G., Kosztra, B., Maucha, G., Pataki, R. 2010. Implementation and achievements of CLC2006. Final Report. Copenhagen: European Environment Agency.

CLC2000 Finland Final Report (2005). Helsinki: Finnish Environment Institute (SYKE).

Dent, D. B. 1996. *Cartography: Thematic map design*, 4th ed. Boston: Wm. C. Brown.

Engberg, A. 2005. Swedish CLC2000 Final Report. Gävle: Lantmäteriet.

Feranec, J., Hazeu, G., Christensen, S., Jaffrain, G. 2007. CORINE land cover change detection in Europe. Case Studies of the Netherlands and Slovakia. *Land Use Policy* 24: 234–247.

Heymann, Y., Steenmans, Ch., Croissille, G., Bossard, M. 1994. *CORINE Land Cover. Technical guide*. Luxembourg: Office for Official Publications of the European Communities.

Nunes de Lima, M. V. (Ed.). 2005. *IMAGE2000 and CLC2000: Products and methods*. Ispra, Italy: European Commission, Joint Research Centre, and European Environment Agency.

Steenmans, Ch., Perdigão, V. 2001. Update of the CORINE land cover database. In G. Groom, and T. Reed (Eds.), *Strategic landscape monitoring for the Nordic countries* (pp. 89–93). Copenhagen: Nordic Council of Ministers.

Section II

Methodology

3

CORINE Land Cover Nomenclature

Jan Feranec, Gerard Hazeu, Barbara Kosztra, and Stephan Arnold

CONTENTS

3.1 Background

The concept of CoORdination of INformation on the Environment (CORINE) Land Cover (CLC) nomenclature is mentioned in the *CLC Technical Guide* (Heymann et al., 1994: 25). The most important concept is the arrangement of landscape objects into groups on the basis of their relationships according to Figure 3.1 (cf. McConnell and Moran, 2001) and a certain number of requirements:

- It must be possible to identify all European Community territory; no "unclassified land" is admissible.
- The heading of classes must correspond to the needs of future users of the CLC database and the state of the environment.
- Heading terminology must be unambiguous and avoid the vague terms often resorted to by many photointerpreters when encountering uncertain areas.

The logical structure of CLC is an example of the hierarchic arrangement of landscape objects that corresponds with the theoretical scheme in Figure 3.1 (proceeding from the smaller number of generalized classes of higher hierarchic level to the large number of more detailed classes at the lower level). Landscape objects on the Earth surface (of the European continent) are arranged in two basic groups: *Water* and *Land,* further specified: *Ocean and Sea, Continental water, Land without plant cover, Land with plant cover,* and so forth. CLC nomenclature groups these sets of objects into 5 classes of the first level, 15 classes of the second level, and 44 classes of the third level (see Table 3.1). The aim of CLC was to characterize the European landscapes (from north to south and from east to west) at a scale of 1:100,000 with the minimum mapping unit (MMU) of 25 ha using satellite images. Ancillary data (particularly black-and-white aerial photographs, thematic and topographic maps, and information obtained by field checking) supported the most possible accurate identification of the CLC classes on satellite images.

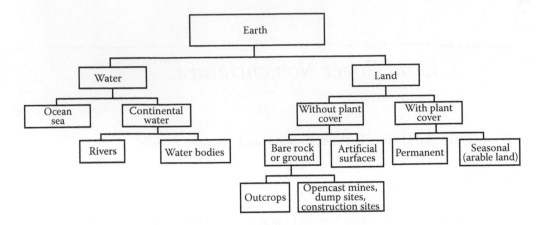

FIGURE 3.1
Theoretical schematic construction of a land cover nomenclature. (From Heymann, Y., Steenmans, Ch., Croissille, G., Bossard, M. 1994. *CORINE land cover. Technical guide*. Luxembourg: Office for Official Publications European Communities, 1994. With permission of European Environment Agency.)

3.2 Relationship between Land Use and Land Cover in the Context of CLC Nomenclature

Humans perceive the landscape as a combination of natural physical objects and objects (re-)created by them (agricultural land, artificial urban objects). LC represents the biophysical state of the real landscape, which means that it consists of natural but also modified (cultivated) and artificial objects (cf. Barnsley et al., 2001; Cihlar and Jansen, 2001; Feranec and Otahel, 2001; McConnell and Moran, 2001; Comber et al., 2005; Di Gregorio, 2005; Giri, 2012), whereas land use (LU) refers to the purpose for which land is used (Feranec et al., 2007: 236). It is important to emphasize that urbanized objects (artificial surfaces) or intensively used agricultural objects (arable land, permanent crops) are LC, but the term also indicates their LU. The nature and appearance of the natural or seminatural objects do not mean that they are not used or that they do not have a function. For instance, a piece of grassland can be used in the framework of agriculture, recreation, and nature conservation or for military purposes (Feranec et al., 2004, 2007).

In the aforementioned context the CLC nomenclature does not strictly separate LC from LU. It is greatly influenced by using satellite images as the primary data source to capture the European landscape situation. Only physiognomic attributes of landscape objects are extracted from satellite images. As the extraction of LU characteristics from satellite images is problematic or even impossible, these are incorporated indirectly into class descriptions (e.g., a class 124 airport is an example of LU; the class includes grassland and various artificial surfaces manifesting themselves by means of LC attributes; their pattern forms an integral class of LU denoted "airport," etc.).

This mix of LC/LU in the description of CLC nomenclature is recognized by several authors. Mücher et al. (2000) believe that CLC classes are heterogeneous and/or are determined by functional LU and consequently consist of various LC types. Jansen (2010) explains this problem as arising from inconsistent application of LC and LU ordering criteria and inconsistent use of these criteria at the different levels of CLC nomenclature. Di Gregorio and O'Brien (2012) note that the CLC nomenclature includes several combinations of LC and LU in 44 classes of the third level.

TABLE 3.1

CLC Nomenclature

---

1 Artificial surfaces

 11 Urban fabric

  111 Continuous urban fabric

   Most of the land is covered by structures and the transport network. Building, roads, and artificially surfaced areas cover more than 80% of the total surface. Nonlinear areas of vegetation and bare soil are exceptional.

  112 Discontinuous urban fabric

   Most of the land is covered by structures. Buildings, roads, and artificially surfaced areas associated with vegetated areas and bare soil, which occupy discontinuous but significant surfaces.

 12 Industrial, commercial, and transport units

  121 Industrial or commercial units

   Artificially surfaced areas (with concrete, asphalt, tarmacadam, or stabilized, e.g., beaten earth) without vegetation occupy most of the area, which also contains buildings and/or vegetation.

  122 Road and rail networks and associated land

   Motorways and railways, including associated installations (stations, platforms, embankments, linear greenery narrower than 100 m). Minimum width for inclusion: 100 m.

  123 Port areas

   Infrastructure of port areas, including quays, dockyards, and marinas.

  124 Airports

   Airport installations: runways, buildings, and associated land.

 13 Mine, dump, and constructions sites

  131 Mineral extraction sites

   Areas with open-pit extraction of construction material (sandpits, quarries) or other minerals (open-cast mines). Includes flooded gravel pits, except for river-bed extraction.

  132 Dump sites

   Public, industrial, or mine dump sites.

  133 Construction sites

   Spaces under construction development, soil or bedrock excavations, earthworks.

 14 Artificial nonagricultural vegetated areas

  141 Green urban areas

   Areas with vegetation within urban fabric; includes parks and cemeteries with vegetation and mansions and their grounds.

  142 Sport and leisure facilities

   Camping grounds, sports grounds, leisure parks, golf courses, racecourses, etc. Includes formal parks not surrounded by urban areas.

2 Agricultural areas

 21 Arable land

  211 Nonirrigated arable land

   Cereals, legumes, fodder crops, root crops, and fallow land. Includes flowers and fruit trees (nurseries cultivation) and vegetables, whether open field, under plastic or glass (include market gardening). Includes aromatic, medicinal, and culinary plants. Does not include permanent pastures.

  212 Permanently irrigated arable land

   Crops irrigated permanently or periodically, using a permanent infrastructure (irrigation channels, drainage network, and additional irrigation facilities). Most of these crops cannot be cultivated without an artificial water supply. Does not include sporadically irrigated land.

  213 Rice fields

   Land prepared for rice cultivation. Flat surfaces with irrigation channels. Surfaces periodically flooded.

 22 Permanent crops

  221 Vineyards

   Areas planted with vines.

  222 Fruit trees and berry plantations

   Parcels planted with fruit trees or shrubs: single or mixed fruit species, fruit trees associated with permanently grassed surfaces. Includes chestnut and walnut groves and hop plantations.

(Continued)

TABLE 3.1 (CONTINUED)

CLC Nomenclature

---

223 **Olive groves**

Areas planted with olive trees, including mixed occurrence of olive trees and vines on the same parcel.

23 **Pastures**

231 **Pastures**

Dense grass cover, of floral composition, dominated by graminacea, not under a rotation system. Mainly for grazing, but the fodder may be harvested mechanically. Includes areas with hedges (bocage).

24 **Heterogeneous agricultural areas**

241 **Annual crops associated with permanent crops**

Nonpermanent crops (arable land) associated with permanent crops on the same parcel.

242 **Complex cultivation patterns**

Juxtaposition of small parcels of diverse annual crops, pasture, and/or permanent crops.

243 **Land principally occupied by agriculture, with significant areas of natural vegetation**

Areas principally occupied by agriculture, interspersed with significant natural areas.

244 **Agroforestry areas**

Annual crops or grazing land under the wooded cover of forestry species.

3 **Forest and semi-natural areas**

31 **Forests**

311 **Broad-leaved forests**

Vegetation formation composed principally of trees, including shrub and bush understoreys, where broad-leaved species predominate.

312 **Coniferous forests**

Vegetation formation composed principally of trees, including shrub and bush understoreys, where coniferous species predominate.

313 **Mixed forests**

Vegetation formation composed principally of trees, including shrub and bush understoreys, where neither broad-leaved nor coniferous species predominate.

32 **Scrub and/or herbaceous vegetation associations**

321 **Natural grasslands**

Low productivity grassland. Often situated in areas of rough, uneven ground. Frequently includes rocky areas, briars, and heathland.

322 **Moors and heathland**

Vegetation with low and closed cover, dominated by bushes, shrubs, and herbaceous plants (heather, briars, broom, gorse, laburnum, etc.).

323 **Sclerophyllous vegetation**

Bushy sclerophyllous vegetation includes maquies and garrigue. In the case of shrub vegetation areas composed of sclerophyllous species such as *Juniperus oxycedrus* and heathland species such as *Buxus* spp. or *Ostrya carpinifolia* with no visible dominance (each species occupy about 50% of the area), priority will be given to sclerophyllous vegetation and the whole area will be assigned class 323.

324 **Transitional woodland/scrub**

Bushy or herbaceous vegetation with scattered trees. Can represent either woodland degradation or forest regeneration/recolonization.

33 **Open spaces with little or no vegetation**

331 **Beaches, dunes, sands**

Beaches, dunes, and expanses of sand or pebbles in coastal or continental location, including beds of stream channels with torrential regime.

332 **Bare rocks**

Scree, cliffs, rock outcrops, including active erosion, rocks and reef flats situated above the high-water mark.

333 **Sparsely vegetated areas**

Includes steppes, tundra, and badlands. Scattered high-altitude vegetation.

334 **Burnt areas**

Areas affected by recent fires, still mainly black.

(Continued)

TABLE 3.1 (CONTINUED)

CLC Nomenclature

---

335 Glaciers and perpetual snow

Land covered by glaciers or permanent snowfields.

4 Wetlands

41 Inland wetlands

411 Inland marshes

Low-lying land usually flooded in winter, and more or less saturated by water all year round.

412 Peat bogs

Peatland consisting mainly of decomposed moss and vegetation matter. May or may not be exploited.

42 Maritime wetlands

421 Salt marshes

Vegetated low-laying areas, above the high-tide line, susceptible to flooding by seawater. Often in the process of filling in, gradually being colonized by halophilic plants.

422 Salines

Salt-pans, active or in process of abandonment. Sections of salt marsh exploited for the production of salt by evaporation. They are clearly distinguishable from the rest of the marsh by their parcelation and embankment systems.

423 Intertidal flats

Generally unvegetated expanses of mud, sand, or rock lying between high and low water marks. 0 m contour on maps.

5 Water bodies

51 Inland waters

511 Water courses

Natural or artificial water courses serving as water drainage channels. Includes canals. Minimum width for inclusion: 100 m.

512 Water bodies

Natural or artificial stretches of water.

52 Marine waters

521 Coastal lagoons

Stretches of salt or brackish water in coastal areas that are separated from the sea by a tongue of land or other similar topography. These water bodies can be connected to the sea at limited points, either permanently or for parts of the year only.

523 Estuaries

The mouth of a river within which the tide ebbs and flows.

523 Sea and ocean

Zone seaward of the lowest tide limit.

Source: Heymann, Y., Steenmans, Ch., Croissille, G., Bossard, M. et al. 1994. *CORINE land cover. Technical guide.* Luxembourg: Office for Official Publications European Communities; Bossard, M. Feranec, J., Otahel, J. 2000. *CORINE land cover. Technical guide.* Addendum 2000, Technical Report 40. Copenhagen: European Environment Agency. Available at http://www.eea.europa.eu/data-and-maps/data/corine-land-cover -clc1990-250-m-version-06-1999/corine-land-cover-technical-guide-volume-2. With permission.

3.3 CLC Nomenclature and the Third-Level CLC Class Definitions

Four main specifications are associated with the CLC nomenclature:

1. The CLC classes were created for a scale 1:100,000.

2. The size of the smallest identified area (MMU) of classes CLC is 25 ha and the minimum width of the linear feature is 100 m.

3. Satellite images (supplemented by various ancillary data) are the primary source used for identification of areas of CLC classes.

TABLE 3.2

Physiognomic Attributes Relevant for Identification of CLC Classes

Main CLC Classes (1st Level)	Description of 1st Level CLC Classes
Urban fabric areas	Size, shape, and density of the buildings, share of supplementing parts of the class (e.g., square, width of the streets, gardens, urban greenery parking lots), character of transport network, size and character of neighboring water bodies, arrangement of infrastructure, size of quays, character of the runway surfaces, state of the dumps, and arrangement and share of playgrounds and sport halls
Agricultural areas	Share of dispersed greenery within agricultural land, arrangement and share of areas of permanent crops, relationships of grasslands with urban fabric, occurrence of dispersed houses (cottages), arrangement and share of agricultural land (arable land), grasslands, permanent crops and natural vegetation (mainly trees and bushes), irrigation channel network
Forest and seminatural areas	Character (composition), developmental stage and arrangement of vegetation (mainly trees and bushes), share of grass and dispersed greenery (composition density)
Wetlands	Character of substrate, water, and vegetation
Water bodies	Character (shape) of water bodies

Source: Feranec, J., Hazeu, G., Christensen, S., Jaffrain, G. 2007. CORINE land cover change detection in Europe. Case Studies of the Netherlands and Slovakia. *Land Use Policy*, 24:234–247. With permission from Elsevier.

4. The structure of the nomenclature consists of 5 items of the first level, 15 items of the second level, and 44 items of the third level (Heymann et al., 1994) (see Table 3.1). The *CLC Technical Guide—Addendum 2000* (Bossard et al., 2000) contains the detailed definitions, characteristics of a set of landscape objects that are their components and graphical sketches of their representative patterns, terrestrial photographs, and rules of generalisation. The CLC nomenclature is not a classification system, but only a legend for identification of LC classes at the scale 1:100,000 using satellite images (cf. McConnell and Moran, 2001). Classification systems should be independent from the spatial scale and from the means used to collect the information recorded (FAO, 1997).

In the conventional procedure of CLC mapping, it is the photointerpreter who manually delimits landscape objects (natural, modified, and human-created; see Table 3.2) by their physiognomic attributes (shape, size, color, texture, and pattern) and attaches CLC classes. The spatial/contextual relationships between landscape objects make it possible for the interpreter to comprehend the content of patterns proper to individual CLC class definitions.

The CLC nomenclature and its illustrated guidelines contain a series of rules that have been used for the production of CLC1990, CLC2000, CLC2006, and CLC2012. This nomenclature contributes to the generation of a harmonized pan-European LC dataset used by various European and national institutions.

3.4 Current Developments of CLC Nomenclature

Owing to its geographical coverage and significant timeline, CLC and its nomenclature is currently considered the quasi-standard for LC/LU mapping in Europe. However, since its inception in 1985, technical circumstances and user requirements of CLC have changed considerably. Meanwhile, an increasing number of countries have been migrating from the conventional visual photointerpretation toward semiautomatic approaches of CLC production to avoid high labor costs of photointerpretation and to improve repeatability and consistency with national datasets.* The methodology of semiautomated data production (described in Section 5.5 of Chapter 5) requires a systematic decomposed description of CLC classes to make it possible to generate them from available datasets in a harmonized way across Europe. Besides that, semantic inconsistencies (gaps and overlaps) in the class definitions have been identified during implementation, causing uncertainties of interpreting and separating CLC classes.[†,‡] In consequence of the aforementioned concerns, the need for revision of the nomenclature has become evident for all stakeholder parties.

Based on conceptual considerations developed by the EAGLE working group (EIONET Action Group on Land Monitoring in Europe)[§] (Arnold et al., 2013), a decompository approach has been applied testwise to CLC classes to tackle the aforementioned issues. In a follow-up step, structural and textual modifications of the CLC Nomenclature Guidelines have been proposed (Kosztra and Arnold, 2013).

3.4.1 Proposed Modification of CLC Class Description Structure

In the current structure of the CLC nomenclature guidelines (Bossard et al., 2000), there are four kinds of information listed under the subheading "This heading includes" of a class description[¶]:

1. Landscape situations or LU types for which the class does apply
2. Landscape elements (can be any piece of landscape visible in the real world, such as a group of trees, a house, a lake, etc.) whose existence makes it mandatory to assign the class (as single elements or in combination with others)
3. Landscape elements that can occur in the class, but are not mandatory
4. Landscape situations/elements that would lead to another class assignment, but because of their size not reaching the 25-ha MMU they are generalized to the given class

This mixture of applicable landscape forms with mandatory and optional elements makes it difficult to distinguish between cases where all listed elements must be present or just a selection of them. Also, the class definitions contain an uneven depth of content detail across the guidelines. Further, defining details are often distributed among several textual positions (class name, extension, subtitle, includes/excludes, generalization rules). All of these issues raise difficulties in the bottom-up compilation of CLC classes through generalization and aggregation or data merging from different data sources.

---

* HELM deliverable 3.1: Commonalities, differences, and gaps in national and sub-national land monitoring systems.
† HELM deliverable 2.3: Practical experiences with bottom-up land monitoring approaches.
‡ HELM deliverable 4.1: Prerequisites and criteria for aligning national/sub-national land monitoring activities.
§ http://ETC/-lusi.eionet.europa.eu/EAGL%20-%20Eionet%20Group%20LC.
¶ It is recommended to have on hand the CLC nomenclature guidelines for better understanding of the following explanations.

The proposed structural enhancement of the CLC nomenclature guidelines aims at giving order to (1) the applicability/nonapplicability of a class for a spatial landscape unit and to (2) the precise elementary content what's in/what's out of the class. In the revised class definitions most subheadings are kept from the original structure, while the following subheadings have been introduced and/or enhanced:

> *This heading is applicable for*: Particular real landscape situation or spatial setting or LU when the given class should be used
>
> *This heading includes*: Landscape elements that make up the LC class as defining features
>
> *This heading is not applicable for*: Real landscape instances and LU types for which the given class is not appropriate
>
> *This heading excludes*: Landscape elements that are excluded from class by definition and must not occur

3.4.2 Modification of Thematic Content of CLC Classes

Besides restructuring, contentwise changes to the text have also been applied based on feedback from national teams during former CLC inventories and on experience gained during training and verification activities of the CLC Technical Team. This first of all included the removal of identified thematic/threshold overlaps and gaps. Second, thematic content was complemented with

- Newly appearing landscape/LU phenomena, for example, solar panel parks, wind power plant, ski pistes with artificial snow making, habitat restoration areas.
- Regional characteristics missing or not emphasized enough, for example, inland salt planes.
- Landscape types and elements frequently misclassified during implementation due to lack of/insufficient representation in class descriptions, for example, abandoned/reclaimed mineral extraction sites, annually sown grass.
- Landscape elements composing a class and LU characteristics typical of a class, but not specifically mentioned before. Many of these are logically evident, and therefore are implicitly understood by a human photointerpreter, but need to be listed for the semiautomated compilation of CLC classes.
- Items found to be omitted at cross-checking of class descriptions. For example, an item mentioned in the "nonapplicable for" list of one class should appear in the "applicable for" list of relevant class.
- Threshold values (minimum/maximum area coverage by class components in area percentage) formerly not added, but logically deductible from definition of other classes.

The results of the guideline enhancement will enable a more consistent harmonized CLC production by both traditional photointerpretation and bottom-up/semiautomated methods during the subsequent inventories.

References

Arnold, S., Kosztra, B., Banko, G., Smith, G., Hazeu, G., Bock, M., Valcarcel Sanz, N. 2013. The EAGLE concept—A vision of a future European Land Monitoring Framework. In R. Lasaponara, L. Masini, and M. Biscione (Eds.), *Towards Horizon 2020: Earth observation and social perspectives. 33th EARSeL Symposium Proceedings* (pp. 551–568). Matera, Italy: European Association of Remote Sensing Laboratories and Consiglio Nazionale delle Richerche (National Research Council of Italy).

Barnsley, M. J., Møller-Jansen, L., Barr, S. L. 2001. Inferring urban land use by spatial and structural pattern recognition. In J. P. Donnay, M. J. Barnsley, and P. A. Longley (Eds.), *Remote sensing and urban analysis* (pp. 115–144). London: Taylor & Francis.

Bossard, M., Feranec, J., Otahel, J. 2000. *CORINE Land Cover technical guide*. Addendum 2000, Technical Report 40. Copenhagen: European Environment Agency.

Cihlar, J., Jansen, L. J. 2001. From land cover to land use: A methodology for efficient land use mapping over large areas. *Professional Geographer* 53:275–289.

Comber, A., Fisher, P., Wadsworth, R. 2005. What is land cover? *Environment and Planning B: Planning and Design* 32:199–209.

Di Gregorio, A. 2005. *Land cover classification system: Classification concepts and user manual.* Software version 2. Rome: Food and Agriculture Organization of the United Nations.

Di Gregorio, A., O'Brien, D. 2012. Overview of land-cover classifications and their interoperability. In Ch. P. Giri (Ed.), *Remote sensing of land use and land cover: Principles and applications* (pp. 37–47). Boca Raton, FL: CRC Press.

FAO. 1997. *AFRICOVER land cover classification*. Remote Sensing Centre Series, 70. Rome: Food and Agriculture Organization of the United Nations.

Feranec, J., Hazeu, G., Christensen, S., Jaffrain, G. 2007. CORINE land cover change detection in Europe. Case Studies of the Netherlands and Slovakia. *Land Use Policy* 24:234–247.

Feranec, J., Otahel, J. 2001. *Land cover of Slovakia*. Bratislava: Veda.

Feranec, J., Otahel, J., Cebecauer, T. 2004. Zmeny krajinnej pokrývky—Zdroj informácií o dynamike krajiny (Land cover changes—Information source about landscape dynamics). *Geografický Časopis* 56:33–47.

Giri, P. Ch. 2012. Brief overview of remote sensing of land cover. In Ch. P. Giri (Ed.), *Remote sensing of land use and land cover: Principles and applications* (pp. 3–12). Boca Raton, FL: CRC Press.

Heymann, Y., Steenmans, Ch., Croissille, G., Bossard, M. 1994. *CORINE land cover. Technical guide.* Luxembourg: Office for Official Publications of the European Communities.

Jansen, L. J. M. 2010. *Analysis of land change with parametrised multi-level class sets: Exploring the semantic dimension*. PhD diss., Wageningen University.

Kosztra, B., Arnold, S. 2013. Deliverable "Proposal for enhancement of CLC nomenclature guidelines" EEA subvention 2013. WA1 Task 261_1_1: Applying EAGLE concept to CLC guidelines enhancement. 2013 Implementation Plan of the European Topic Center on Spatial Information and Analysis. Copenhagen: European Topic Centre for Spatial Information and Analysis.

McConnell, W. J., Moran, E. F. 2001. Meeting in the middle: The challenge of meso-level integration: An international workshop on the harmonisation of land use and land cover classification. LUCC Report Series No. 5. Louvain-la-Neuve: LUCC International Project Office.

Mücher, C. A., Steinnocher, K., Kressler, F., Heunks, C. 2000. Land cover characterization and change detection for environmental monitoring of pan-Europe. *International Journal of Remote Sensing* 21(6/7):1159–1181.

4

Satellite Data Used

Tomas Soukup, Jan Feranec, and Gerard Hazeu

CONTENTS

4.1 Background

The concept of European CoORdination of Information on the Environment (CORINE) Land Cover (CLC) mapping has been from the very beginning tightly linked to the use of satellite images as the fundamental input data (Heymann et al., 1994: 40). While during a feasibility study back in 1985 some alternative approaches for a LC inventory were tested (e.g., ground survey, aerial photographs), based on a cost–benefit analysis results, finally, an approach using high-resolution Earth observation satellite data were selected. The added value of the remote sensing approach is clear, as it provides a synoptic view of the Earth's surface by capturing at once information for much larger areas than aerial survey data. In addition, passing regularly over the same area, the changes in land use/land cover (LU/LC) also can be periodically monitored. LC mapping using satellite imagery has a long tradition, and remote sensing image archives provide inputs for LC/LU, and its changes for a period of more than 45 years. Use of satellite data in the CLC context has benefited from previous activities, in particular stimulated by the US Landsat program, but also evolved in time as outlined in the following paragraphs.

4.2 Criteria for Satellite Data Selection

To understand the CLC product characteristics, it is necessary to understand the basic characteristics of the satellite data used in the CLC mapping process. Right from the beginning, selection of satellite images for CLC mapping activities has been driven by the following main criteria:

- Data are available on a regular basis for the whole European continent (subject to there being no cloud cover), so that all of the countries can be covered once or twice a year.

- Data can capture sufficient detail (target scale 1:50,000–1:100,000).
- Data are inexpensive.
- Data are objective, so the sensor–transmission–reception system involves no human intervention.
- Data are in digital form, which has a number of advantages in data handling and processing profiting from continuous advances in image processing techniques.

On an operational level, the preceding requirements have been implemented for each CLC mapping campaign by selecting a particular sensor as the source of the base satellite data. The selection of image acquisition dates was based on the national technical team requirements to cover vegetation period in particular country.*

4.3 Sensor Characteristics

A sensor is a device that gathers energy and converts it into a signal, presenting it in a form suitable for further analysis of the target under investigation (Lillesand and Kiefer, 1979; Mather, 2004). Sensors used for CLC mapping are those operating in the optical-infrared spectrum. Use of microwave sensors has not been utilized yet in the CLC context although there is a discussion about their potential for heavily clouded regions. Different satellite sensors can capture different objects on the Earth's surface up to the limit based on particular sensor technical specifications. The most important characteristics that determine the capability of the satellite sensor to detect such landscape objects relate to resolution: spatial, spectral, radiometric, and temporal.

Spatial resolution is the projection of a satellite sensor's detection element onto the ground, that is, the minimum segment observed, often called pixel size (Lillesand and Kiefer, 1979; Mather, 2004). In general, landscape objects can be distinguished (detected and outlined) only if larger than the sensor's pixel size. Images in which only large features are visible are said to have coarse or low resolution. In fine or (very) high resolution images, much smaller objects can be detected. For CLC mapping purposes, high-resolution sensors (10–30 m pixel size) are utilized.

Spectral resolution describes the ability of the sensor to define fine wavelength intervals, that is, sampling the spatially segmented image in different spectral intervals. Satellite sensors measure reflected/emitted energy within several specific sections (also called image bands) of the electromagnetic spectrum. Different classes of features and details in an image can be distinguished by comparing their responses over distinct wavelength ranges (Lillesand and Kiefer, 1979; Mather, 2004). In general, landscape objects can be distinguished only if spectrally different from surrounding objects. Spectral emissivity curves characterize the reflectance and/or emitance of a feature or object over a variety of wavelengths. Multispectral sensors used in the CLC mapping context usually have up to 10 different band measurements in each pixel of the images they produce. Most of the sensors used for the CLC mapping typically include visible green, visible red, and near infrared spectral bands.*

The temporal resolution of satellite sensor indicates the revisit time of the satellite for each particular area and determines the potential frequency of satellite data acquisition. The temporal resolution of a sensor depends on a variety of factors, including the satellite/sensor capabilities to change instrument orientation, the swath overlap, and latitude (Lillesand and Kiefer, 1979; Mather, 2004). The higher the temporal resolution, the shorter is the distance of time between the acquisitions of images. Nevertheless, actual acquisition frequency may be compromised by meteorological conditions (e.g., cloud and haze presence) during the acquisition period—the period in which the satellite passes over a certain place to acquire data. Temporal resolution is very important characteristic, as acquisition of cloud-free wall-to-wall European coverage (or coverages[†]) still represents a challenging task. In the CLC mapping context, most of the satellites used have a temporal resolution of about 2 weeks and better.

The main characteristics of the high-resolution satellite images used are summarized in Table 4.1.

4.4 Summary of Data Used

An overview of particular satellite data used in different CLC mapping campaigns is presented in Table 4.2. As can be seen, the resolution of satellite sensors used within the context of CLC mapping has evolved. First, methodological testing was conducted with a range of spatial resolution from of 10 m (Satellite Pour l'Observation de la Terre [SPOT]) to 80 m (Landsat Multispectral Scanner System [MSS]) data to evaluate the technical, but also the organizational and financial feasibility of continental LC mapping. The initial mapping for the reference year 1990 in the first countries was even performed in 80-m resolution (Landsat MSS), but later, with availability of the Landsat 5 satellite, 25-m resolution has been established as the common base for further mapping. CLC2000 mapping followed the same line. As the Landsat 7 satellite was not available for CLC2006, a new source of satellite imagery was introduced and used. A collaboration with the European Space Agency (ESA) within the Global Monitoring for Environment and Security (GMES)/Copernicus framework provided as additional resources—SPOT 4 and 5 and Indian Remote Sensing (IRS) P6 satellites. The IMAGE2006[‡] is for the first time composed of two multitemporal coverages of more than 3800 SPOT 4/5 and IRS-P6 Linear Imaging Self Scanning Sensor (LISS)-III/ResourceSat 2 scenes. Each of the scenes is also provided in 20-m resolution. The same combination of satellite imagery has also been used in the 2012 mapping campaign with the addition of RapidEye constellation imagery for the second

---

* In addition, one has to consider also *radiometric resolution,* which is a rather technical measure of the sensor describing its ability to discriminate very slight differences in energy. The finer the radiometric resolution of a sensor, the more sensitive it is to detecting small differences in reflected or emitted energy. In the CLC context, the radiometric resolution of the used imagery ranges from 6 bits (Landsat MSS) to 16 bits (Landsat 8).

† For the CLC2006 mapping campaign, dual coverage has been introduced to support better temporal variation based interpretation.

‡ Orthorectification and preprocessing of images is organized centrally. Individual ortho-rectified images and image mosaics (called IMAGE<year>) are provided both in national and European coordinate systems. From 2006 on, image data are considered as a multipurpose product, serving also other Copernicus services beside CLC, and are distributed via Copernicus data warehouse (DWH) operated by the European Space Agency (ESA) (Lima, 2009; Büttner, 2014).

TABLE 4.1

Overview of EO Satellites Used (and Potentially to Be Used) in CLC Contexts

Satellites	Landsat 1-2-3	Landsat 4-5	Landsat 7	Landsat 8	SPOT 1-2-3	SPOT 4-5	IRS-P6/ResourceSat 2[a]	RapidEye	Sentinel 2
Sensor	MSS	TM	ETM+	OLI	HRV	HRVIR/HRG	LISS-III	MS	MSI
Launching date	1972, 1975, 1978	1982, 1987	1999	2013	1986, 1990, 1993	1998, 2002	2003, 2011	2008	2015
Spectral bands (μm):[b]									
PAN			0.52–0.90*	0.50–0.68*	0.50–0.73*	0.48–0.71[c]	0.62–0.68*[d]		
COASTAL/AEROSOL				0.43–0.45					0.43–0.45
BLUE		0.45–0.52	0.45–0.52	0.45–0.52				0.44–0.51	0.46–0.52*
GREEN	0.50–0.60	0.52–0.60	0.52–0.60	0.52–0.60	0.50–0.59	0.50–0.59	0.52–0.59	0.52–0.59	0.54–0.58*
RED	0.60–0.70	0.63–0.69	0.63–0.69	0.63–0.68	0.61–0.68	0.61–0.68	0.62–0.68	0.63–0.69	0.65–0.68*
"RED EDGE"								0.69–0.73	0.69–0.71+
NIR	0.70–0.80	0.76–0.90	0.76–0.90	0.85–0.89	0.79–0.89	0.79–0.89	0.77–0.86	0.76–0.85	0.73–0.75+[e]
NIR	0.80–1.10								0.78–0.90*
WATER VAPOUR									0.93–0.95
CIRRUS				1.36–1.39					1.36–1.39
SWIR		1.55–1.75	1.55–1.75	1.56–1.66		1.58–1.75	1.55–1.7		1.56–1.65+
SWIR		2.08–2.35	2.08–2.35	2.10–2.30					2.10–2.28+
Spatial resolution (m)[f]	80	30	30, 15*	30, 15*	20, 10*	20, 10, 5	23.5, 5.8*	6.5	60, 20+, 10*
Radiometric resolution (bits)	8	8	8	12/16[g]	8	8	7	12	12
Temporal resolution (days)	18	16	16	16	26	26	24	1[h]	10
Altitude (km)	920	705	705	705	822	830	817	630	786
Swath width (km)	183	185	185	185	60	2 × 60[i]	140	77	290

a ResourceSat 2 successor of IRS-P6 (ResourceSat 1).
b Thermal (TIR) bands of sensors are not listed.
c SPOT 4 PAN range is 0.61–0.68 μm with resolution 10 m, SPOT 5 PAN range is 0.48–0.71 μm with resolution 5 m.
d Panchromatic sensor LISS-IV on board, but not used in the CLC context.
e There are two additional NIR bands available in 20 m: B7 (0.77–0.79 μm), B8b (0.85–0.87 μm).
f In general, different bands can have different spatial resolution. Please see * and + marks for particular spectral bands to see their resolution.
g Landsat 8 imagery has a radiometric resolution of 12 bits (16 bits when processed into Level 1 data products) compared to 8 bits for its predecessor.
h RapidEye constellation consists of five identical satellites.
i SPOT 5 satellite has on board two HRG sensors that can capture data simultaneously, that is, capture the territory with a total width of 120 km.

TABLE 4.2

Evolution of Satellite Data Used for CLC Mapping

Campaign	CLC1990	CLC2000	CLC2006	CLC2012
Name	IMAGE1990	IMAGE2000	IMAGE2006	IMAGE2012
Satellite data	Landsat 5 MSS/TM	Landsat 7 ETM	SPOT 4/5 and IRS P6 LISS III	IRS P6 LISS III and RapidEye
Coverage(s)	Single date	Single date	Dual date	Dual date
Time consistency	1986–1998	2000 ± 1 year	2006 ± 1 year	2011–2012
Spatial resolution	25 m	25 m	25 and 20 m	25, 20, and 5 m

Source: http://land.copernicus.eu/pan-european/corine-land-cover/view.

coverage. A spatial resolution of 20 m became now the baseline for production. Future CLC mapping campaigns (CLC2016+) shall be based mainly on the capacity of new Copernicus Sentinel 2 satellites. Owing to the huge acquisition capacity of the Sentinel2 constellation, more temporal information is foreseen to better support seasonal variability detection.

As seen, resolution specification of the sensors used during the different mapping campaigns is quite similar. Slight technical differences occurred that were compensated for CLC purposes by the application of specific band combinations and preprocessing. The satellite sensors determine the approximate scale of work, which for CLC is set to 1:50,000–1:100,000, facilitating the detection of essential landscape features by means of their representation in image.

References

Büttner, G. 2014. CORINE land cover and land cover change products. In I. Manakos, and M. Braun (Eds.), *Land use and land cover mapping in Europe: Practices and trends* (pp. 55–74). Remote Sensing and Digital Image Processing, 18. Dordrecht: Springer.

Heymann, Y., Steenmans, Ch., Croissille, G., Bossard, M. 1994. *CORINE land cover. Technical guide.* Luxembourg: Office for Official Publications of the European Communities.

Hoersch, B., Amans, V. 2015. Copernicus space component data access portfolio: Data warehouse 2014–2020. Issue Date 15/05/2015, Ref COPE-PMAN-EOPG-TN-15-0004, ESA-ESRIN.

Lillesand, M. T., Kiefer, W. R. 1979. *Remote sensing and image interpretation.* New York: John Wiley & Sons.

Lima, V. 2009. Report on progress WPI.3, 3rd steering committee meeting, European Environment Agency, April 22, 2009.

Mather, M. P. 2004. *Computer processing of remotely-sensed images: An introduction.* Chichester: John Wiley & Sons.

5

Interpretation of Satellite Images

Jan Feranec, Gerard Hazeu, and Tomas Soukup

CONTENTS

5.1 Background

Various methods contribute to discernment of land cover (LC) and its changes. They are classified into two groups:

- Methods based on field research and analysis of maps
- Methods based on remote sensing data

The first group makes use of topographic maps at large scales, city maps, or maps of agricultural crops, which can be verified and updated by field mapping. Spatial precision and verity of the mapped subject correspond to the technological possibilities of the geodetic instruments used and interpretation capacities of field workers (Feranec and Otahel, 2001).

Advances offered by the second group include the development of remote sensing techniques (providing different types of images, principally digital ones with increasing spatial, spectral, and temporal resolution), as well as a means of widening the approaches to image interpretation. Numerous authors dealt with the possibilities using aerial photographs in the process of identification satellite and map presentation of LC classes, for instance, Schneider (1974), Baker et al. (1979), Befort (1986), Seger (1989), Lo and Noble (1990), Schott (1997), and others. Giri (2012) summarizes and provides the most recent survey of contributions of satellite images to identification of LC classes and their changes.

The aim of this chapter is to explain briefly the visual interpretation, computer-assisted photointerpretation (CAPI), and semiautomated methods of satellite images interpretation used for generations of CLC1990, CLC2000, CLC2006, and CLC2012 data and $CLC_{1990-2000}$, $CLC_{2000-2006}$, and $CLC_{2006-2012}$ data regarding the changes.

5.2 Why Are Visual Photointerpretation and CAPI Still Used in CLC Methodology?

Although the position of methods of digital processing of satellite images in identification of LC classes and their changes (Coppin et al., 2004; Rogan and Chen, 2004; Treitz and Rogan, 2004; Chen et al., 2012; Pouliot et al., 2012) is strong, visual photointerpretation and CAPI methods remain important as well (Steenmans and Perdigao, 2001; Büttner et al., 2004; Feranec et al., 2007). As Feranec et al. (2007: 235) explains

- CoORdination of Information on the Environment (CORINE) Land Cover (CLC) classes are very heterogeneous in terms of their spectral characteristics. Objects that fall under one class can be fairly different (e.g., artificial surfaces, urban greens, small water bodies, etc.) and such varied objects cannot be classified via a computer approach into one class only on the basis of the spectral signature or texture.

- *Association* as the interpretation sign (Feranec, 1999) used as a classification criterion in the framework of the corresponding algorithm of image processing is difficult (especially for CLC classes such as 242, 243, 324, but also others); sorting out such classes requires awareness of spatial relationships between the corresponding objects in the landscape, which an experienced interpreter can accomplish using visual interpretation or CAPI.

- Natural conditions can modify the spectral properties of objects. The identification of LC classes is affected by these conditions. The same LC class can have different spectral signatures due to, for example, a different level of groundwater.

5.3 Visual Photointerpretation

Visual photointerpretation is best described as the record on transparent overlays of objects visible on satellite images (hardcopies in CLC1990). The interpreter uses interpretation signs (characteristics) such as shape, size, color, texture, pattern, and association for the visual analysis of satellite imagery. Signs are described in detail in the Addendum (Bossard et al., 2000) and partially in the *CORINE Land Cover Technical Guide* (Heymann et al., 1994). Individual classes are identified and recorded by drawing polygons on a transparent overlay fixed on top of a satellite image hardcopy printed at a scale of 1:100,000. Polygons (also called faces in topological terminology) describe areas bounded by lists of arcs sometimes called links (Jones, 1997: 34). Interpretation is facilitated by using various ancillary data (principally aerial photographs, topographic and thematic maps, and field knowledge). The result of the interpreter's work is interpretation schemes at a scale of 1:100,000. They contain polygons of CLC classes marked by a three-digit code and corners of the concerned sheet of the topographic map at a scale of 1:100,000 marked by crosses (see Figure 5.1). To avoid problems when superimposing the sheets of interpretation schemes, the interpreted contents exceed the size of the map sheet by about 1 cm. The final step of this method is digitization of interpretation schemes. Visual photointerpretation was used until end of the 1990s in all countries participating in the CLC1990 project (Feranec and Otahel, 2001).

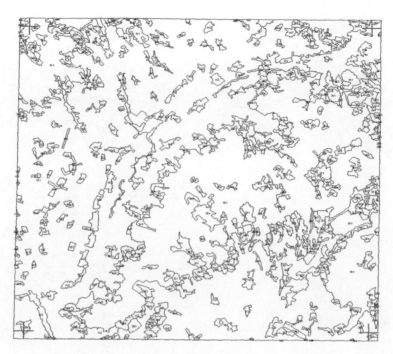

FIGURE 5.1
Example of interpretation scheme. (From Feranec, J., Otahel, J. 2001. *Land cover of Slovakia*. Bratislava: VEDA. With permission of VEDA.)

5.4 CAPI

CAPI is a method of interpretation of satellite images in which the interpreter also uses interpretation signs and identifies LC classes. However, this time he or she does not use the hardcopy of satellite images, but rather digital images on the computer screen, where he or she marks—digitizes—the borders of CLC classes (see Figure 5.2). If the interpretation process is part of the Geographical Information System (GIS), the interpreter can readily and more precisely use the ancillary data in digital or analog format. As in the preceding method, the skills of the interpreter distinctly influence the quality of interpretation results.

According Nunes de Lima (2005: 92), the advantages of CAPI are

- Better resolution of satellite images on the computer screen than on the hard copy images at a constant scale of 1:100,000 overlaid by a transparency
- The possibility of operative enlargement and comparison with ancillary data in digital or analog format (mainly aerial photographs, topographic and thematic maps)
- The possibility of more precise delineation of CLC classes

In CLC2006, 33 out of 38 participating countries applied this method.

Differences in the CAPI method stem from four main sources (Ben-Asher et al., 2013: 39):

- Software used, whether standard or bespoke
- In situ data varying according to the monitoring programs applied in the member states

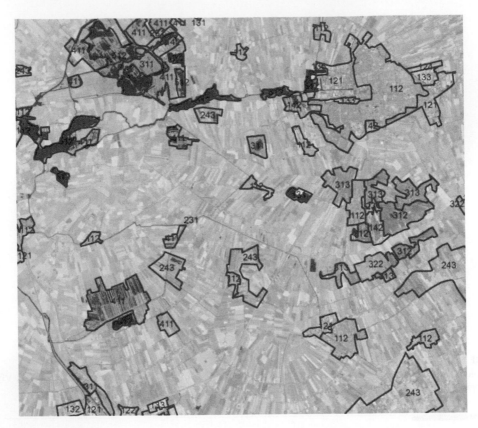

FIGURE 5.2
Screenshot of digitized polygons of CLC classes. (With permission of Gerard Hazeu, Altera—WUR.)

- Human factors, for example, differences in training, background, experience, and understanding
- Inconsistencies, gaps, and insufficient descriptions in the CLC nomenclature

5.5 Semiautomated Methods

Meanwhile, an increasing number of countries are migrating from conventional visual photointerpretation or CAPI toward semiautomatic approaches of CLC data production. They do so in particular to avoid high labor costs of photointerpretation and to improve repeatability and consistency with national datasets.* Methodological key elements of semiautomated data production are GIS generalization of national LC datasets and GIS integration of elementary in situ data (data merging). Both of these require a systematic decomposed description of CLC classes to be able to generate them from available datasets in a harmonized way across Europe. These methods were initially applied in Finland,

---

* HELM deliverable 3.1: Commonalities, differences and gaps in national and sub-national land monitoring systems. 2013. Lasmanova, S., Büttner, G., Kosztra, B. (Eds.). Available at http://www.fp7helm.eu/fileadmin /site/fp7helm/HELM_3_1_Commonalities_and_Differences.pdf.

Iceland, Norway, Sweden, Switzerland, and the United Kingdom; in CLC2012 and later, also in Germany, Ireland, and Spain. The United Kingdom, which does not have a corresponding national dataset for CLC2012, is now reverting to the CAPI approach.

Common practices for most semiautomated methods are

- Classification of satellite data into LC information
- GIS integration of various LU/LC data from the national database (data merging)
- Generalization to aggregate to larger spatial resolution than national monitoring data
- Production of individual thematic layers for each CLC class and subsequent merging of the layers according to a priority list (Ben-Asher et al., 2013: 39)

5.6 CLC Change Identification Methodology

A CLC change (CLCC) is interpreted as a categorical change in which one LC class or its part(s) is replaced by another LC class(es) (cf. Coppin et al., 2004). An example of conversion is the change of an arable land area into a discontinuous urban fabric area or a pasture area into a mineral extraction site area.

The basic condition for identification of LC changes by application of satellite images is the existence of changes in spectral reflectance. Such changes are manifest on images as a result of changes of characteristics of interpretation signs (shape, color, texture, pattern, etc.). From the methodological point of view it means that images acquired in two or more time horizons are used in the identification of LC changes (Feranec et al., 2007).

Two basic methods are used for identification of LC changes (a contiguous changed area of at least 5 ha having a width of at least 100 m) (EEA-ETC/TE, 2002) in the context of CLC projects: (1) CAPI and (2) semiautomated methods (see Sections 5.4 and 5.5). These methods are based on the updating or backdating of a LC data layer (see Figure 5.3; Feranec et al., 2005). The updating or backdating can take place in two ways: LC changes are the results of an overlay of two CLC datasets (approach A; see Figure 5.4) or the "changes first" method in which the interpreted and identified LC changes are merged with a status layer (approach B; see Figure 5.4) (Büttner et al., 2004; Feranec et al., 2007).

Detailed characteristics of approaches A and B are included in the study of Feranec et al. (2007). According to these authors the updating method minimizes the time needed

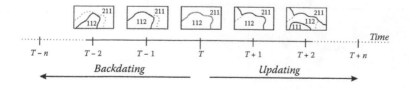

FIGURE 5.3

Basic principles of updating and backdating. The referential layer, a copy of which—the template—is modified in updating or backdating subject to the changes of LC shape, for instance, in the $T + 1$ time horizon or $T - 1$, and so forth is in red. (Feranec, J., Cebecauer, T., Otahel, J. 2005. *Photo-to-photo interpretation manual* (revised). Institute of Geography, Slovak Academy of Sciences, Bratislava, BIOPRESS document, biopress-d-13-1.3. 104 pp. With permission of Elsevier.)

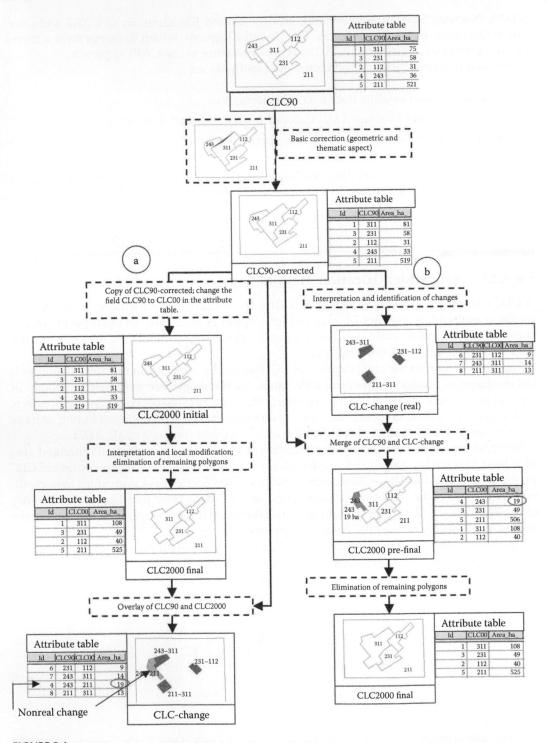

FIGURE 5.4
Generation of the CLC 2000 by the computer-assisted photointerpretation method: Two different approaches—(a), (b). (Feranec, J., Hazeu, G., Christensen, S., Jaffrain, G. 2007. Corine Land Cover change detection in Europe. Case Studies of the Netherlands and Slovakia. *Land Use Policy* 24: 234–247. With permission.)

by an interpreter to generate a new data layer. This approach (Figure 5.4a) makes it possible to check the minimum mapping unit and amalgamation of the residual parts of polygons smaller than 25 ha with the neighboring polygons (by using the priority table; see EEA-ETC/TE, 2002). The approach "changes first" (Figure 5.4b) in which, for example, the CLC2000 data layer is completed according to the formula CLC2000 = CLC1990 + CLCC2000 change, is characterized in *CORINE Land Cover update, I&CLC2000 project* (EEA-ETC/TE, 2002).

As the European Environment Agency (EEA) is most of all interested in the layer of changes, approach B, which principally identifies LC changes, dominates in the pan-European context. Subsequently, by joining the layer of (e.g., CLCC2012) changes with the preceding CLC2006 layer, CLC2012 = CLC2006 + CLCC2012 is produced.

References

Baker, R. D., DeSteiguer, J. E., Grant, D. E., Newton, M. J. 1979. Land use/land cover mapping from aerial photographs. *Photogrammetric Engineering and Remote Sensing* 45:661–668.

Befort, W. 1986. Large-scale sampling photography for forest habitat: Type identification. *Photogrammetric Engineering and Remote Sensing* 52:101–108.

Ben-Asher, Z., Gilbert, H., Haubold, H., Smith, G., Strand, G.-H. 2013. *HELM—Harmonised European Land Monitoring: Findings and recommendations of the HELM Project*. Tel-Aviv: The HELM Project.

Bossard, M., Feranec, J., Otahel, J. 2000. *CORINE land cover technical guide*. (p. 105), Addendum 2000, Technical Report 40. Copenhagen: European Environment Agency, Available at: http://terrestrial .eionet.eea.int.

Büttner, G., Feranec, J., Jaffrain, G., Mari, L., Maucha, G., Soukup, T. 2004. The CORINE Land Cover 2000 Project. In R. Reuter (Ed.), *EARSeL eProceedings*, (Vol. 3, pp. 331–346). Paris: European Association of Remote Sensing Laboratories.

Chen, X., Giri, Ch. P., Vogelman, J. E. 2012. Land-cover change detection. In Ch. P. Giri (Ed.), *Remote sensing of land use and land cover: Principles and applications* (pp. 153–175). Boca Raton, FL: CRC Press.

Coppin, P., Jonckheere, J., Nackaerts, K., Muys, B., Lambin, E. 2004. Digital change detection methods in ecosystem monitoring: A review. *International Journal of Remote Sensing* 25:1565–1596.

EEA-ETC/TE. 2002. CORINE Land Cover update. I&CLC2000 project. Technical guidelines, Available at: http://terrestrial.eionet.eu.int.

Feranec, J. 1999. Interpretation element "association": Analysis and definition. *International Journal of Applied Earth Observation and Geoinformation* 1:64–67.

Feranec, J., Cebecauer, T., Otahel, J. 2005. *Photo-to-photo interpretation manual* (revised). Institute of Geography, Slovak Academy of Sciences, Bratislava, BIOPRESS document, biopress-d-13-1.3. 104 pp.

Feranec, J., Hazeu, G., Christensen, S., Jaffrain, G. 2007. Corine Land Cover change detection in Europe. Case Studies of the Netherlands and Slovakia. *Land Use Policy* 24:234–247.

Feranec, J., Otahel, J. 2001. *Land cover of Slovakia*. Bratislava: VEDA.

Giri, Ch. 2012. Brief overview of remote sensing of land cover. In Ch. P. Giri (Ed.), *Remote sensing of land use and land cover: Principles and applications* (pp. 3–12). Boca Raton, FL: CRC Press.

Heymann, Y., Steenmans, Ch., Croissille, G., Bossard, M. 1994. *CORINE Land Cover. Technical guide*. Luxembourg: Office for Official Publications of the European Communities.

Jones, B. Ch. 1997. *Geographical information systems and computer cartography*. Essex: Longman.

Lo, C. P., Noble, W. E. 1990. Detailed urban land-use and land-cover mapping using large format camera photographs: An evaluation. *Photogrammetric Engineering and Remote Sensing* 52:197–206.

Nunes de Lima, M. V. (Ed.). 2005. *IMAGE2000 and CLC2000: Products and methods*. Ispra, Italy: European Commission and Joint Research Centre.

Pouliot, D., Latifovic, R., Olthof, I., Fraser, R. 2012. Supervised classification approaches for the development of land-cover time series. In Ch. P. Giri (Ed.), *Remote sensing of land use and land cover: Principles and applications* (pp. 177–190). Boca Raton, FL: CRC Press.

Rogan, J., Chen, D. 2004. Remote sensing technology for mapping and monitoring land-cover and land-use change. *Progress in Planning* 61:301–325.

Schneider, S. 1974. *Luftbild und Luftbildinterpretation*. Berlin: Walter de Gruyter.

Schott, R. J. 1997. *Remote sensing: The image chain approach*. New York: Oxford University Press.

Seger, M. 1989. Landnutzungsanalyse aufgrund einer Farbinfrarot-Orthophotokarte. *Mitteilungen der Österreichischen Geographischen Gesellschaft* 131:5–26.

Steenmans, Ch., Perdigao, V. 2001. Update of the CORINE Land Cover database. In G. Groom and T. Reed (Eds.), *Strategic landscape monitoring for the Nordic countries* (pp. 101–107). Copenhagen: Nordic Council of Ministers.

Treitz, P., Rogan, J. 2004. Remote sensing for mapping and monitoring land-cover and land-use change—An introduction. *Progress in Planning* 61:269–279.

6

Accuracy Assessment of CLC Data

György Büttner, Barbara Kosztra, and Gergely Maucha

CONTENTS

6.1 Introduction

Quality control (QC) and quality assurance (QA) have been key elements of the European CoORdination of Information on the Environment (CORINE) Land Cover (CLC) projects since the early days of CLC inventories (Heymann et al., 1994; Bossard et al., 2000). We distinguish two different QA/QC activities:

1. *Verification* has a corrective purpose and is implemented as part of production. Through its feedback loops, verification is a tool for geometric standardization and thematic harmonization of CLC throughout Europe.

2. *Validation* aims to assess the accuracy of the final database and is implemented after completion of production. Validation should be fully independent from production and rely on data with higher spatial resolution than those used in production.

The most important documents providing guidance on CLC mapping and CLC change (CLCC) mapping are CLC nomenclature guidelines (Bossard et al., 2000), CLC2006 technical guidelines (Büttner and Kosztra, 2007), and the *Manual of CORINE Land Cover Changes* (Büttner and Kosztra, 2011).

The objective of this chapter is to present the practice of the European verification of CLC data, focusing on current practice in the CLC2012 project (Büttner et al., 2014), and to summarize the results of existing European validation studies.

6.2 Verification of CLC Data

6.2.1 Organizing Verification

Since the start of CLC inventories, the number of participating countries grew to 39 in CLC2012. An important element of CLC project management on behalf of the European Environment Agency (EEA) is quality control of the work done by national teams. Usually two verifications by EEA and/or European Topic Centre on Spatial Information and Analysis (ETC/SIA*) experts are organized for each country during a CLC inventory. In previous inventories, two on-site verification visits were organized for each participating country. However, owing to improved availability of digital in situ data and the short (1.5 years) implementation time of CLC2012, many of these visits have been replaced by remote verification actions (data sent by countries for remote checking). In countries working with regional teams (e.g., Italy and Spain) all regions are verified separately.

- The first verification is done when 10%–20% of the country area is interpreted. The main purpose is to reveal problems and provide feedback in the early phase of implementation to influence further production.
- The second verification is due when approximately 75% of the country area is interpreted. The main purpose is to check the database close to completion and suggest overall improvements if still needed. Working units for checking are selected by the EEA/ETC.

Verification (on-site or remote) of CLC and CLCC (in the CLC2012 project meaning usually the coverage of revised CLC2006 and CLCC$_{2006-2012}$ databases) is done visually, based on the following materials:

- Satellite images (IMAGE2006, IMAGE2012) used to derive CLC and CLCC
- Optionally topographic maps or links to the Web Map Service (WMS) implementations

During the last two CLC inventories, frequently Google Earth (GE) imagery has been the only ancillary/in situ data available. GE coverage has improved recently in most parts of Europe and it supports verification of CLC and also CLCC in an efficient way.

A custom-made software tool, InterCheck (Büttner et al., 2002a; Taracsák, 2012), serves as a software platform to carry out the verification. Quality remarks based on visual control are provided in point coverages linking the thematic comment to a specific location.

---

* Formerly European Topic Centre for Land Use and Spatial Information (ETC/LUSI) and European Topic Centre on Terrestrial Environment (ETC/TE).

Remarks related to the revised CLC and CLCC datasets are provided in separate files. The result of the verification of a working unit is expressed in qualitative terms:

- **A** (accepted): Only minor problems were found.
- **CA** (conditionally accepted): There are more problems but they are relatively easy to correct; after corrections the working unit is accepted.
- **R** (rejected): There are many mistakes in the database (incorrect application of the nomenclature, omitted changes, false changes, etc.), which require considerable work to correct.

Usually, altogether about 5%–25% of the area of countries is checked in the course of the two verification missions, depending on the area of the country, number of photointerpreters, and the quality of the work.

6.2.2 Implementing Verification

Because of the proposed "change mapping first" updating methodology (Büttner and Kosztra, 2007) the revised CLC2006 and the $CLCC_{2006-2012}$ layers are verified as two outputs of CLC2012 production. In some countries (e.g., Finland and Germany) where the revised CLC2006 is not produced because of the specific methodology, as an exception the CLC2012 layer is verified instead of the revised CLC2006. Verification is carried out by sporadic visual checking of the revised CLC2006 (or CLC2012) layer and detailed checking of the $CLCC_{2006-2012}$ layer.

- Checking of the revised CLC2006 concentrates on thematic and geometric properties of the CLC2006 layer, that is, if the nomenclature is applied properly and if delineation is precise enough. Basically, the revised CLC2006 will determine the quality of CLC2012, because $CLCC_{2006-2012}$ covers only a small percentage of the area.
- Checking $CLCC_{2006-2012}$ concentrates on two issues: (1) Do the mapped changes represent real change processes in the environment? (2) Are there any omitted changes?

Verification does not provide any quantitative accuracy figures, just a qualitative evaluation to influence the further production process. The verification steps are

- Automated technical conformity checking of both layers to reveal technical mistakes (e.g., smaller than the minimum mapping unit [MMU] polygons, invalid codes) and topological mistakes (e.g., a hole in the revised CLC2006 database, overlapping and multipart polygons).
- Visual checking of revised CLC2006 layer, including (1) checking CLC2006 statistics (to reveal nonrelevant codes) and (2) visual evaluation of "critical" classes (sampled according to CLC code, polygon size range, etc.).
- Visual checking of the $CLCC_{2006-2012}$ layer means
  - Checking CLC change statistics (to reveal nonrelevant changes; see Figure 6.1).
  - Depending on the amount of changes visual checking of (1) all changes or (2) just a subset of changes. Owing to the large number of potential change

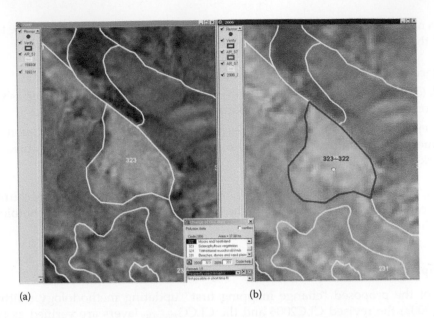

(a)   (b)

FIGURE 6.1
Example of verification of CLC2006. (a) CLC2000 (yellow lines) and IMAGE2000. (b) $CLCC_{2000-2006}$ (magenta) and IMAGE2006. The change polygon is proposed to delete. Moors and heathland (322) is characteristic of Atlantic climate, while sclerophyllous shrub (323) is typical in the Mediterranean. Change between them in 6 years is not possible.

types changes are grouped into major change process types (e.g., urban sprawl, changes of agriculture plantations, forest management changes [Büttner and Kosztra, 2011]). Change process types with the largest area are always checked. For each verified change polygon both the delineation and the coding is checked. Coding should reflect real change processes as visible on images.

- Looking for omitted changes.

6.3 Validation of CLC Data

The objective of the validation of CLC data is to derive thematic accuracy figures based on independent, very high resolution in situ data as reference. The date of in situ data should coincide as much as possible with the acquisition date of satellite images used to derive CLC. In situ data—especially thematic datasets—used in validation must not have been used in compilation of the CLC databases.

Validation is carried out either on the national or the European level. Some countries have performed a national validation of their CLC database, using country-specific approaches. In most cases spatially and thematically more detailed national data were used for that purpose (Hazeu, 2003; Caetano et al., 2006; Aune-Lundberg and Strand, 2010).

Validation in the CLC context means thematic validation. The aim of validating European CLC is to assess its thematic accuracy by means of a statistical method. Major challenges related to validation of European CLC are (1) to gain access to appropriate in situ (reference) data with European coverage to support the validation and (2) to validate CLC classes covering small area in Europe. In Section 6.3.1 two European validation exercises are presented:

1. Validation of CLC2000 by means of Eurostat Land Use/Cover Area Frame Statistical Survey (LUCAS) data (Büttner and Maucha, 2006; Maucha and Büttner, 2006)
2. Validation of CLCC$_{2000-2006}$ data by means of GE imagery (Büttner et al., 2010)

6.3.1 Validation of CLC2000 by Means of LUCAS Data

The aim of the exercise was to assess thematic quality of CLC2000 data.

6.3.1.1 Databases Used

Two European databases played a role in validating CLC2000:

- IMAGE2000 data meant Landsat 7 satellite imagery, taken in the period 1999–2001 covering countries participating in the CLC2000 project. The ortho-rectified satellite images in national projection provided the basis of deriving CLC2000.
- LUCAS data: Eurostat coordinated the national LUCAS surveys in 2001–2003 covering 18 countries (Avikainen et al., 2003; Duhamel et al., 2003). The total land area was divided into grids of 18 km (Primary Sampling Units [PSUs]). LUCAS land cover (LLC) and LUCAS land use (LLU) data were collected on the Secondary Sampling Units (SSUs) distributed along PSUs (Figure 6.2). Four field photographs (N, S, E, and W directions) were taken around the central SSU. The most important characteristics of CLC and LUCAS are compared in Table 6.1. LUCAS fulfilled most of the criteria for validation data because of the following characteristics: (1) higher geometric accuracy than used in CLC mapping; (2) higher thematic accuracy because LC/LU data were derived usually by field observations; (3) nearly coincident data acquisition with IMAGE2000; and (4) independent, as not used during production of CLC2000.

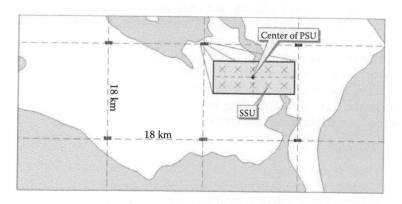

FIGURE 6.2
Two-stage sampling of the LUCAS (Avikainen et al., 2003). The sampling scheme of LUCAS 2001–2002 was not continued by EUROSTAT in later LUCAS surveys.

TABLE 6.1

Main Parameters of CLC2000 and LUCAS 2001–2002

	CORINE Land Cover	LUCAS
Coverage	EU25, Bulgaria, Croatia, Liechtenstein, and Romania	EU15, Estonia, Hungary, and Slovenia
Database characteristics	Single land cover with land use elements	Separate LU and LC; field photographs
Method of database production	Mapping based on satellite images and in situ data	Area frame sampling in 18 × 18 km grid (PSU); 10 sampling points (SSU) in each PSU
Nomenclature	LC (with some land use elements); 44 classes in 5 major groups	LC: 57 classes in 7 groups, emphasis on agriculture Land use: 14 classes in 4 groups
Observation unit	25-ha MMU, minimum width 100 m	7 m^2 (0.1 ha = 1000 m^2 in exceptional cases)
Coding	Single Level 3 code in each polygon	One (or two) LC code; one (or two) LU code in each SSU plus additional environmental information
Geometric accuracy specification	Better than 100 m	1–3 m
Thematic accuracy specification	Better than 85%	Na

Source: Büttner, G. and Maucha, G. *The thematic accuracy of CORINE Land Cover 2000. Assessment using LUCAS.* Technical Report 7. Copenhagen: European Environment Agency, 2006.

6.3.1.2 *Methodology*

For validation of CLC2000 two different methods were applied, which measure different kinds of accuracy:

1. Automatic comparison of CLC2000 and LUCAS LC/LU codes
2. Reinterpretation of IMAGE2000 with the help of LUCAS data (LLU and LLC codes and photographs) following interpretation rules of CLC

6.3.1.2.1 *Ad 1: Automatic Comparison*

The nomenclatures of CLC and LUCAS, although similar, are not directly comparable. Therefore as first step a correspondence table between CLC and LLC classes was created. As CLC is not a pure LC nomenclature, not only the LLC, but also the LLU information needed to be considered during comparison. Therefore similarity tables were constructed between CLC and LLU classes. The automatic comparison in these tables resulted in the degree of agreement between the two databases. Each SSU was checked, whether the CLC code corresponded to the appropriate LLC and LLU classes according to the correspondence table. The agreement was considered positive if both the LC and LU values corresponded to the CLC data according to the tables. Finally, the agreement cases were summed and the percent total agreement (PTA) was computed.

6.3.1.2.2 *Ad 2: Reinterpretation*

Validation was based on a reinterpretation of IMAGE2000 around the LUCAS PSU, using also the reference information provided by LUCAS landscape photographs as well as the LLC and LLU information. The main advantages of this method are: (1) CLC generalization rules and complex class definitions can be considered; (2) LLC and LLU codes in the 1200-meter × 300 meter area surrounding the central SSU (where the photos were taken) provide the spatial context; and (3) field photographs provide in situ data for reinterpretation.

Results of this method can be considered as a thematic reliability estimation of CLC2000, as CLC2000 data are compared to an independent CLC interpretation. The reinterpretation included three steps.

1. First, the validation expert visually interpreted the location around the central LUCAS sampling point by using IMAGE2000 data, LLU and LLC data, and landscape photographs. No information from the CLC database was shown at this time. CLC interpretation rules had to be respected when selecting the CLC code (first CLC control code). After confirmation the code could not be changed.

2. Second, the CLC polygon boundaries were also displayed on top of IMAGE2000 and LUCAS points, but still without CLC class information. This situation was interpreted again by assigning the second CLC control code. After confirmation the code could not be changed.

3. In the third step the actual CLC code was also displayed. By comparing the actual CLC code and the control codes the validating expert had to evaluate the situation as one of the following cases: (a) clear agreement, (b) another interpretation is possible (the interpretation was still acceptable), (c) wrong interpretation, or (d) not enough information is available (e.g., cloudy IMAGE2000, poor quality LUCAS photos, etc.).

6.3.1.3 Estimating the Uncertainty of Accuracy Values

A certain number of samples are needed to be able to provide representative estimation of the parameters of accuracy. The binomial distribution as a statistical model is used to calculate the standard deviations for error values. Uncertainty of accuracy values is expressed by the confidence interval of the binomial distribution. The confidence interval may be approximated by the confidence interval of the normal distribution.

$$\hat{p} \pm z \cdot \sqrt{\frac{1}{n} \cdot \hat{p} \cdot (1-\hat{p})}$$

where n is the number of all samples and $\hat{p}$ is the probability of the true accuracy (approximated in the practice as the number of agreements divided by the number of all valid samples). In the case of a 95% confidence interval the coefficient $z = 1.96$ (Wallis, 2013). Note that the binomial distribution function is not symmetrical (unlike the normal distribution); therefore the preceding expression is a simple but not perfect approximation of the confidence interval of a binomial distribution.

Although the binomial distribution may be a good approximation for the PSU level LUCAS sampling (see Section 6.3.1.2.2), it is biased on the SSU level (see Section 6.3.1.2.1), because the SSU samples cannot be considered as totally independent.

6.3.1.4 Results

6.3.1.4.1 Ad 1: Automatic Comparison

CLC2000 and LUCAS LU&LC were compared for 18 countries and analyzed using 99,936 SSUs by means of the correspondence table. The benefit of this method is the fast, automatic computation by using a large number of samples.

TABLE 6.2

Results of the Automatic Comparison of CLC2000 and LUCAS

SSUs	LC Agreed	LU Agreed	PTA (CLC and LUCAS Agreed)
99,936	80,067	88,626	75,598
100%	80.1% ± 0.2%	88.7% ± 0.2%	75.6% ± 0.2%

TABLE 6.3

Results of the Reinterpretation of IMAGE2000 Using LUCAS

PSUs	Clear Agreements	Wrong Interpretation	Another Interpretation Possible	Acceptable CLC2000 Classification
8115	6681	1057	377	7058
100%	82.3% ± 0.8%	13.0% ± 0.7%	4.6% ± 0.5%	87.0% ± 0.7%

The PTA between CLC2000 and LUCAS LC&LU (meaning how CLC2000 approximates the reality) is 75.6% ± 0.2% (Table 6.2). With regard to the generalization effect of the 25-ha MMU of CLC this result can be considered rather satisfactory.

6.3.1.4.2 Ad 2: Reinterpretation

Using the methodology of reinterpretation altogether 8115 samples were used for the calculation of accuracy figures (Table 6.3).

The main results of the reinterpretation are as follows:

- The reliability of CLC2000 is 87.0% ± 0.7%. This value is based on an independent CLC interpretation performed around LUCAS PSUs. It can be concluded that the 85% accuracy requirement specified in *CORINE Land Cover Technical Guide* (Büttner et al., 2002b) has been fulfilled.

- The highest Level 3 reliability (>95%) was obtained for Water courses (511), Water bodies (512), Industrial and commercial units (121), and Discontinuous urban fabric (112).

- The two largest CLC classes—Nonirrigated arable land (211) and Coniferous forest (312)—were estimated with high reliability (between 90% and 95%). Similar reliabilities were found for Agroforestry areas (244) and Permanently irrigated arable land (212).

- The lowest reliabilities were obtained for Sparsely vegetated areas (333), Fruit trees and berry plantations (222), and Land principally occupied by agriculture, with significant areas of natural vegetation (class 243), highlighting the difficulties in interpreting these classes.

- The majority (78%) of classification errors occurred on Level 3 and Level 2 of the CLC nomenclature. Level 1 misclassifications occurred mostly between "agriculture" and "forest and semi-natural" classes.

- About half of the 44 CLC classes could not be validated either because of low representativity* or because of intentional omission by LUCAS (glaciers and coastal water areas).

---

* Small CLC classes having occurrences around 0.1%. Many of the classes in the artificial surfaces group belong here.

6.3.2 Validation of CLC Changes by Means of GE Data

The aim of this exercise was to assess the thematic quality of the European $CLCC_{2000-2006}$ database, through independent interpretation and comparison with the CLC change layer. The challenge was how to handle the low rate of changes (see Section 6.3.2.2).

6.3.2.1 Databases Used

In the optimal case the materials supporting validation would be very high resolution (VHR) satellite imagery or ortho-photos taken in the years 1999–2001 and 2005–2007 at a resolution better than that of IMAGE2000 and IMAGE2006, respectively and topographic maps at scale 1:50,000 or better. However, the large number of participating countries (39) and access right issues made it unrealistic to collect VHR imagery and topographic maps. Therefore reinterpretation of IMAGE2000 and IMAGE2006 was done supported by the use of GE imagery. GE was a useful support, especially because of its time-series feature (Figure 6.3).

6.3.2.2 Methodology

Stratified random point sampling was selected to compile a list of samples for independent interpretation and comparison with the $CLCC_{2000-2006}$ layer. Owing to a very large number of change types in the European $CLCC_{2000-2006}$ database—more than 900 different Level 3 change types (Büttner et al., 2010)—it was not possible to test all of them. Therefore some kind of selection or grouping of Level 3 changes had to be done.

Two sampling schemes were implemented:

- Maximum 100 randomly placed sampling points were selected for each of the 25 CLC Level 1 change types, to estimate the commission error for $CLCC_{2000-2006}$, grouped according to Level 1 change types. In this exercise the entire population of CLCC polygons was sampled.
- Additional samples were selected (about 100 samples for each case) for a number of Level 3 change types of special interests. The concept of Land Cover Flows (LCFs) (Haines-Young and Weber, 2006) and the European $CLCC_{2000-2006}$ statistics were used to select the change types to be considered. In this exercise, 10 Level 3 change types covering about half of all $CLCC_{2000-2006}$ polygons were sampled.

(a)

(b)

FIGURE 6.3
High-resolution GE images provided useful basis for validation. (a) September 2003. (b) September 2006. The class arable land was changed to construction site (Turkey).

Although the estimation of commission error is a feasible task, the statistically representative estimation of the amount of omitted changes is not feasible without appropriate stratification. Between 2000 and 2006, 1.24% of the area of participating countries has changed in Europe. This is distributed among the 25 Level 1 change types and among the more than 900 Level 3 change types (Büttner et al., 2010). Without stratification, 98.76% of Europe's territory, that is, the nonchanged area, would have had to be examined in search for omitted changes via random sampling, requiring an extraordinarily high number of samples. Lacking a method for appropriate stratification, the omission error of the $CLCC_{2000-2006}$ database was therefore not estimated.

Validation points were interpreted in a way similar to that described in Section 6.3.1.2. The reinterpretation of each sample concluded with the evaluation of the mapped change as "correct" or "not correct." The response "other" could be chosen if it was not possible to make a decision (i.e., missing or poor quality images). For the "not correct" case one of the following standard explanations had to be provided: (1) no change; (2) no change, temporal difference only; (3) no change because CLC2000 code was not corrected; (4) no change and CLC2006 code is not correct; and (5) change exists, but with different attributes.

6.3.2.3 Accuracy of CLC Level 1 Changes

Accuracy figures for the 25 Level 1 changes were calculated together with a standard deviation of estimation. The accuracy refers to cases when the change was found by the original photointerpreter and the given attributes were correct. The overall accuracy of $CLCC_{2000-2006}$ database (commission errors only) is 87.8%, that is, exceeds the target value of 85%.

- 17 of the 25 Level 1 change types have an accuracy higher than 85%. The far largest change type is the *internal changes in Forest/semi-natural Level 1 class* (dominated by forest clear-cuts and forest growth). Almost three-fourths of all changes belong to this type; therefore its 89.5% accuracy is an important finding.

- Another important change type with >85% accuracy: *Agriculture changed to Forest/ semi-natural area* and *internal changes within Artificial areas* (e.g., construction site turned to residential area).

- There are eight Level 1 change types on the lower end of the accuracy list (accuracy below 80%). Important change types included *Forest/semi-natural area changed to Artificial surface* (e.g., new constructions in forest) and *Forest/semi-natural area changed to Agriculture*.

6.3.2.4 Accuracy of Dominant Level 3 Changes

Considering the 10 Level 3 change types selected as "flagships" of major LCFs it was found that all but two change types were above the 85% limit. The following five change types have extra high (above 90%) accuracy (the name of the relevant LCF is given in parentheses):

- *Arable land changed to Construction site* (part of "urban residential sprawl")
- *Arable land converted to Water body* ("creation of new water bodies")
- *Coniferous forest burnt* ("changes of LC due to natural and multiple causes")

TABLE 6.4

Accuracy Figures for Selected Level 3 CLC Changes (Commission Error Only)

CLC2000 Class	CLC2006 Class	No. of Samples	Accuracy (%)	Uncertainty (%)	Size of Change Class (% of total changes)
211	133	101	96.04	3.80	1.08
211	512	100	96.00	3.84	0.20
312	334	96	93.75	4.84	0.50
133	112	100	93.00	5.00	0.71
312	324	110	92.73	4.85	34.21
211	324	100	83.00	7.36	1.11
211	112	96	82.29	7.64	0.97
211	231	100	82.00	7.53	1.30
324	312	99	76.77	8.32	8.25
231	211	97	76.29	8.46	1.34
Total:	–	999	–	–	49.67

- *Construction of residential area finished* ("internal transformation of urban areas")
- *Coniferous forest changed to Transitional woodland-shrub* ("forest creation and management"). This is the largest Level 3 change, providing more than one-third of area of all CLC changes

Two change types were mapped with an accuracy lower than 85%:

- *Growth of Coniferous forests* ("forest creation and management"). This change type concerns rather significant areas in Europe. Mapping it consistently is difficult without proper in situ data.
- *Pasture/set-aside land changed to Arable land* ("agriculture internal conversions"). Mapping it consistently is difficult without proper in situ data, for example, agricultural subsidy control datasets (IACS*).

Level 3 changes presented in Table 6.4 represent almost 50% of area of all CLCC. Because not the whole CLC change polygon population was sampled in this exercise, overall accuracy was not calculated.

6.3.3 Conclusions

Validation of CLC and CLCC data is not an easy endeavour. Finding appropriate reference data with European coverage poses a serious difficulty. In the two examples presented LUCAS data and GE imagery were used, being optimal for validation. LUCAS, on the one hand, gives preference to agriculture areas in sampling design, causing a shift in thematic focus. Moreover, LUCAS is available only for European Union member states. GE imagery, on the other hand, is not equally available for all types of landscapes. VHR images are usually rarely provided for areas with few inhabitants (e.g., above the Polar Circle). The potential alternative solution would be to collect ortho-photos with European coverage,

---

* Integrated Administration and Control System. IACS is the most important system for the management and control of payments to farmers made by the member states in application of the Common Agricultural Policy (IACS, 2014).

but currently it is time consuming and costly and therefore not feasible. In the future an appropriate European VHR coverage (acquired during the vegetation season) could solve this problem.

Currently automating the validation process is possible only to a limited extent (see Section 6.3.1.2). Owing to the complex class definitions and the lack of European-wide in situ data, sampling points have to be interpreted visually. This makes validation time-consuming and expensive. The improvement foreseen in availability of in situ data in the participating countries resulting from the Infrastructure for Spatial Information in the European Community (INSPIRE) process might in the longer term allow replacing the expensive photointerpretation with an automated procedure in validating CLC and CLCC.

Although at present the estimation of commission error is feasible for CLC as well as for CLCC, the statistically representative estimation of omission errors for the small CLC classes as well as for CLCC is hindered by the large sample size required. The anticipated future improved availability of proper in situ information will help to overcome this problem because the sample size will no longer be a limiting factor.

CLC2000 having been validated on the European level, validation was not repeated for the next update (CLC2006) because of the difficulties mentioned in the preceding text. It must, however, be noted that owing to low rate of CLCC, CLC2006 is very similar to CLC2000, and therefore its validation probably would have provided results similar to those of CLC2000. CLC2012 datasets (CLC2012 as well as $CLCC_{2006-2012}$) produced under GIO land has been validated in 2015.

References

Aune-Lundberg, L., Strand, G.-H. 2010. CORINE Land Cover 2006. The Norwegian CLC2006 project. Report from the Norwegian Forest and Landscape Institute, 11. Ås: Norwegian Forest and Landscape Institute.

Avikainen, J., Delincé, J., Croi, W., Kayadjanian, W., Bettio, M. Mariano, A. 2003. LUCAS Land Use/Cover Area Frame Statistical Survey. Technical Document No. 1: Sampling plan. Luxembourg: Eurostat.

Bossard, M., Feranec, J., Otahel, J. 2000. *CORINE Land cover technical guide*. Addendum 2000. Technical Report 40. Copenhagen: European Environment Agency.

Büttner, G., Feranec, J., Jaffrain, G. 2002b. *CORINE Land Cover update 2000. Technical guidelines*. Technical Report 89. Copenhagen: European Environment Agency.

Büttner, G., Kosztra, B. 2007. *CLC2006 technical guidelines*. Technical Report 17. Copenhagen: European Environment Agency.

Büttner, G., Kosztra, B. 2011. *Manual of CORINE Land Cover changes*. Copenhagen: European Environment Agency, European Topic Centre on Spatial Information and Analysis.

Büttner, G., Kosztra, B., Maucha, G., Pataki, R. 2010. Implementation and achievements of CLC2006. Final report to EEA. Copenhagen: European Topic Centre on Spatial Information and Analysis.

Büttner, G., Maucha, G. 2006. *The thematic accuracy of CORINE Land Cover 2000. Assessment using LUCAS*. Technical Report 7. Copenhagen: European Environment Agency.

Büttner, G., Maucha, G., Taracsák, G. 2002a. Inter Change: A software support for interpreting land cover changes. In T. Benes (Ed.), *Geoinformation for European-wide interpretation. Proceedings of the 22nd EARSeL Symposium* (pp. 93–98). Rotterdam: Millpress.

Büttner, G., Soukup, T., Kosztra, B. 2014. *CLC2012—Addendum to CLC2006 technical guidelines.* Copenhagen: European Environment Agency, European Topic Centre on Spatial Information and Analysis.

Caetano, M., Mata, F., Freire, S. 2006. Accuracy assessment of the Portuguese CORINE Land Cover map. In A. Marcal (Ed.), *Global developments in environmental Earth observation from Space. Proceedings of the 25th EARSeL Symposium* (pp. 459–467). Rotterdam: Millpress.

Duhamel, C., Eiden, G., Aifantopoulou, D., Croi, W. 2003. LUCAS Land Use/Cover Area Frame Statistical Survey. Technical Document No. 2: The Nomenclature. Luxembourg: Eurostat.

Haines-Young, R., Weber, J.-L. 2006. Land accounts for Europe 1990–2000. Towards integrated land and ecosystem accounting. EEA Report 11. Copenhagen: European Environment Agency.

Hazeu, G. W. 2003. CLC2000. Land Cover database of the Netherlands. Alterra Rapport, 775. Wageningen: Alterra.

Heymann, Y., Steenmans, C., Croissille, G., Bossard, M. 1994. *CORINE Land Cover. Technical guide.* Luxembourg: Office for Official Publications of the European Communities.

IACS. 2014. Integrated Administration and Control System (IACS). Available at http://ec.europa.eu /agriculture/direct-support/iacs/index_en.htm.

Maucha, G., Büttner, G. 2006. Validation of the European CORINE Land Cover 2000 database. In A. Marcal (Ed.), *Global developments in environmental Earth observation from space. Proceedings of the 25th EARSeL Symposium* (pp. 449–457). Rotterdam: Millpress.

Taracsák, G. 2012. CLC 2012 support package. Available at http://clc2012.taracsak.hu/.

Wallis, S. A. 2013. Binomial confidence intervals and contingency tests: Mathematical fundamentals and the evaluation of alternative methods. *Journal of Quantitative Linguistics* 20(3):178–208.

7

CORINE Land Cover Products

György Büttner

CONTENTS

7.1 Introduction

In accordance with legal provisions (see Chapter 8), CoORdination of INformation on the Environment (CORINE) Land Cover (CLC) data are disseminated both by the European Environment Agency (EEA) and the member states. As national products might vary in geometric and thematic resolution, the following introduction lists only the standard CLC products disseminated by the EEA (EEA, 2014).

7.2 European CLC Products

CLC data are produced by participating countries (39 in CLC2012) in vector format in a national projection and with a minimum mapping unit (MMU) of 25 ha. The CLC Change (CLCC) product has higher resolution: MMU = 5 ha. In the course of the European data integration national data are converted to the European Terrestrial Reference System 1989 (ETRS89), the Lambert Azimuthal Equal Area (LAEA) projection (epsg: 3035), and merged to seamless European vector database (Büttner et al., 2004) (see list below).

Technical Characteristics of Standard, Primary CLC Data

MMU, CLC	125 ha
MMU, CLCC	15 ha
Minimum mapping width, CLC and CLCC	100 m
Nomenclature	Standard European Level 3, 44 classes (Bossard et al., 2000)
Projection	National (country level), ETRS89 LAEA
Type	Vector, polygon topology

TABLE 7.1

European CORINE Land Cover Products

Products	Type	Characteristics
CLC1990, CLC2000, CLC2006, CLC2012	Vector	44 classes; simplified border matching along country boundaries
CLC1990, CLC2000, CLC2006, CLC2012	Raster	Rasterized vector product. 44 classes, 100-m and 250-m grid
CLC-Change$_{1990-2000}$ CLC-Change$_{2000-2006}$ CLC-Change$_{2006-2012}$	Vector	Change polygons have two attributes: CLC code for year$_1$ and CLC code for year$_2$
CLC-Change$_{1990-2000}$ CLC-Change$_{2000-2006}$ CLC-Change$_{2006-2012}$	Raster	Two 100-m grids and two 250-m grids (formation code and consumption code)

CLC data in raster format are derived products. Primary European vector data are rasterized with 100-m and 250-m grid sizes. All of the previous CLC inventories are available for distribution in vector as well as in raster format, including the so-called status layers (CLC1990, CLC2000, CLC2006, and CLC2012) and the respective change layers (Table 7.1). CLC products are identified by a version number (as of January 2016 the latest is Version 18). A new version number indicates the increase of the coverage (new areas are mapped or a new inventory is produced*). Table 7.1 presents the European CLC data disseminated by the EEA Data Service (Büttner, 2014) and the Copernicus Land programme (Copernicus, 2015). The full time series (historical, actual, and future) are disseminated via land.copernicus .eu (Copernicus, 2015), while historical time series are downloadable also via eea.europa.eu (EEA, 2014). All datasets are accompanied by extensive metadata.

7.3 Using CLCC Products

Users interested in CLC changes should always rely on the corresponding CLCC product and never on the difference (intersect) of the two status layers. The CLCC product includes the real LC changes mapped at the higher resolution of 5 ha MMU, while the intersect includes the difference of two generalized lower resolution (25-ha MMU) datasets. The consequence of the "change mapping first" methodology (Büttner and Kosztra, 2007)— and eventually that of the different MMUs—is that the difference (intersect) between two consecutive status layers (e.g., CLC2000 and CLC2006) will differ from the corresponding CLCC layer (e.g., CLCC$_{2000-2006}$). The magnitude of difference depends on the size distribution of change polygons. If there are many changes in the size range of 5–25 ha, the difference can be significant. If all changes were larger than 25 ha, then there would be no difference. In addition, new CLC status layers include revisions (correction) of the previous status layer, which then cannot be distinguished from actual LC changes when intersecting two status layers.

---

* For example, the reason for producing the V17 was the inclusion of the new CLC1990, CLC2000, and CLC2006 data produced for the Azores Islands (PT). The next CLC version (V18) includes results of the CLC2012 inventory.

References

Bossard, M., Feranec, J., Otahel, J. 2000. *CORINE Land Cover technical guide*. Addendum 2000. Technical Report 40. Copenhagen: European Environment Agency.

Büttner, G. 2014. CORINE land cover and land cover change products. In I. Manakos and M. Braun (Eds.), *Land use and land cover mapping in Europe: Practices and trends* (pp. 55–74). Remote Sensing and Digital Image Processing 18. Dordrecht, the Netherlands: Springer.

Büttner, G., Feranec, J., Jaffrain, G., Mari, L., Maucha, G., Soukup, T. 2004. The CORINE Land Cover 2000 project. Available at: http://www.eproceedings.org/static/vol03_3/03_3_buttner2.pdf.

Büttner, G., Kosztra, B. 2007. *CLC2006 technical guidelines*. Technical Report 17. Copenhagen: European Environment Agency.

Copernicus. 2015. Copernicus Land Monitoring Services. Available at: http://land.copernicus.eu.

EEA. 2014. Global search on data, maps and indicators. Available at: http://www.eea.europa.eu /data-and-maps/find/global#c12=Corine+Land+Cover.

8

Product Dissemination Policy

György Büttner

CONTENTS

8.1 Introduction

Three documents are presented regulating the right of use of European Land Monitoring data managed by the European Environment Agency (EEA): (1) the EEA standard reuse policy (EEA, 2014); (2) legal provisions of the Grant Agreements between EEA and countries participating in Global Monitoring for Environment and Security (GMES) Initial Operations (GIO) Land (Sousa, 2014); and (3) Regulation (EU) No. 1159/2013 regarding the establishment of registration and licensing conditions for GMES users and defining criteria for restricting access to GMES dedicated data and GMES service information (EU, 2013).

8.2 General Product Dissemination Policy of the EEA

Rights of use of the European CoORdination of Information on the Environment (CORINE) Land Cover (CLC) products introduced in Chapter 7 are governed by the EEA standard reuse policy. The following is the citation of the EEA website (EEA, 2014): "Unless otherwise indicated, reuse of content on the EEA website for commercial or non-commercial purposes is permitted free of charge, provided that the source is acknowledged." The copyright holder is the EEA.

The EEA reuse policy follows Directive 2003/98/EC of the European Parliament and the Council on the reuse of public sector information throughout the European Union and Commission Decision 2006/291/EC, Euratom on the reuse of Commission documents. "The EEA accepts no responsibility or liability whatsoever for the reuse of content accessible on its website."

Any inquiries about reuse content on the EEA website should be addressed to copyright @eea.europa.eu.

8.3 Right of Use of GMES/Copernicus Products

Ownership and use of the results of the GIO Land monitoring 2011–2013 in the framework of regulation (EU) No. 911/2010—Pan-EU component action is regulated as follows in the Grant Agreements between the EEA and the countries (A. Sousa, personal communication, 2014):

- Any results of the action, including information and data (in particular $CLCC_{2006-2012}$ and CLC2012, including metadata; and the enhanced high-resolution layers, if applicable), produced in the framework of the action shall be the property of the European Union (EU).
- […] the EU grants the beneficiary the right to make free use of the results of the action, including information and data, free of charge and as it deems fit, and, in particular, to display, reproduce by any technical procedure, translate, or communicate the results of the action by any medium, provided it does not thereby breach the publicity obligations […].

8.4 Free Use of GMES/Copernicus Products

Further aspects of the open dissemination of the GMES/Copernicus products are found in the Regulation (EU) 1159/2013 (EU, 2013):

- Access to Sentinel* data should be free, full and open […].
- […] GMES should be considered as a European contribution to building the Global Earth Observation System of Systems (GEOSS). Therefore, the GMES open dissemination should be fully compatible with GEOSS data sharing principles.
- GMES dedicated data and GMES service information should be free of charge for the users to capitalize on the social benefits arising from an increased use of GMES dedicated data and GMES service information.

References

EEA. 2014. Legal notice. http://www.eea.europa.eu/legal/copyright.
EU. 2013. Commission Delegated Regulation (EU) No. 1159/2013.

---

* Sentinel 2 is a dedicated Copernicus satellite (Sentinel 2A was successfully launched on 23 June 2015). It is the first European satellite providing high-resolution Earth Observation data for the pan-European component of Copernicus Land Monitoring.

9

High-Resolution Layers

György Büttner, Gergely Maucha, and Barbara Kosztra

CONTENTS

9.1 Introduction

The most frequently expressed user criticisms regarding CoORdination of Information on the Environment (CORINE) Land Cover (CLC) are that (1) databases are not detailed enough; (2) classes of nomenclature contain a mixture of land cover (LC) and land use (LU) information (Arnold and Kosztra, 2013); (3) some of the classes (e.g., mixed agriculture) are difficult to translate to other systems, for example, the Food and Agriculture Organization of the United Nations Land Cover Classification System (FAO LCCS) (Herold et al., 2009); and (4) CLC data cannot be used for statistical comparisons with other surveys (Sjolbørg Flo Heggem et al., 2010).

The High-Resolution Layers (HRLs), as part of the pan-European component of the Global Monitoring for Environment and Security (GMES)/Copernicus Land (GIO Land) Project (European Union [EU], 2010; European Environment Agency [EEA], 2013a), are used in an attempt to overcome part of the aforementioned limitations by providing detailed information on LC instead of the landscape-level information given by CLC. This will afford the possibility of, for example, populating CLC polygons with more precise LC information (see Chapter 26). The strategy was to produce high-resolution LC information for all Level 1 layers of CLC, a goal that is partly fulfilled, as the thematic correspondence table shows (Table 9.1). Further HRLs will be needed to observe those LC classes that are not covered by the current HRLs, for example, arable land, shrub land, bare surfaces, and permanent ice. Other parameters of CLC and HRLs are compared in Table 9.2. The GIO Land Project includes also two other mapping components:

- Global (under the Joint Research Centre [JRC] in Ispra, Italy)
- Local (under the European Environment Agency [EEA]): Urban Atlas, Riparian Zones, Natura 2000, (these are ongoing), and Coastal Zone (under planning)

TABLE 9.1

Thematic Correspondence between CLC and HRLs

CLC Level 1 Code	CLC Level 1 Class Name	Corresponding HRL (as Part of the Level 1 CLC Class)
1	Artificial surfaces	Imperviousness (IMD)
2	Agricultural areas	Permanent grassland (GRA)
3	Forests and semi-natural areas	Tree cover density (TCD) and Forest type (FTY)
4	Wetlands	Wetlands (WET)
5	Water bodies	Permanent (inland) water bodies (PWB)

TABLE 9.2

Comparison of CLC and HRLs

	CLC	HRL
Resolution	25 ha (e.g., 500 m × 500 m)	20 m × 20 m (intermediate product) 100 m × 100 m (final product)
Means of service provision	Bottom-up (by national teams)	Top-down (by service providers with contribution by national teams)
Classification	Computer-assisted photointerpretation (in majority of countries)	Computer classification followed by manual postprocessing
Basic input data used	IMAGE2006, IMAGE2012, and national in situ data	IMAGE2006 IMAGE2009 and IMAGE2012 (GRA,WET, PWB) VHR imagery (TCD,FTY) AWiFS imagery (GRA,WET, PWB)
Planned update cycle	6 years	3 years
Primary data format	Vector	Raster

9.2 Derivation of HRLs

Service Providers (SPs) produce the six intermediate thematic layers (see Table 9.3) at 20 m pixel resolution in the national projection (Figure 9.1). Depending on the country's choice, either the national team verifies the intermediate product by using national databases, or in the absence of the country's willingness the SP in charge carries out the verification (EEA, 2013b).

In the next step the countries are encouraged to enhance the intermediate layers (EEA, 2013c); otherwise, it is carried out by the SP. Enhancement means correcting deficiencies of the automatic satellite image classification by removing major commission errors and adding information to the HRLs in the case of important omissions. Enhancement should remain at the scale of classification provided by the satellite imagery used (20 m pixel size).

To derive final products, enhanced intermediate data are transformed from national to European projection (ETRS89 Lambert Azimuthal Equal-Area [LAEA]), mosaicked, and aggregated to 100 m resolution. Final European mosaics are available on the Copernicus land portal (Copernicus Land, 2015). Final data in the national projection are accessible from the countries.

TABLE 9.3

Summary of HRL Products

Product Name	Short Name	Intermediate Product (20 m) National Projection	Final Product (100 m) European Projection	Expected Accuracy, Final Product (%)
Degree of Imperviousness 2012	IMD	Imperviousness 0–100%	Imperviousness 0–100%	90
Tree Cover Density 2012	TCD	Tree cover density 0–100%	Tree cover density 0–100%	90
Forest Type 2012	FTY	Dominant leaf type: Broadleaved, coniferous forests and nonforest	Dominant leaf type: Broadleaved, coniferous, mixed forests and nonforest	90
Permanent Grassland 2006–2009–2012	GRA	Permanent grassland and no permanent grassland (binary map)	Occurrence of permanent grassland 0–100%	80
Wetlands 2006–2009–2012	WET	Wetland and non-wetland (binary map)	Occurrence of wetlands 0–100%	80
Permanent Water Bodies 2006–2009–2012	PWB	Water and non-water (binary map)	Occurrence of water bodies 0–100%	90

Source: EEA (European Environment Agency). GIO land High Resolution Layers. Available at: http://land .copernicus.eu/user-corner/publications/gio-land-high-resolution-layers/view. 2013a.

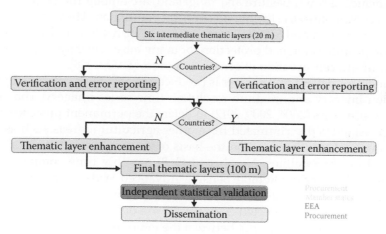

FIGURE 9.1

Flowchart of the derivation of high-resolution layers. (Dufourmont, personal communication, 2011.)

9.2.1 Degree of Imperviousness 2012

The Degree of Imperviousness 2012 (IMD) layer aims to map surfaces where the original (semi-) natural cover or water surface is replaced by an artificial, often impervious cover (Di Gregorio and Jansen, 2005). The high-resolution imperviousness dataset representing the artificially sealed areas is produced using automated processing of calibrated Normalized Difference Vegetation Index (NDVI) data derived from IMAGE2012 data. A per-pixel estimate of sealed soil is derived as the index for the degree of imperviousness (0–100%). Similar data for 2006 and 2009 were produced in the frame of GMES precursor activities and Geoland 2 (Geoland 2, 2010), respectively. These are used to achieve full consistency of the 100-m resolution time series products (J. Weichselbaum, personal communication, 2012).

9.2.2 Tree Cover Density and Forest Type

The Tree Cover Density (TCD) product includes information on all kinds of trees (forest trees and nonforest trees). The Forest Type (FTY) product includes only forest trees (coniferous, deciduous, mixed), and nonforest trees (trees in urban context, trees in agriculture plantations) are excluded by following the FAO forest definition (FAO, 1998).* The primary forest product is the TCD (density range: 0–100%).

FTY/dominant leaf type is produced from TCD layer by filtering and classification. The definition of FTY largely follows the FAO definition (FAO, 1998):

- TCD is higher than 10%, meaning that a low TCD does not mean forest.
- Minimum mapping unit (MMU) = 0.5 ha applicable both for tree-covered areas and for non-tree-covered areas ("holes"), meaning that a very small patch of trees is not considered as forest.
- Minimum mapping width (MMW) = 20 m.

9.2.3 Permanent Grassland

Grasslands, including sown pasture and rangeland, are among the largest ecosystems in the world and contribute to the livelihoods of rural population. They are a source of goods and services such as food and forage, energy, and wildlife habitat, and also provide carbon and water storage and watershed protection for many major river systems. Grasslands are important for in situ conservation of genetic resources (FAO, 2008).

The Permanent Grassland (GRA) layer is produced by means of combined processing of high-resolution imagery and medium resolution time-series imagery. The analysis uses the three reference years (2006, 2009, 2012) to detect the permanent presence of grassland. Permanent grassland is discriminated from other agricultural areas such as arable land, temporary grassland, and bare soil on the basis of seasonal variations in spectral reflectance observed by the satellites. Number of (3–8) monthly time composite of medium resolution (60-m pixel) Indian Remote Sensing Advanced Wide Field Sensor (IRS AWiFS) data (reference year 2012) is used as additional information for the classification process. Isolated pixels (up to four connected 20-m pixels) are deleted as "salt-and-pepper" noise in intermediate product. A difference between the intermediate and final grassland products is that the final product will not include grasslands in the urban context, airports, and sports and recreation areas. Quality problems and general dissatisfaction with the GRA product emerged from the verification reports by the countries. An analysis and discussions led to a suggested new grassland product (NGR) as a mitigation option, with a re-defined scope and specification. NGR focuses especially on natural grasslands. The original definition of GRA will be reconsidered in further evolution of a Copernicus grassland product.

---

* In intermediate (20 m × 20 m) FTY product nonforest trees (trees in urban context and trees in agriculture plantations) are included. They are, however, excluded in the final (100 m × 100 m) product.

9.2.4 Wetlands

Wetlands are areas where water is the primary factor controlling the environment and the associated plant and animal life. They occur where the water table is at or near the surface of the land, or where the land is covered by shallow water.

A high-resolution dataset of wetlands is produced by means of using the three reference years (2006, 2009, 2012) to detect the permanent presence of surface wetness. The discrimination of wetlands from other LC classes such as water, grassland, or forest seasonal variations is performed with IRS AWiFS monthly time series (as described for grassland under Section 9.2.3). As a consequence of the applied definition of wetlands, dry wetlands (e.g., raised bogs) are not included.

9.2.5 Permanent Water Bodies

A water body is a body of water forming a physiographical feature, for example, a sea or a reservoir (*Oxford Dictionaries*). Permanent presence of surface water is mapped based on

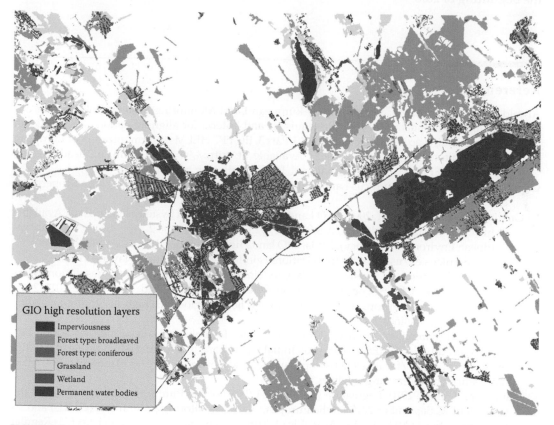

FIGURE 9.2
Székesfehérvár and surroundings (HU) as mapped by the intermediate High-Resolution Layers (HRLs). Imperviousness indicates a larger town and several smaller settlements. A highway and major roads are easy to recognize as linear impervious surfaces. Broadleaved forests are mostly visible around the largest lake. A few small coniferous stands can also be seen. It is expected that broadleaved patches inside built-up areas (trees in urban context) will be eliminated from final product. Permanent grasslands represent agriculture areas (pastures, hayfields). Lake Velencei is the largest water body on the area. Its SW part is protected wetland. Other smaller lakes are fishponds and water reservoirs. The extended white area is arable land, which is not included in current HRLs.

the analysis of three reference years (2006, 2009, 2012). Similarly to GRA and WET products, IRS medium resolution AWiFS monthly time series are used to separate temporary water from permanent water bodies for the reference year 2012 and exclude effects of seasonal changes in water coverage. The PWB layer should include rivers, channels, permanent lakes, ponds, coastal lagoons, and estuaries. But PWB should not include liquid dump sites, areas with temporary inundation, and water-logging and sea and ocean.

9.3 Example

An example of a composite map of high-resolution layers is presented for a rural area in Hungary. The 20 m × 20 m resolution intermediate products are shown in the Figure 9.2. This kind of information is available for 39 European countries (about 6 million km^2) from the beginning of 2015.

References

Arnold, S., Kosztra, B. 2013. Enhancing the European Land Monitoring System: Collection of criteria for a future data model. In Prerequisites and criteria for aligning national/sub-national land monitoring activities. Deliverable 4.1 Part 3, EU FP7 HELM—Harmonization of European Land Monitoring (grant no. 261562). Available at: http://www.fp7helm.eu/fileadmin/site/fp7helm/HELM_4_1_Aligning_national_land_monitoring_activities.pdf.

Copernicus Land. 2015. Copernicus Land Monitoring Services. Available at: http://land.copernicus.eu/.

Di Gregorio, A., Jansen, L. J. M. 2005. FAO Land Cover Classification System. Rome: FAO. Available at: http://www.fao.org/docrep/008/y7220e/y7220e06.htm#bm06.2.

EEA (European Environment Agency). 2013a. GIO land High Resolution Layers. Available at: http://land.copernicus.eu/user-corner/publications/gio-land-high-resolution-layers/view.

EEA (European Environment Agency). 2013b. Guidelines for verification of high-resolution layers produced under GMES/Copernicus Initial Operations (GIO) Land Monitoring 2011–2013. http://forum.eionet.europa.eu/nrc_land_covers/library/gio-land/high-resolution-layers-hrl/verification-guidelines.

EEA (European Environment Agency). 2013c. Guidelines for enhancement of high-resolution layers produced under GMES/Copernicus Initial Operations (GIO) Land Monitoring 2011–2013. Available at: http://forum.eionet.europa.eu/nrc_land_covers/library/gio-land/high-resolution-layers-hrl/enhancement-guidelines.

EU (European Union). 2010. Regulation (EU) No 911/2010 of the European Parliament and of the Council of 22 September 2010 on the European Earth monitoring programme (GMES) and its initial operations (2011 to 2013). Available at: http://www.google.com/search?q=EU+Regulation+%28EU%29+n%C2%B0911%2F2010.

FAO (Food and Agriculture Organization of the United Nations). 1998. FRA2000 on definitions of forest and forest change. Available at: http://www.fao.org/docrep/006/ad665e/ad665e06.htm.

Food and Agriculture Organization of the United Nations. 2008. Are grasslands under threat? Available at: http://www.fao.org/ag/agp/agpc/doc/grass_stats/grass-stats.htm.

Geoland 2. 2010. Geoland 2 Project. Available at: http://www.geoland2.eu/portal/service/List Service.do?serviceCategoryId=CA80C581.

Herold, M. R., Hubald, A., Di Gregorio, A. 2009. Translating and evaluating land cover legends using the UN Land Cover Classification System. Available at: http://nofc.cfs.nrcan.gc.ca/gofc-gold /Report%20Series/GOLD_43.pdf.

Oxford Dictionaries. Available at: http://www.oxforddictionaries.com/definition/english/waterbody? q=waterbodies.

Sjolbørg Flo Heggem, E., Strand, G.-H. 2010. CORINE Land Cover 2000. The Norwegian CLC2000 project. Report from the Norwegian Forest and Landscape Institute, 10. As: Norwegian Forest and Landscape Institute.

Wikipedia. 2014. http://en.wikipedia.org/wiki/Grassland.

Section III

European CLC Data Layers

10

CORINE Land Cover 1990 (CLC1990): Analysis and Assessment

Tomas Soukup, Jan Feranec, Gerard Hazeu, Gabriel Jaffrain,
Marketa Jindrova, Miroslav Kopecky, and Erika Orlitova

CONTENTS

10.1 Background

Completion of the CLC1990 data layer in the context of the eponymous project (produced under the European Commission) took 10 years (Feranec et al., 2012), the longest time compared to the three CLC data layers that followed. The time range of the interpreted images even covered 12 years (1986–1998) (Feranec et al., 2012). This fact should be borne in mind especially when comparing the results of land cover (LC) change analysis from different parts of Europe for the period 1990–2000. The progressive inclusion of the partner European Union (EU) countries from Central and Eastern Europe into the project after 1993 had significantly influenced the completion of the CLC1990 data layer.

Image maps produced from satellite data without ortho-correction (Büttner, 2014) were used in the process of visual interpretation. Furthermore, the spatial resolutions of satellite images used for the period 1986–1988 were not equal (e.g., for the south of France Landsat MSS images with 80-m resolution were used while Satellite Pour l'Observation de la Terre, Satellite for observation of Earth [SPOT 2] images with 20-m resolution were used for the north of France), which led to variation in the geometric accuracy of the interpretation. A description of the complete interpretation procedure is provided in Section 5.3 of Chapter 5. It should be noted that the missing ortho-correction of image maps and deformation of the plastic sheet used by interpreters for manual drawing of the identified CoORdination of INformation on the Environment (CORINE) Land Cover (CLC) class areas often caused geometric distortion of the resulting CLC1990 data (Büttner, 2014). Büttner reports that the quality control of CLC1990 data was questionable, as the results of interpreters' work existed only on the plastic sheet.

According to Table 2.1 in Chapter 2 the geometric accuracy of satellite images was ≤50 m. The geometric accuracy of CLC data is 100 m and their thematic accuracy is ≥85%. The size of the smallest identified area or minimum mapping unit (MMU) is 25 ha and its minimum width is 100 m.

10.2 Statistical Characteristics

The seamless CLC1990 data layer provides information about the spatial distribution of the LC classes in 26 European countries (27 countries with Turkey; see Table 2.2, Chapter 2). From the chronological point of view, it is the first harmonized set of environmental data from a considerable part of Europe that is available for a broad public.

Tables 10.1 and 10.2 and Figures 10.1 and 10.2 present the statistical characteristics of CLC1990 classes. Table 10.1 offers an overview of surface areas of CLC1990 classes. The largest part of Europe's surface in that period is covered by *Agricultural areas* (classes 211–244) with an area of 1,899,481.2 km², in total 55.1% of the area of LC in Europe. Classes 311–324, which define the *Forest landscape*, cover the second largest area, that is, 1,266,258.2 km², equating to 36.8%. *Artificial surfaces* (classes 111–142) cover 146,552.8 km², or 4.3%. *Open space with little or no vegetation* (classes 331–335) covers only 49,179.9 km², or 1.4%; *Water bodies* (classes 511–522; *Sea and ocean* were not taken into account as they include a buffer zone around Europe's land mass of 15 km) cover an area of 43,442.5 km², or 1.3%, and *Wetlands* (classes 411–423) cover an area of 39,575.3 km², that is, 1.1% (see Table 10.1).

Figures 10.1 (relative values in %) and 10.2 (absolute values in km²) show the extent (by means of graphs) of CLC Level 1 classes for the 26 European countries included in the CLC1990 data layer. *Artificial surfaces* have a relatively greatest representation in small countries: Malta (ca. 30%), Belgium (ca. 20%), and the Netherlands (ca. 10%). *Agricultural areas* take a relatively large share of the country surface area in Denmark (ca. 76%) and Hungary, Ireland, Lithuania, the Netherlands, and Poland (between 60% and 67%). *Forest and semi-natural areas* cover about 80% of the total surface of Montenegro, and for Austria, Estonia, Greece, Croatia, and Slovenia the share of the total surface covered by forest and semi-natural areas is between 54% and 64%. Low percentages in terms of relative representations in pan-European context are the figures for *Wetlands* and *Water bodies*. *Wetlands* are relatively important by surface area in Ireland (ca. 19%) and the Netherlands (ca. 7%). The largest shares for *Water bodies* are encountered in the Netherlands (ca. 8%) and Estonia (ca. 5%) (see Figure 10.1).

Figure 10.2 makes it possible to compare the CLC1990 Level 1 classes (highest hierarchical level). The area of *Artificial surfaces* is, naturally, largest in big European countries such as France and Germany. Likewise, *Agricultural areas* are also largest in France, Germany, Italy, Poland, and Spain. The largest *Forest and semi-natural areas* are in France, Germany, Greece, Italy, Poland, Romania, and Spain. The largest areas occupied by *Wetlands* are in Ireland and the largest *Water bodies* are in Germany, France, the Netherlands, Poland, and Romania (see Figure 10.2).

Table 10.2 provides insight into the largest CLC1990 classes (CLC Level 3 classes as explained in Table 3.1) in terms of extent. Class 112 leads among *Artificial surfaces* in all countries (besides 111 in Spain). Class 211 is prominent among *Agricultural areas* except for Ireland (231), Latvia (231), Liechtenstein (231), the Netherlands (231), Croatia (242), Luxembourg (242), Slovenia (242), Montenegro (243), and Malta (243). The greatest representation in *Forest and semi-natural areas* is that of class 311 (11 countries), 312 (8 countries), 313 (4 countries), 323 (3 countries), 321 (one country), and 324 (one country). Among *Wetlands*, class 411 is prominent in 13 countries. The number of countries in which the other *Wetlands* classes prevail is as follows: 412 (5 countries), 423 (4 countries), 421 (3 countries), and 422 (one country). In the case of *Water bodies*, class 512 prevails in 21 countries, 511 prevails in 4 countries, and 521 and 522 prevail each in one country. CLC class 112 definitely dominates the urbanized landscape of all countries that participate in the CLC1990 Project. The

TABLE 10.1

Statistical Characteristics of the CLC1990, CLC2000, CLC2006, and CLC2012 (in Percentage and km²) Data Layers of Europe

CLC Classes Third Level	1990 Total Area (in km²)	Share (in %)	2000 Total Area (in km²)	Share (in %)	2006 Total Area (in km²)	Share (in %)	2012 Total Area (in km²)	Share (in %)
111	5493.65	0.16	6106.33	0.12	6206.33	0.12	5708.59	0.11
112	109,047.48	3.17	138,914.82	2.74	147,164.52	2.90	157,047.22	3.09
121	14,978.15	0.43	20,616.23	0.41	22,852.84	0.45	27,102.11	0.53
122	1261.33	0.04	1950.34	0.04	2521.22	0.05	3147.87	0.06
123	780.56	0.02	1054.62	0.02	1127.90	0.02	1126.75	0.02
124	2176.25	0.06	3077.32	0.06	3230.03	0.06	3235.56	0.06
131	4686.84	0.14	6354.55	0.13	6757.65	0.13	6914.30	0.14
132	916.33	0.03	1102.62	0.02	1101.63	0.02	1160.83	0.02
133	942.13	0.03	1197.64	0.02	1654.22	0.03	1908.44	0.04
141	1883.49	0.05	3009.91	0.06	3002.98	0.06	3453.85	0.07
142	4386.57	0.13	8671.01	0.17	10,259.92	0.20	11,684.66	0.23
211	987,335.46	28.66	1,106,114.04	21.79	1,107,356.04	21.82	1,105,438.29	21.78
212	28,218.32	0.82	31,683.54	0.62	32,994.65	0.65	35,678.41	0.70
213	5684.27	0.17	5702.22	0.11	5982.37	0.12	6189.91	0.12
221	39,214.16	1.14	39,696.84	0.78	38,555.62	0.76	39,491.48	0.78
222	24,593.58	0.71	25,281.78	0.50	26,012.88	0.51	29,056.09	0.57
223	39,005.72	1.13	39,967.19	0.79	41,469.34	0.82	45,927.54	0.90
231	302,703.44	8.79	385,003.28	7.59	382,372.19	7.53	404,030.70	7.96
241	9918.32	0.29	9902.34	0.20	9584.44	0.19	5999.89	0.12
242	247,453.06	7.18	275,204.60	5.42	262,329.08	5.17	209,148.40	4.12
243	183,895.16	5.34	229,594.25	4.52	224,129.46	4.42	203,359.50	4.01
244	31,459.70	0.91	31,895.89	0.63	32,938.72	0.65	31,347.92	0.62
311	410,817.01	11.93	527,269.22	10.39	522,307.31	10.29	547,761.02	10.79
312	318,825.98	9.26	717,940.84	14.15	701,549.52	13.82	770,960.46	15.19
313	182,222.96	5.29	302,550.65	5.96	311,480.70	6.14	275,279.63	5.42
321	94,041.88	2.73	129,504.69	2.55	129,311.48	2.55	142,592.83	2.81

(Continued)

TABLE 10.1 (CONTINUED)

Statistical Characteristics of the CLC1990, CLC2000, CLC2006, and CLC2012 (in Percentage and km²) Data Layers of Europe

CLC Classes Third Level	1990 Total Area (in km²)	1990 Share (in %)	2000 Total Area (in km²)	2000 Share (in %)	2006 Total Area (in km²)	2006 Share (in %)	2012 Total Area (in km²)	2012 Share (in %)
322	28,134.52	0.82	174,706.98	3.44	163,815.14	3.23	173,408.79	3.42
323	96,932.95	2.81	101,631.92	2.00	101,313.75	2.00	98,951.76	1.95
324	135,282.86	3.93	253,349.41	4.99	274,289.05	5.40	219,042.17	4.32
331	3285.77	0.10	7120.68	0.14	7023.45	0.14	7004.78	0.14
332	15,628.34	0.45	70,374.21	1.39	68,834.65	1.36	70,508.65	1.39
333	27,045.78	0.79	137,293.32	2.71	138,899.56	2.74	133,641.43	2.63
334	1733.05	0.05	1512.67	0.03	1227.11	0.02	1073.65	0.02
335	1486.97	0.04	16,903.91	0.33	16,475.66	0.32	15,970.48	0.31
411	10,985.56	0.32	12,562.80	0.25	11,880.81	0.23	11,713.59	0.23
412	17,022.49	0.49	99,323.66	1.96	104,568.07	2.06	115,475.05	2.27
421	2513.40	0.07	3133.61	0.06	3384.80	0.07	3604.39	0.07
422	726.32	0.02	756.99	0.01	593.78	0.01	553.75	0.01
423	8327.49	0.24	11,343.42	0.22	12,246.14	0.24	12,120.91	0.24
511	8725.97	0.25	12,177.86	0.24	12,639.00	0.25	12,299.77	0.24
512	27441.37	0.80	114,869.82	2.26	115,644.60	2.28	116,560.01	2.30
521	5360.41	0.16	5824.23	0.11	5826.78	0.11	5893.30	0.12
522	1914.77	0.06	3102.38	0.06	2646.14	0.05	3308.56	0.07
Total	**3,444,489.82**	**100.00**	**5,075,354.63**	**100.00**	**5,075,561.53**	**100.00**	**5,075,883.29**	**100.00**

Source: EEA. 2014. Land accounts data viewer 1990, 2000, 2006. Available at: http://www.eea.europa.eu/data-and-maps/data/data-viewers/land-accounts.

Note: Values are aggregations of 26 countries (except Turkey).

TABLE 10.2

Statistical Characteristics of the Dominating CLC1990 Level 3 Classes within the Groups of the First Hierarchical Level in European Countries and Turkey

Country	Artificial Surfaces			Agricultural Areas			Forest and Semi-Natural Areas			Wetlands			Water Bodies		
	Dominating CLC Class	Area (in Km²)	% of Total Country's Area	Dominating CLC Class	Area (in Km²)	% of Total Country's Area	Dominating CLC Class	Area (in Km²)	% of Total Country's Area	Dominating CLC Class	Area (in Km²)	% of Total Country's Area	Dominating CLC Class	Area (in Km²)	% of Total Country's Area
Albania[a]															
Austria	112	3057.97	3.64	211	11,053.54	13.17	312	21,409.29	25.51	411	220.78	0.26	512	453.32	0.54
Belgium	112	5058.63	16.49	211	6758.70	22.04	313	2624.93	8.56	412	47.44	0.15	512	94.79	0.31
Bosnia/Herzegovina[a]															
Bulgaria	112	4092.43	3.69	211	38,863.09	35.01	311	23,327.06	21.01	411	91.84	0.08	512	630.13	0.57
Croatia	112	1346.32	2.37	242	10,250.43	18.04	311	16,976.52	29.88	411	174.43	0.31	512	299.85	0.53
Cyprus[a]															
Czech Republic	112	3568.74	4.52	211	35,555.35	45.07	312	16,577.80	21.01	411	52.74	0.07	512	492.17	0.62
Denmark	112	1877.03	4.26	211	28,188.58	63.97	312	1990.47	4.52	423	582.21	1.32	521	460.39	1.04
Estonia	112	495.42	1.09	211	6633.68	14.58	313	8557.06	18.81	412	1222.94	2.69	512	2096.66	4.61
Finland[a]															
France	112	19,260.78	3.49	211	153,426.90	27.80	311	88,865.10	16.10	423	2130.14	0.39	512	1965.56	0.36
Germany	112	21,229.92	5.87	211	139,544.44	38.59	312	56,643.93	15.66	423	2785.01	0.77	512	2967.83	0.82
Greece	112	1589.13	1.20	211	15,640.69	11.84	323	23,564.45	17.84	421	344.98	0.26	512	862.08	0.65
Hungary	112	4122.96	4.43	211	49,606.56	53.33	311	14,341.46	15.42	411	909.96	0.98	512	1226.27	1.32
Iceland[a]															
Ireland	112	703.82	1.00	231	38,226.67	54.12	312	2482.19	3.51	412	12,399.22	17.55	512	1222.83	1.73
Italy	112	8835.84	2.93	211	80,605.86	26.75	311	54,689.22	18.15	421	432.61	0.14	512	1677.32	0.56
Kosovo[a]															
Latvia	112	518.26	0.80	231	9331.43	14.44	313	12,686.13	19.63	412	1310.62	2.03	512	1052.83	1.63
Liechtenstein[a]															

(Continued)

TABLE 10.2 (CONTINUED)

Statistical Characteristics of the Dominating CLC1990 Level 3 Classes within the Groups of the First Hierarchical Level in European Countries and Turkey

Country	Artificial Surfaces			Agricultural Areas			Forest and Semi-Natural Areas			Wetlands			Water Bodies		
	Dominating CLC Class	Area (in km²)	% of Total Country's Area	Dominating CLC Class	Area (in km²)	% of Total Country's Area	Dominating CLC Class	Area (in km²)	% of Total Country's Area	Dominating CLC Class	Area (in km²)	% of Total Country's Area	Dominating CLC Class	Area (in km²)	% of Total Country's Area
Lithuania	112	1469.08	2.24	211	21,881.79	33.39	312	7562.55	11.54	412	389.34	0.59	512	1068.92	1.63
Luxembourg	112	164.67	6.35	242	617.34	23.80	311	640.59	24.70	0	0	0	512	7.18	0.28
Macedonia FYR[a]															
Malta	112	66.09	20.95	243	151.63	48.06	323	49.78	15.78	422	0.26	0.08	0	0	0
Montenegro	112	107.76	0.78	243	1831.59	13.25	311	3669.41	26.55	411	118.62	0.86	512	253.41	1.83
Netherlands	112	2541.56	6.37	231	11,380.32	28.54	312	1628.64	4.08	423	2289.23	5.74	512	2642.60	6.63
Norway[a]															
Poland	112	7712.05	2.47	211	139,954.43	44.77	312	55,517.30	17.76	411	1084.28	0.35	512	3742.01	1.20
Portugal	112	1335.31	1.45	211	11,733.95	12.71	311	11,868.75	12.85	421	191.13	0.21	522	454.13	0.49
Romania	112	12,849.64	5.39	211	81,151.09	34.04	311	48,157.31	20.20	411	3832.98	1.61	511	1660.64	0.70
Serbia	112	2100.91	2.71	211	20,034.48	25.80	311	20,717.21	26.68	411	211.92	0.27	511	603.78	0.78
Slovakia	112	2222.96	4.53	211	16,744.43	34.16	311	10,415.46	21.25	411	57.70	0.12	512	151.82	0.31
Slovenia	112	416.59	2.05	242	2778.98	13.71	313	4475.80	22.07	411	25.19	0.12	511	53.57	0.26
Spain	111	2518.30	0.50	211	103,744.00	20.49	323	55,120.94	10.89	411	541.75	0.11	512	2037.18	0.40
Sweden[a]															
Switzerland[a]															
Turkey	112	7052.00	0.90	211	123,092.27	15.79	333	102,617.89	13.16	411	2254.47	0.29	512	10,450.59	1.34
United Kingdom[a]															

[a] Country did not participate in the CLC1990 project.

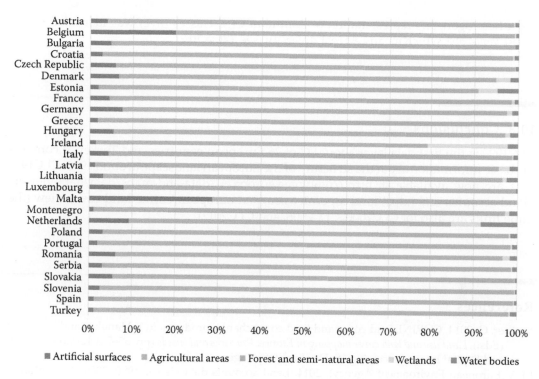

FIGURE 10.1
Extent of CLC1990 classes of the first hierarchic level in European countries (relative values in %).

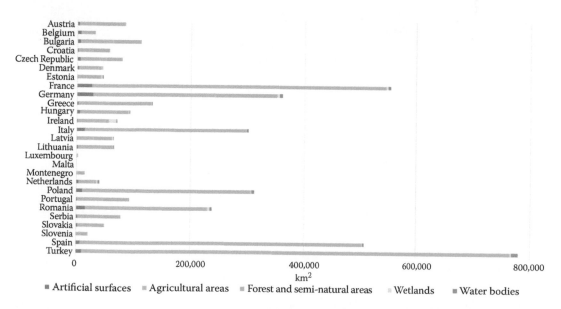

FIGURE 10.2
Extent of CLC1990 classes of the first hierarchic level in European countries (absolute values in km²).

most heterogeneous representation of dominant CLC classes (five) was identified in the *Forest landscape*, which means a great diversity (see Table 10.2).

10.3 Conclusions

CLC1990 is the first data layer about LC of 26 European countries and Turkey. It contains the basic information about the structure of the European landscape around 1990. CLC1990 data are now available at: http://www.eea.europa.eu/data-and-maps/data/data-viewers /land-accounts and http://land.copernicus.eu/pan-european/corine-land-cover/view (the same links are also for CLC2000, CLC2006, and CLC2012).

References

Büttner, G. 2014. CORINE land cover and land cover change products. In I. Manakos and M. Braun (Eds.), *Land use and land cover mapping in Europe: Practices and trends* (pp. 55–74). Remote Sensing and Digital Image Processing 18. Dordrecht: Springer.

EEA (European Environment Agency). 2014. Land accounts data viewer 1990, 2000, 2006. http:// www.eea.europa.eu/data-and-maps/data/data-viewers/land-accounts.

Feranec, J., Soukup, T., Hazeu, G. W., Jaffrain, G. 2012. Land cover and its change in Europe: 1990–2006. In G. Chandra (Ed.), *Remote sensing of land use and land cover: Principles and applications* (pp. 285–301). Boca Raton, FL: CRC Press.

11

CORINE Land Cover 2000 (CLC2000): Analysis and Assessment

Tomas Soukup, Jan Feranec, Gerard Hazeu, Gabriel Jaffrain, Marketa Jindrova, Miroslav Kopecky, and Erika Orlitova

CONTENTS

11.1 Background

The CLC2000 data layer is the product of the IMAGE and CLC2000 Project managed by the European Environment Agency (EEA) and the Joint Research Centre (JRC). The time consistency of applied satellite images was 2000 ± 1 year, hence much better than for the images used in CLC1990 (Feranec et al., 2012). The computer-assisted photointerpretation (CAPI) method, its variants A—updating of the revised CLC1990 vector layer on the basis of satellite images from 2000 ± 1 year (changes are derived by overlaying the revised CLC1990 and CLC2000) or B—computer derivation of CLC2000 layer on the basis of $CLC_{1990-2000}$ changes identified by interpretation of satellite images coupled with the revised CLC1990 data layer (see Sections 5.4 and 5.6 in Chapter 5). Before the CAPI method could be applied the original CLC1990 was converted into digital format. With the help of the CAPI, methodological errors caused by drawing on transparencies during the production of CLC1990 were eliminated. Polynomial transformation or rubber-sheeting was used to eliminate the bulk geometric distortion. Residual geometric errors exceeding 100 m and coding mistakes were removed by editing (Büttner, 2014: 59). Data were also revised and generalized by removing the areas smaller than 25 ha. A copy of this revised CLC1990 was the basis for the CLC2000 data layer. The production time for CLC2000 data was reduced compared to the previous data layer from 10 to 4 years. All CLC1990 and CLC2000 data are today freely accessible from the EEA to any person or legal entity (Büttner, 2014). Geometric accuracy of used satellite images improved to ≤25 m. Likewise, the geometric accuracy of CLC data is better than 100 m and the thematic accuracy is ≥85%. The size of the smallest identified area (minimum mapping unit [MMU]) and its width (25 ha and 100 m; see Table 2.1 in Chapter 2) remained unchanged compared to the CLC1990 layer.

11.2 Statistical Characteristics

The second seamless data layer CLC2000 provides information about LC classes (their occurrence and areas) in as many as 38 European countries plus Turkey (total 39 countries; see Table 2.2 in Chapter 2). The increasing demand from users in the environmental domain has led to a distinct expansion of countries producing CLC2000. CoORdination of Information on the Environment (CORINE) Land Cover (CLC) data gained an important place at the European level and in various countries in the context of various environmental assessments.

Tables 10.1* and 11.1 and Figures 11.1 and 11.2 present the statistical characteristics of CLC2000 classes. An overview of surface areas of CLC2000 classes in Europe is presented in Table 10.1. In this data layer the *Forest landscape* (CLC classes 311–324) covers 2,206,953.7 km², corresponding to 43.5% of the area of the European LC. The *Agricultural landscape* (CLC classes 211–244) stretches over 2,180,045.9 km², which equates to 42.9%. As such both landscapes have surface areas of the same order of magnitude. As far as size is concerned, they are followed by *Open space with little or no vegetation* (CLC classes 331–335) with an area of 233,204.8 km², equating to 4.6%. The area of *Artificial surfaces* (CLC classes 111–142) is in total 192,055.4 km² or 3.8%. *Water bodies* (CLC classes 511–522) occupy 135,974.3 km², or 2.7%. The *Wetlands* (CLC classes 411–423), with an area of 127,120.5 km², or 2.5%, cover the smallest area.

Figures 11.1 (relative values in %) and 11.2 (absolute values in km²) demonstrate the extent of CLC classes on the first hierarchic level for 38 European countries and Turkey. The largest relative representation is that of *Artificial surfaces* in small countries: Malta (ca. 30%), Belgium (ca. 20%), and Liechtenstein and the Netherlands (ca. 12%, see Figure 11.1). *Agricultural areas* dominate in Denmark (ca. 75%), and their dominance in countries such as France, Hungary, Ireland, Lithuania, the Netherlands, and Poland is between 60% and 65%. For Belgium, the Czech Republic, Romania, Serbia, and the United Kingdom the relative share of the country surface is around 57%–58%. *Forest and semi-natural areas* occupy the largest areas in Iceland, Montenegro, and Norway (80%–90%); Finland and Sweden (ca. 75%); and Albania, Austria, Bosnia/Hercegovina, Liechtenstein, Macedonia FYR, and Switzerland (59%–70%). The relative representation of *Wetlands* in all European countries is much lower: Ireland (ca. 16%); Iceland, Finland, the Netherlands, Norway, and Sweden (6%–7%); and Estonia (ca. 4%). *Water bodies* cover almost 5%–8% in Finland, Sweden, and the Netherlands. Around 3%–4% of Norway, Estonia, and Switzerland is covered with *Water bodies*.

Absolute values of CLC2000 classes (in km²) of the first hierarchic level are presented in Figure 11.2, with an additional option to compare their sizes in individual countries. *Artificial surfaces* have the largest size (total km²) in the large countries, Germany and France, but they are small(er) in other big countries such as Italy, Poland, Romania, Spain, Sweden, and the United Kingdom. Likewise, *Agricultural areas* are largest in France, Germany, Italy, Poland, Romania, Spain, and the United Kingdom. Large areas of *Forest and semi-natural areas* are in Finland, Norway, and Sweden, but also in Spain and France. *Wetlands* are largest in Finland, Iceland, Ireland, Norway, Sweden, and United Kingdom. *Water bodies* are largest in Sweden, Finland, and Norway.

---

* Values for CLC2000 are based on data from 38 countries (except Turkey).

TABLE 11.1

Statistical Characteristics of the Dominant CLC2000 Third Level Classes within the Groups of the First Hierarchical Level in European Countries and Turkey

Country	Artificial Surfaces			Agricultural Areas			Forest and Semi-Natural Areas			Wetlands			Water Bodies		
	Dominant CLC Class	Area (in km²)	% of Total Country's Area	Dominant CLC Class	Area (in km²)	% of Total Country's Area	Dominant CLC Class	Area (in km²)	% of Total Country's Area	Dominant CLC Class	Area (in km²)	% of Total Country's Area	Dominant CLC Class	Area (in km²)	% of Total Country's Area
Albania	112	312.60	1.09	242	2700.88	9.39	311	6375.21	22.16	411	56.85	0.20	512	439.37	1.53
Austria	112	3090.71	3.68	211	10,981.06	13.09	312	21,419.26	25.53	411	220.32	0.26	512	458.48	0.55
Belgium	112	5113.37	16.67	211	6719.89	21.91	313	2654.12	8.65	412	48.46	0.16	512	102.16	0.33
Bosnia/Herzegovina	112	488.14	0.96	242	7869.40	15.47	311	16,238.05	31.92	411	52.91	0.10	512	181.59	0.36
Bulgaria	112	4094.49	3.69	211	39,109.99	35.23	311	23,405.80	21.08	411	94.94	0.09	512	628.00	0.57
Croatia	112	1373.55	2.42	242	10,173.55	17.91	311	16,899.16	29.74	411	178.54	0.31	512	296.53	0.52
Cyprus	112	436.42	4.72	211	2447.51	26.46	323	1606.15	17.36	421	19.63	0.21	512	14.06	0.15
Czech Republic	112	3616.18	4.58	211	32,640.95	41.38	312	17,015.28	21.57	411	52.53	0.07	512	508.91	0.65
Denmark	112	1915.75	4.35	211	27,981.66	63.49	312	1797.92	4.08	423	583.30	1.32	521	460.31	1.04
Estonia	112	509.47	1.12	211	6641.73	14.60	313	8374.44	18.41	412	1230.53	2.70	512	2095.82	4.61
Finland	112	3588.23	1.06	211	16,069.98	4.75	312	100,357.63	29.64	412	22,214.45	6.56	512	30,914.82	9.13
France	112	19,796.68	3.59	211	153,548.13	27.82	311	88,975.41	16.12	423	2127.70	0.39	512	2087.61	0.38
Germany	112	22,184.07	6.13	211	136,850.21	37.84	312	56,281.94	15.56	423	2826.72	0.78	512	3148.91	0.87
Greece	112	1645.95	1.25	211	15,359.30	11.62	323	23,323.42	17.65	421	340.18	0.26	512	860.89	0.65
Hungary	112	4138.05	4.45	211	49,515.07	53.23	311	14,813.38	15.93	411	920.21	0.99	512	1261.61	1.36
Iceland	142	117.98	0.11	231	2456.61	2.37	322	36,065.86	34.87	412	6507.81	6.29	512	1216.34	1.18
Ireland	112	867.22	1.23	231	36,393.58	51.52	324	3396.71	4.81	412	11,384.35	16.12	512	1212.14	1.72
Italy	112	9336.81	3.10	211	79,928.36	26.52	311	55,297.33	18.35	421	432.11	0.14	512	1694.10	0.56
Kosovo	112	203.44	1.86	242	1818.98	16.67	311	3959.02	36.29	0	0	0	512	15.26	0.14
Latvia	112	518.26	0.80	231	9282.11	14.36	313	11,974.51	18.52	412	1308.37	2.02	512	1055.16	1.63
Liechtenstein	112	17.01	10.57	211	19.63	12.19	313	37.86	23.52	0	0	0	511	2.29	1.42

(Continued)

TABLE 11.1 (CONTINUED)

Statistical Characteristics of the Dominant CLC2000 Third Level Classes within the Groups of the First Hierarchical Level in European Countries and Turkey

Country	Artificial Surfaces			Agricultural Areas			Forest and Semi-Natural Areas			Wetlands			Water Bodies		
	Dominant CLC Class	Area (in km²)	% of Total Country's Area	Dominant CLC Class	Area (in km²)	% of Total Country's Area	Dominant CLC Class	Area (in km²)	% of Total Country's Area	Dominant CLC Class	Area (in km²)	% of Total Country's Area	Dominant CLC Class	Area (in km²)	% of Total Country's Area
Lithuania	112	1476.49	2.25	211	22,277.37	34.00	312	7314.88	11.16	412	389.53	0.59	512	1071.46	1.64
Luxembourg	112	174.82	6.74	242	613.48	23.65	311	633.56	24.42	0	0	0	512	7.18	0.28
Macedonia FYR	112	297.55	1.17	211	2543.88	10.01	311	7691.19	30.25	411	18.52	0.07	512	515.02	2.03
Malta	112	66.87	21.19	243	150.72	47.76	323	49.02	15.53	422	0.26	0.08	0	0	0
Montenegro	112	107.96	0.78	243	1828.82	13.23	311	3681.20	26.64	411	119.21	0.86	512	252.98	1.83
Netherlands	112	2983.48	7.48	231	10,715.18	26.87	312	1619.16	4.06	423	2277.34	5.71	512	2676.77	6.71
Norway	112	1796.11	0.55	243	9693.44	2.99	333	81,387.80	25.14	412	21,307.05	6.58	512	13,339.04	4.12
Poland	112	7793.80	2.49	211	139,790.76	44.72	312	55,135.65	17.64	411	1019.40	0.33	512	3823.53	1.22
Portugal	112	1766.26	1.91	211	11,013.70	11.93	311	12,582.72	13.63	421	189.17	0.20	522	452.48	0.49
Romania	112	12,882.08	5.40	211	81,525.55	34.20	311	48,629.05	20.40	411	3825.20	1.60	511	1658.54	0.70
Serbia	112	2134.70	2.75	211	19,902.97	25.63	311	20,687.10	26.64	411	211.56	0.27	511	594.12	0.77
Slovakia	112	2253.73	4.60	211	16,682.78	34.03	311	10,649.70	21.72	411	42.86	0.09	512	202.02	0.41
Slovenia	112	416.71	2.06	242	2778.88	13.71	313	4474.77	22.07	411	25.60	0.13	511	53.57	0.26
Spain	112	2843.96	0.56	211	100,064.09	19.77	323	53,531.61	10.57	411	551.19	0.11	512	2412.19	0.48
Sweden	112	4040.75	0.90	211	30,056.36	6.68	312	216,938.74	48.21	412	28,155.92	6.26	512	36,235.52	8.05
Switzerland	112	2292.12	5.55	211	7149.03	17.32	312	6334.99	15.35	411	34.15	0.08	512	1385.66	3.36
Turkey	112	8247.03	1.06	211	122,335.56	15.69	333	102,545.73	13.15	411	2506.55	0.32	512	11,501.67	1.48
United Kingdom	112	12,250.09	4.94	231	67,761.27	27.32	322	29,392.26	11.85	412	5142.05	2.07	512	2171.45	0.88

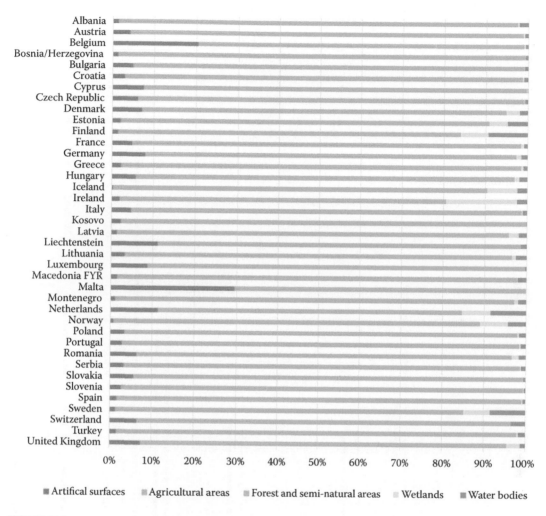

FIGURE 11.1

Extent of CLC2000 classes of the first hierarchic level in European countries (relative values in %).

Table 11.1 contains an overview of the dominating CLC2000 classes (CLC class Level 3 as explained in Table 3.1) in terms of areas in individual countries. Class 112 of *Artificial surfaces* was the largest in all countries, besides Iceland. Class 211 of *Agricultural areas* was the largest in 25 countries, followed by 242 (5 countries), 231 (4 countries), and 243 (4 countries). The dominant class within the group of *Forest and semi-natural areas* was class 311 (15 countries) along with classes 312 (10 countries), 313 (5 countries), 323 (4 countries), 322 (2 countries), and 324 and 333 (both one country). Among *Wetlands* class 411 is dominant in 18 countries. The number of countries in which the other *Wetlands* classes prevail is as follows: class 412 (10 countries), class 421 (5 countries), class 423 (4 countries), and class 422 (in one country). Class 512 of *Water bodies* prevails in 31 countries. In 5 countries class 511 prevails and classes 521 and 522 both dominate in one country. CLC class 112 almost completely dominates in the urbanized landscape of all countries (apart from Iceland) that participate in the CLC2000 Project. The most heterogeneous representation of dominant CLC classes (7) was identified in the *Forest landscape* (see Table 11.1), as in CLC1990.

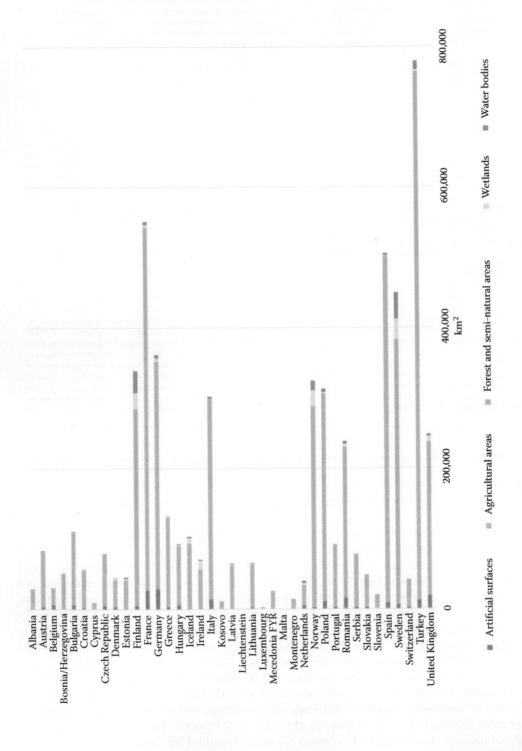

FIGURE 11.2

Extent of CLC2000 classes of the first hierarchic level in European countries (absolute values in km²).

11.3 Conclusions

The CLC2000 data layer provides spatial information on 44 CLC classes, their occurrence, and surface area in 38 European countries and Turkey. Existence of this data layer compatible with the 1990 data layer has created the first option of identifying LC changes in Europe in 1990–2000.

References

Büttner, G. 2014. CORINE land cover and land cover change products. In I. Manakos and M. Braun (Eds.), *Land use and land cover mapping in Europe: Practices and trends* (pp. 55–74). Remote Sensing and Digital Image Processing 18. Dordrecht: Springer.

Feranec, J., Soukup, T., Hazeu, G.W., Jaffrain, G. 2012. Land cover and its change in Europe: 1990–2006. In G. Chandra (Ed.), *Remote sensing of land use and land cover: Principles and applications* (pp. 285–301). Boca Raton, FL: CRC Press.

12

CORINE Land Cover 2006 (CLC2006): Analysis and Assessment

Tomas Soukup, Jan Feranec, Gerard Hazeu, Gabriel Jaffrain,
Marketa Jindrova, Miroslav Kopecky, and Erika Orlitova

CONTENTS

12.1 Background

The CLC2006 data layer is the eponymous product coordinated by the European Environ-ment Agency (EEA) in cooperation with the European Commission (EC) and European Space Agency. The project became part of the Fast Track Service on Land Monitoring (FTS-LM) under the Global Monitoring for Environment and Security (GMES) program since 2012 referred to as Copernicus. Time consistency of interpreted satellite images was 2006 ± 1 year and CLC2006 data were available after 3 years. The applied interpretation of the computer-assisted photointerpretation (CAPI) method was complemented by ancillary data (scanned topographic maps, digital color aerial photographs—ortho-photos, results of the Land Parcel Identification System [LPIS]; Büttner, 2014), which enhanced the quality of interpretation results. Geometric accuracy of the satellite images used, geometric accu-racy of CoORdination of Information on the Environment (CORINE) Land Cover (CLC) data, and the size of the smallest identified area and its minimum width did not change compared with the CLC2000 layer (see Table 2.1 in Chapter 2). Thirty-eight European countries and Turkey participated in the CLC2006 Project (see Table 2.2).

12.2 Statistical Characteristics

Tables 10.1 and 12.1 and Figures 12.1 and 12.2 present the statistical characteristics of CLC2006 classes. An overview of areas of CLC2006 classes in Europe is provided in Table 10.1. In this data layer the *Forest landscape*, again the largest (CLC classes 311–324), cov-ers 2,204,067.0 km^2 corresponding to 43.4% of the area of the European land cover (see Table 10.1). The *Agricultural landscape* (CLC classes 211–244) slightly diminished compared

TABLE 12.1

Statistical Characteristics of the Dominant CLC2006 Third-Level Classes within the Groups of the First Hierarchical Level in European Countries and Turkey

Country	Artificial Surfaces			Agricultural Areas			Forest and Semi-Natural Areas			Wetlands			Water Bodies		
	Dominant CLC Class	Area (in Km²)	% of Total Country's Area	Dominant CLC Class	Area (in Km²)	% of Total Country's Area	Dominant CLC Class	Area (in Km²)	% of Total Country's Area	Dominant CLC Class	Area (in Km²)	% of Total Country's Area	Dominant CLC Class	Area (in Km²)	% of Total Country's Area
Albania	112	699.82	2.43	243	3036.07	10.55	311	6284.08	21.84	421	46.63	0.16	512	443.17	1.54
Austria	112	3459.04	4.12	211	11,633.14	13.86	312	21,877.34	26.07	411	208.05	0.25	512	466.68	0.56
Belgium	112	5154.75	16.81	211	6701.20	21.85	313	2659.15	8.67	412	50.33	0.16	512	105.57	0.34
Bosnia/Herzegovina	112	562.87	1.11	242	7707.10	15.15	311	16,039.04	31.53	411	50.82	0.10	512	191.28	0.38
Bulgaria	112	4115.54	3.71	211	38,910.15	35.05	311	23,082.03	20.79	411	89.80	0.08	512	649.21	0.58
Croatia	112	1397.60	2.46	242	10,226.53	18.00	311	16,850.92	29.66	411	186.85	0.33	512	300.28	0.53
Cyprus	112	485.76	5.25	211	2387.68	25.82	323	1577.38	17.06	421	19.64	0.21	512	15.55	0.17
Czech Republic	112	3765.26	4.77	211	30,140.24	38.21	312	17,241.10	21.86	411	59.75	0.08	512	519.35	0.66
Denmark	112	1977.66	4.49	211	27,868.26	63.23	312	1799.68	4.08	423	588.22	1.33	521	465.87	1.06
Estonia	112	536.14	1.18	211	6726.09	14.79	313	8322.36	18.29	412	1237.40	2.72	512	2098.35	4.61
Finland	112	3643.56	1.08	211	17,189.45	5.08	312	96,848.84	28.61	412	21,993.00	6.50	512	30,911.40	9.13
France	112	20,919.46	3.79	211	154,132.29	27.93	311	87,970.57	15.94	423	2128.50	0.39	512	2143.31	0.39
Germany	112	22,956.70	6.35	211	135,418.49	37.44	312	56,173.46	15.53	423	2845.88	0.79	512	3321.63	0.92
Greece	112	2147.17	1.62	211	12,929.57	9.78	323	23,656.65	17.90	421	326.40	0.25	512	912.34	0.69
Hungary	112	4315.50	4.64	211	49,358.02	53.06	311	15,020.65	16.15	411	761.93	0.82	512	1293.30	1.39
Iceland	142	135.21	0.13	231	2448.77	2.37	322	35,995.84	34.80	412	6501.79	6.29	512	1220.82	1.18
Ireland	112	1067.17	1.51	231	35,801.65	50.60	324	4164.46	5.89	412	10,873.60	15.37	512	1175.41	1.66
Italy	112	9557.17	3.17	211	80,762.30	26.80	311	54,707.83	18.15	421	413.19	0.14	512	1732.84	0.57
Kosovo	112	212.38	1.95	242	1822.53	16.71	311	3988.21	36.56	0	0	0	512	15.28	0.14
Latvia	112	534.92	0.83	211	9993.10	15.46	313	11,617.34	17.97	412	1342.48	2.08	512	1044.90	1.62
Liechtenstein	112	17.79	11.05	211	21.53	13.37	312	45.55	28.29	411	1.53	0.95	511	2.19	1.36

(Continued)

TABLE 12.1 (CONTINUED)

Statistical Characteristics of the Dominant CLC2006 Third-Level Classes within the Groups of the First Hierarchical Level in European Countries and Turkey

Country	Artificial Surfaces			Agricultural Areas			Forest and Semi-Natural Areas			Wetlands			Water Bodies		
	Dominant CLC Class	Area (in Km²)	% of Total Country's Area	Dominant CLC Class	Area (in Km²)	% of Total Country's Area	Dominant CLC Class	Area (in Km²)	% of Total Country's Area	Dominant CLC Class	Area (in Km²)	% of Total Country's Area	Dominant CLC Class	Area (in Km²)	% of Total Country's Area
Lithuania	112	1484.75	2.27	211	22,235.37	33.96	313	7297.84	11.15	412	392.96	0.60	512	1064.70	1.63
Luxembourg	112	182.61	7.04	242	469.24	18.09	311	639.01	24.63	0	0	0	512	6.84	0.26
Macedonia FYR	112	319.65	1.26	211	2514.56	9.89	311	7331.55	28.84	411	19.52	0.08	512	529.74	2.08
Malta	112	66.89	21.19	243	149.75	47.45	323	49.89	15.81	422	0.26	0.08	0	0	0
Montenegro	112	107.74	0.78	243	1841.92	13.33	311	3625.37	26.24	411	113.75	0.82	512	253.17	1.83
Netherlands	112	3202.99	8.03	231	10,275.58	25.77	312	1610.48	4.04	423	2282.19	5.72	512	2690.95	6.75
Norway	112	1810.89	0.56	243	9692.41	2.99	333	81,382.74	25.13	412	21,301.88	6.58	512	13,338.41	4.12
Poland	112	9817.96	3.14	211	139,094.64	44.51	312	55,851.06	17.87	411	988.25	0.32	512	3867.12	1.24
Portugal	112	2313.68	2.51	211	9906.45	10.73	324	14,265.63	15.45	421	184.44	0.20	512	540.76	0.59
Romania	112	13,045.32	5.47	211	82,880.47	34.76	311	48,147.15	20.20	411	3362.34	1.41	511	2086.74	0.88
Serbia	112	2213.09	2.85	211	20,210.03	26.03	311	20,772.09	26.75	411	222.46	0.29	511	600.26	0.77
Slovakia	112	2177.58	4.44	211	16,771.98	34.21	311	10,730.90	21.89	411	26.79	0.05	512	211.83	0.43
Slovenia	112	417.34	2.06	242	2781.84	13.72	313	4485.79	22.12	411	24.41	0.12	511	51.93	0.26
Spain	112	3376.11	0.67	211	97,559.10	19.27	323	52,160.36	10.30	411	543.21	0.11	512	2483.22	0.49
Sweden	112	4105.56	0.91	211	29,967.34	6.66	312	210,143.52	46.70	412	28,229.13	6.27	512	36,218.64	8.05
Switzerland	112	2316.25	5.61	211	7301.17	17.69	312	6234.07	15.10	411	36.47	0.09	512	1386.23	3.36
Turkey	112	8352.96	1.07	211	122,135.62	15.66	333	102,532.33	13.15	411	2473.39	0.32	512	11,557.82	1.48
United Kingdom	112	12,579.40	5.07	231	70,413.21	28.38	322	22,771.56	9.18	412	11,074.08	4.46	512	2155.64	0.87

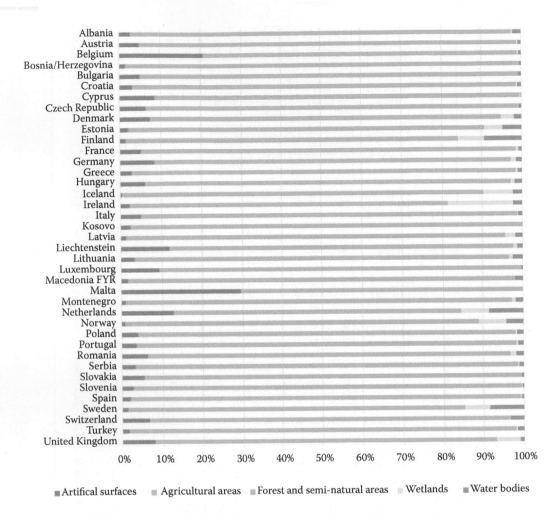

FIGURE 12.1
Extent of CLC2006 classes of the first hierarchic level in European countries (relative values in %).

to the previous year to 2,163,724.8 km², which equates to 42.6%. As far as size is concerned, it is followed by *Open space with little or no vegetation* (CLC classes 331–335), with an area of 232,460.4 km² equating to 4.6%. The area of *Artificial surfaces* (CLC classes 111–142) is in total 205,879.2 km², or 4.0%. *Water bodies* (CLC classes 511–522) occupy 136,756.5 km², or 2.7%. The *Wetlands* (CLC classes 411–423), with an area of 132,673.6 km², or 2.6%, are the smallest.

Figures 12.1 (relative values in %) and 12.2 (absolute values in km²) contain an overview of the extent of CLC classes on the first hierarchic level for 38 European countries and Turkey that participated in the CLC2006 Project. The largest relative representation of *Artificial surfaces* in CLC2006 is in small countries: Malta (ca. 30%), Belgium (ca. 20%), and Liechtenstein and the Netherlands (>10%) (see Figure 12.1). *Agricultural areas* prevail in Denmark (ca. 75%); Hungary (ca. 70%); Ireland, Lithuania, the Netherlands, and Poland (>60%). *Forest and semi-natural areas* distinctly prevail in Norway and Montenegro (ca. 80%); Finland and Sweden (ca. 75%); Iceland (almost 90%, classes 322, 332, 333, and 335 are predominantly represented, not the forest); Albania (ca. 70%); and Switzerland,

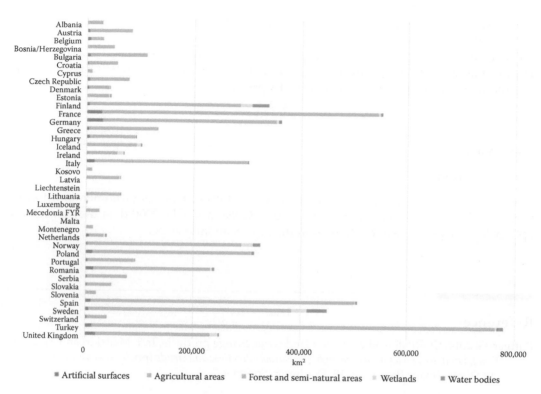

FIGURE 12.2

Extent of CLC2006 classes of the first hierarchic level in European countries (absolute values in km²).

Liechtenstein, Macedonia FYR, Bosnia/Herzegovina, and Austria (>60%). Relative representation of *Wetlands* is low with the exception of Ireland, where they cover 15%, and the Netherlands, Poland, Norway, and Iceland (below 10%). *Water bodies* are also scarcely represented: Finland (ca. 9%), the Netherlands and Sweden (ca. 8%), and Estonia and Norway (ca. 4%–5%).

Absolute areas of CLC classes on the first hierarchic levels allow for the comparison of their real size between the European countries (see Figure 12.2). *Artificial surfaces* have their largest size (total km²) in France, Germany, and the United Kingdom. A distinct representation of *Agricultural landscape* is typical for large countries such as France, Germany, Italy, Poland, and Spain. Large areas of *Forest and semi-natural areas* are in Finland, Norway, Spain, and Sweden. Areas of *Wetlands* stand out only in Finland, Iceland, Ireland, Norway, Sweden, and United Kingdom and those of *Water bodies* are largest in Finland, Norway, and Sweden (see Figure 12.2).

Table 12.1 contains information about the dominant representation of CLC2006 classes. As in the two preceding data layers, nothing changed in *Artificial surfaces* as far as the dominant and almost univocal representation of class 112 (except for Iceland) is concerned. In the case of *Agricultural areas* classes 211 prevailed in 26 countries, followed by 242 (5 countries), 243, and 231 (both of the latter in 4 countries). Classes 311, 312, 313, 323, 322, 324, and 333 are present in *Forest and semi-natural areas* as in the CLC2000 data layer. The prevailing representation in *Forest and semi-natural areas* corresponds to classes 311 (14 countries), 312 (10 countries), and 313 (5 countries). This CLC2006 data layer shows within the

Forest landscape also the largest heterogeneity in representation of dominant CLC classes (7). As far as *Wetlands* is concerned, the dominant representation has been attributed to classes 411 (17 countries), 412 (10 countries), 421 (5 countries), 423 (4 countries), and 422 (1 country). The most dominant class regarding *Water bodies* is class 512 (32 countries). Classes 511 and 521 dominate in 4 countries and 1 country, respectively.

12.3 Conclusion

CLC2006 data broadened the inputs for the generation of fresh environmental information. Thanks to their compatibility with the CLC1990 and CLC2000 data they enrich the options for the assessment of changes in the European landscape.

Reference

Büttner, G. 2014. CORINE land cover and land cover change products. In I. Manakos and M. Braun (Eds.), *Land use and land cover mapping in Europe: Practices and trends* (pp. 55–74). Remote Sensing and Digital Image Processing 18. Dordrecht: Springer.

13

CORINE Land Cover 2012 (CLC2012): Analysis and Assessment

Tomas Soukup, György Büttner, Jan Feranec, Gerard Hazeu, Gabriel Jaffrain, Marketa Jindrova, Miroslav Kopecky, and Erika Orlitova

CONTENTS

13.1 Background

The fourth CLC2012 data layer is the product of an eponymous project initiated by the Directorate-General for Enterprise Industry of the EC (Büttner, 2014) and implemented under the Global Monitoring for Environment and Security (GMES) Copernicus Initial Operations (GIO). Interpreted satellite images are from the period 2011–2012. CLC2012 data were accessible in September 2015. Their production time was more than 2 years, the shortest compared with the previous data layers. Computer-assisted photointerpretation (CAPI) was the prevailing method applied in interpreting of satellite images. Finland, Germany, Iceland, Ireland, Norway, Spain, and Sweden applied a semiautomatic methodology based on integration of the existing land use data, satellite image processing, and generalization (Büttner, 2014). Geometric accuracy of the satellite images used, and the geometric accuracy of CLC data, size of the smallest identified area, and its minimum width did not change compared with the CLC2000 and CLC2006 layers (see Table 2.1 in Chapter 2). Thirty-eight European countries and Turkey participated in the CLC2012 project (see Table 2.2).

13.2 Statistical Characteristics

Tables 10.1 and 13.1 and Figures 13.1 and 13.2 present the statistical characteristics of CLC2012 classes. An overview of areas of CLC2012 classes in Europe is provided in Table 10.1. In this data layer the *Forest landscape* area again increased and (CLC classes 311–324) covers 2,227,996.7 km², corresponding to 43.9% of the area of the European land cover (see Table 10.1). The *Agricultural landscape* area (CLC classes 211–244) diminished

TABLE 13.1

Statistical Characteristics of the Dominant CLC2012 Third Level Classes within the Groups of the First Hierarchic Level in European Countries and Turkey

Country	Artificial Surfaces			Agricultural Areas			Forest and Semi-Natural Areas			Wetlands			Water Bodies		
	Dominant CLC Class	Area (in Km²)	% of Total Country's Area	Dominant CLC Class	Area (in Km²)	% of Total Country's Area	Dominant CLC Class	Area (in Km²)	% of Total Country's Area	Dominant CLC Class	Area (in Km²)	% of Total Country's Area	Dominant CLC Class	Area (in Km²)	% of Total Country's Area
Albania	112	664.05	2.31	243	2807.11	9.76	311	6253.91	21.74	421	48.90	0.17	512	460.57	1.60
Austria	112	3763.14	4.48	211	12,887.14	15.36	312	22,405.78	26.70	411	199.35	0.24	512	481.71	0.57
Belgium	112	5165.53	16.84	211	6685.93	21.80	313	2669.04	8.70	412	49.09	0.16	512	107.03	0.35
Bosnia/Herzegovina	112	609.05	1.20	242	7173.38	14.10	311	16,202.63	31.85	411	45.18	0.09	512	191.29	0.38
Bulgaria	112	3872.35	3.49	211	38,491.96	34.67	311	22,988.88	20.71	411	90.90	0.08	512	653.73	0.59
Croatia	112	1502.01	2.64	242	10,071.18	17.72	311	16,570.43	29.16	411	192.71	0.34	512	294.09	0.52
Cyprus	112	496.27	5.37	211	2287.35	24.73	323	1545.69	16.71	421	19.76	0.21	512	21.26	0.23
Czech Republic	112	3817.77	4.84	211	29,013.02	36.78	312	17,149.89	21.74	411	60.09	0.08	512	529.30	0.67
Denmark	112	2020.58	4.58	211	27,897.80	63.23	312	1697.49	3.85	423	590.28	1.34	521	494.90	1.12
Estonia	112	585.88	1.29	211	6866.12	15.09	313	9115.93	20.04	412	1314.42	2.89	512	2106.83	4.63
Finland	112	3264.09	0.96	211	15,473.64	4.57	312	141,660.75	41.88	412	21,250.05	6.28	512	31,345.27	9.27
France	112	22,538.88	4.08	211	153,909.22	27.89	311	88,390.58	16.02	423	2123.93	0.38	512	2177.54	0.39
Germany	112	24,760.05	6.84	211	135,872.79	37.54	312	59,267.38	16.38	423	2928.15	0.81	512	3271.13	0.90
Greece	112	2171.30	1.64	211	12,904.40	9.76	323	23,475.60	17.76	421	325.28	0.25	512	958.98	0.73
Hungary	112	4368.72	4.70	211	48,014.00	51.62	311	14,782.43	15.89	411	767.00	0.82	512	1299.70	1.40
Iceland	142	140.25	0.14	231	2526.35	2.44	322	35,474.63	34.28	412	6624.89	6.40	512	1295.71	1.25
Ireland	112	1109.99	1.57	231	38,654.96	54.71	324	2886.55	4.09	412	10,268.03	14.53	512	1124.77	1.59
Italy	112	10,188.84	3.38	211	80,769.88	26.80	311	55,467.30	18.40	421	442.06	0.15	512	1737.44	0.58
Kosovo	112	281.36	2.58	242	1797.32	16.47	311	4009.27	36.75	411	9.76	0.09	512	23.18	0.21
Latvia	112	774.27	1.20	211	10,786.09	16.69	313	10,570.03	16.35	412	1469.39	2.27	512	1134.36	1.75

(Continued)

TABLE 13.1 (CONTINUED)

Statistical Characteristics of the Dominant CLC2012 Third Level Classes within the Groups of the First Hierarchic Level in European Countries and Turkey

Country	Artificial Surfaces			Agricultural Areas			Forest and Semi-Natural Areas			Wetlands			Water Bodies		
	Dominant CLC Class	Area (in km²)	% of Total Country's Area	Dominant CLC Class	Area (in km²)	% of Total Country's Area	Dominant CLC Class	Area (in km²)	% of Total Country's Area	Dominant CLC Class	Area (in km²)	% of Total Country's Area	Dominant CLC Class	Area (in km²)	% of Total Country's Area
Liechtenstein	112	17.07	10.60	211	21.38	13.28	312	45.22	28.09	411	1.45	0.90	511	2.17	1.35
Lithuania	112	1511.05	2.31	211	21,375.55	32.65	313	7439.71	11.36	412	406.92	0.62	512	1085.33	1.66
Luxembourg	112	192.79	7.43	242	460.26	17.74	311	638.22	24.60	411	0.28	0.01	512	6.55	0.25
Macedonia FYR	112	326.76	1.29	211	2459.92	9.68	311	7274.86	28.62	411	19.73	0.08	512	530.26	2.09
Malta	112	66.93	21.21	243	149.70	47.43	323	49.71	15.75	422	0.26	0.08	0	0	0
Montenegro	112	186.33	1.35	243	1620.40	11.73	311	3690.55	26.71	411	111.78	0.81	512	253.20	1.83
Netherlands	112	3326.87	8.34	231	10,138.09	25.41	312	1595.89	4.00	423	2271.90	5.69	512	2703.17	6.77
Norway	112	1838.13	0.57	243	9685.13	2.99	333	81,372.37	25.13	412	21,294.94	6.58	512	13,340.76	4.12
Poland	112	14,479.62	4.63	211	136,121.09	43.55	312	56,350.34	18.03	411	1020.36	0.33	512	3915.82	1.25
Portugal	112	2492.62	2.70	211	8622.88	9.34	324	14,999.99	16.24	421	179.25	0.19	512	615.56	0.67
Romania	112	10,959.25	4.60	211	86,873.08	36.44	311	49,204.71	20.64	411	3151.98	1.32	511	1814.98	0.76
Serbia	112	2358.31	3.04	211	21,711.64	27.96	311	20,814.87	26.81	411	246.34	0.32	511	628.41	0.81
Slovakia	112	2314.07	4.72	211	16,137.37	32.92	311	10,828.00	22.09	411	38.35	0.08	512	216.19	0.44
Slovenia	112	421.47	2.08	242	2782.52	13.72	313	4526.87	22.33	411	24.40	0.12	511	51.12	0.25
Spain[a]	112	4873.87	0.96	211	100,111.66	19.78	311	50,859.63	10.05	411	443.35	0.09	512	2582.64	0.51
Sweden	112	4168.56	0.93	211	29,954.09	6.66	312	224,639.63	49.92	412	28,377.82	6.31	512	36,219.65	8.05
Switzerland	112	2325.09	5.63	211	7263.48	17.59	312	6151.60	14.90	411	36.49	0.09	512	1386.70	3.36
Turkey[b]															
United Kingdom	112	13,130.63	5.28	231	70,327.42	28.28	322	18,420.93	7.41	412	22,967.89	9.24	512	2243.42	0.90

[a] Preliminary data were used in the CLC2012 pan-European layer.

[b] Country participated in CLC 2012 Project; results will be available in the future.

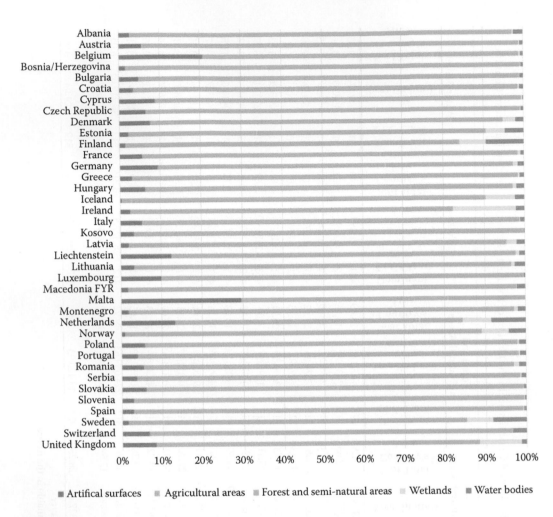

FIGURE 13.1
Extent of CLC2012 classes of the first hierarchic level in European countries (relative values in %).

to 2,114,473.6 km², which equates to 41.7%. As in CLC2006, as far as size is concerned, they are followed by *Open space with little or no vegetation* (CLC classes 331–335), with an area of 228,199.0 km² equating to 4.5%. The area of *Artificial surfaces* (CLC classes 111–142) increased to 223,684.7 km², or 4.4%. The *Wetlands* (CLC classes 411–423) occupy an area of 143,467.7 km², or 2.8%. *Water bodies* (CLC classes 511–522), with an area 138,061.6 km², or 2.7%, are the smallest.

Figures 13.1 and 13.2 show the relative (%) and absolute (km²) representation of classes on the first hierarchic level for individual countries as identified from the most recent CLC2012 data layer. Relative values of *Artificial surfaces* were highest (>10%) in small countries, that is, Belgium, Liechtenstein, Malta, and the Netherlands (see Figure 13.1). *Agricultural areas*, as in CLC2006 data layer, dominate in Denmark (ca. 75%), followed by Hungary (<70%) and France, Germany, Ireland, Lithuania, the Netherlands and Poland (>60%). *Forest and semi-natural areas* dominate in Finland, Iceland, Montenegro, Norway, and Sweden (>70%). In Albania, Austria, Liechtenstein, Slovenia, and Switzerland *Forest*

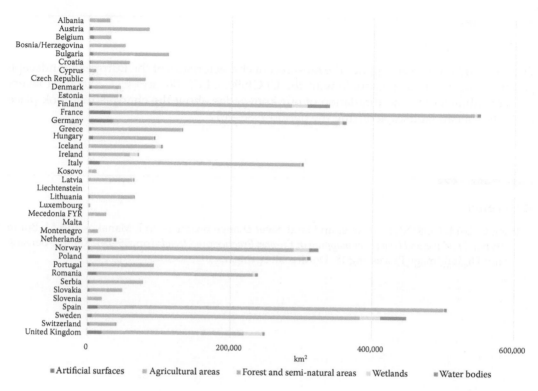

FIGURE 13.2

Extent of CLC2012 classes of the first hierarchic level in European countries (absolute values in km²).

and semi-natural areas occupy between 60% and 70% of the country surface area. *Wetlands* occurred most abundantly in Ireland (ca. 15%) and the United Kingdom (ca. 10%), followed by Finland, Iceland, the Netherlands, Norway, and Sweden (ca. 5%). The area covered by *Water bodies* was even smaller. In Finland, the Netherlands, and Sweden, 7% to 8% of the country surface was covered by *Water bodies*. In the other countries it was below 5%.

Figure 13.2 makes it possible to compare areas of CLC2012 classes between the European countries. Surface areas covered by *Artificial surfaces* are of similar size in France, Germany, and the United Kingdom. *Agricultural areas* are largest in France, Germany, Poland, and Spain. *Forest and semi-natural areas* cover the largest surfaces in Finland, Norway, Spain, and Sweden. Areas of *Wetlands* dominate in Finland, Norway, Sweden, and the United Kingdom. *Water bodies* are largest in Sweden and Finland.

Class 112 is the dominant class within the Level 1 *Artificial surfaces* in 38 countries with the exception of Iceland (see Table 13.1). The dominant representation within the *Agricultural landscape* is class 211 (25 countries), within the *Forest landscape* it is class 311 (15 countries), within the *Wetlands* it is class 411 (18 countries), and within the Level 1 class *Water bodies* it is 512 (in 32 countries). As in CLC1990, CLC2000, and CLC2006 data layers, the CLC2012 data layer also displays the greatest diversity in representation of CLC classes within the *Forest landscape*. On the other side, the representation of class 112 definitely dominates the *Artificial surfaces* (see Table 13.1).

13.3 Conclusion

The CLC2012 data layer supplies the most recent characteristics of the European landscape pattern. Its content is compatible with the CLC1990, CLC2000, and CLC2006 data layers and contributes to the accumulation of our knowledge about the changes that took place in the European landscape during the period 1986–2012.

Reference

Büttner, G. 2014. CORINE land cover and land cover change products. In I. Manakos and M. Braun (Eds.), *Land use and land cover mapping in Europe: Practices and trends* (pp. 55–74). Remote Sensing and Digital Image Processing 18. Dordrecht: Springer.

14

CORINE Land Cover 1990–2000 Changes: Analysis and Assessment

Tomas Soukup, Jan Feranec, Gerard Hazeu, Gabriel Jaffrain,
Marketa Jindrova, Miroslav Kopecky, and Erika Orlitova

CONTENTS

14.1 Background

The option to obtain information about land cover (LC) changes from the CLC1990, CLC2000, CLC2006, and CLC2012 data for the whole of Europe or for individual countries is one of their most important benefits. The compatible contents of CoORdination of Information on the Environment (CORINE) Land Cover (CLC) data from individual time horizons, as well as the applied interpretation procedures, offer a reliable guarantee of this benefit (Feranec et al., 2012; Büttner, 2014). Data concerning $CLC_{1990-2000}$ changes are available for 28 countries (with Turkey; see Table 2.2). Their time interval is not exactly 10 years for all countries, as the generation of data took place in 1986–1998 (see Table 2.1).

The aim of this chapter is to provide an overview of the size of all changes and their percentage of country's total area for the period 1990–2000. The three largest LC change types on the third level of *Artificial surfaces, Agricultural areas, Forest and semi-natural areas*, and *Wetlands and water bodies* are shown, and the area of all changes identified in individual countries are documented as well.

14.2 Percentages of Countries' Areas Affected by LC Changes

The percentages of the countries' surfaces affected by LC changes are shown in Figures 14.1 and 14.2 (Turkey is not included in this analysis, nor in Chapters 15 and 16). It is obvious from the figure and table that more than 10% of the country's surface changed in Portugal, followed by Ireland (8%) and the Czech Republic (>6%). Hungary, Latvia, the Netherlands, Slovakia, and Spain, with changes of about 4%, are also classified in the group of countries

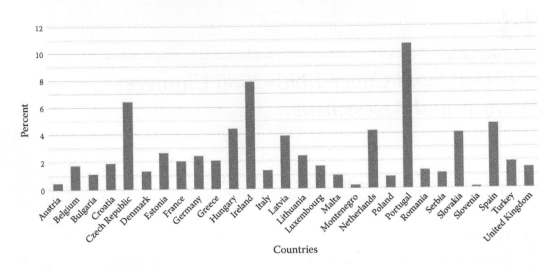

FIGURE 14.1
Percentage of a country's surface affected by LC changes in 1990–2000.

with the largest percentage of LC changes. The largest group of countries is that with 1%–2% changes while Austria, Montenegro, and Slovenia are the countries, with less than 0.5% changes.

The sum of LC changes per country in square kilometers was largest in Spain (23,706.7 km^2), France (11,071.5 km^2), and Portugal (9821.5 km^2) (see Figure 14.2).

The mean annual LC change percentage of the country's overall area reached 0.80% in the Czech Republic. The dominant changes were 211* in favor of 112, 211 in favor of 231, 324 in favor of 312, and 243 in favor of 512. Other countries with high mean annual change rates (>0.70%) were Ireland with 0.79% (dominant changes were 231 in favor of 112, 231 in favor of 211, and 412 in favor of 324 twice—for *Forest and semi-natural areas* and also *Wetlands and water bodies*), Latvia with 0.77% (dominant changes were 312 in favor of 131, 231 in favor of 211, 313 in favor of 324, and 412 in favor of 512), and Portugal with 0.76% (dominating changes were 242 in favor of 112, 211 in favor of 212, 312 in favor of 324, and 211 in favor of 512) (see Figure 14.2).

14.3 Overview of Dominant LC Changes in Individual Countries

Figure 14.2 contains an overview of four quantitative characteristics of LC changes in individual countries and three dominating changes within *Artificial surfaces, Agricultural areas, Forest and semi-natural areas,* and *Wetlands and water bodies* (these are types of classes on the first hierarchic level of CLC nomenclature while *Wetlands and water bodies* were brought together to make one class for purposes of this assessment; see Table 3.1). There were two types of changes: those between classes on the first hierarchic level denoted "interclass changes" (e.g., from *Agricultural areas* into *Artificial surfaces*) and between classes of one

---

* CLC class codes quoted in this chapter and Chapters 15 and 16 are explained in Table 3.1.

Country	Country area (km²)	Artificial surfaces	Agricultural areas	Forest and semi-natural areas	Wetlands and water bodies	Area of all changes (km²)	All changes (% of total country area)	Time difference (CLC 1990–2000), years	Yearly changes (% of total country area)
Albania[a]									
Austria	83,908.99	211 – 142 211 – 112 231 – 142	211 – 142 211 – 112 231 – 142	335 – 332 324 – 312 312 – 324	131 – 512 211 – 512 333 – 512	357.84	0.43	15	0.03
Belgium	30,668.97	211 – 121 242 – 121 242 – 112	211 – 121 242 – 121 242 – 112	324 – 312 312 – 324 324 – 313	211 – 512 312 – 412 131 – 512	527.85	1.72	10	0.17
Bosnia/Herzegovina[a]									
Bulgaria	111,012.95	211 – 131 132 – 211 231 – 131	213 – 211 222 – 211 231 – 211	324 – 311 311 – 324 324 – 313	512 – 411 243 – 512 512 – 211	1216.29	1.10	10	0.11
Croatia	56,813.84	242 – 112 311 – 133 242 – 121	211 – 243 242 – 231 324 – 231	311 – 324 324 – 311 324 – 231	512 – 411 242 – 512 511 – 311	1058.95	1.86	10	0.19
Cyprus[a]									
Czech Republic	78,885.69	211 – 112 132 – 324 131 – 132	211 – 231 231 – 211 211 – 243	324 – 312 312 – 324 324 – 313	243 – 512 211 – 512 131 – 512	5053.85	6.41	8	0.80
Denmark	44,064.24	211 – 112 211 – 142 211 – 121	211 – 324 211 – 112 211 – 142	312 – 324 211 – 324 313 – 324	211 – 411 523 – 331 523 – 123	563.61	1.28	10	0.13
Estonia	45,492.42	243 – 112 324 – 131 131 – 333	231 – 211 211 – 231 231 – 242	313 – 324 312 – 324 311 – 324	324 – 412 312 – 412 412 – 324	1189.52	2.61	6	0.44
Finland[a]									
France	551,899.86	242 – 112 211 – 121 211 – 112	231 – 211 211 – 231 242 – 112	324 – 312 312 – 324 324 – 311	131 – 512 231 – 512 211 – 512	11,071.49	2.01	10	0.20
Germany	361,635.07	211 – 112 211 – 121 211 – 131	211 – 231 211 – 112 211 – 242	312 – 324 321 – 324 324 – 313	523 – 423 131 – 512 211 – 512	8640.58	2.39	10	0.24
Greece	132,066.01	211 – 131 242 – 112 242 – 121	211 – 212 212 – 213 213 – 212	312 – 324 311 – 324 324 – 311	512 – 331 512 – 211 512 – 411	2709.11	2.05	10	0.21
Hungary	93,017.63	211 – 121 211 – 131 211 – 122	231 – 211 211 – 231 211 – 324	324 – 311 311 – 324 211 – 324	411 – 512 211 – 512 231 – 512	4096.37	4.40	9	0.49
Iceland[a]									
Ireland	70,632.15	231 – 112 231 – 142 211 – 112	231 – 211 211 – 231 231 – 242	412 – 324 312 – 324 324 – 312	412 – 324 412 – 312 512 – 321	5575.28	7.89	10	0.79
Italy	301,357.81	211 – 112 242 – 112 211 – 121	243 – 323 211 – 112 242 – 112	324 – 311 243 – 323 321 – 324	211 – 512 411 – 512 512 – 331	3903.64	1.30	10	0.13
Kosovo[a]									
Latvia[b]	64,642.23	312 – 131 311 – 131 313 – 131	231 – 211 211 – 231 222 – 211	313 – 324 312 – 324 311 – 324	412 – 512	2490.91	3.85	5	0.77

FIGURE 14.2

Three largest LC changes on the third level within the first hierarchic level CLC for 1990–2000.

(Continued)

Country	Country area (km²)	Artificial surfaces	Agricultural areas	Forest and semi-natural areas	Wetlands and water bodies	Area of all changes (km²)	All changes (% of total country area)	Time difference (CLC 1990–2000), years	Yearly changes (% of total country area)
Liechtenstein[a]									
Lithuania	65,524.70	133 – 112 211 – 131 133 – 121	231 – 211 211 – 242 231 – 242	313 – 324 312 – 324 311 – 324	243 – 512 242 – 512 512 – 231	1556.70	2.38	5	0.48
Luxembourg[b]	2593.98	242 – 112 231 – 112 243 – 112	242 – 112 231 – 112 243 – 112	312 – 324 311 – 324 313 – 324		42.12	1.62	11	0.15
Macedonia FYR[a]									
Malta[b]	315.49	243 – 112 133 – 121 323 – 131	243 – 112 243 – 131 243 – 121	323 – 131 333 – 131 323 – 132	523 – 123 123 – 523	2.98	0.94	10	0.09
Montenegro	13,818.21	324 – 131 231 – 132 243 – 131	231 – 324 231 – 132 243 – 131	311 – 324 312 – 324 324 – 311	512 – 411 132 – 512 411 – 512	26.99	0.20	10	0.02
Netherlands	39,869.81	231 – 112 211 – 112 242 – 112	231 – 211 231 – 112 211 – 112	211 – 311 211 – 321 231 – 311	231 – 411 331 – 423 231 – 512	1672.01	4.19	14	0.30
Norway[a]									
Poland	312,594.34	211 – 131 133 – 112 211 – 133	231 – 211 211 – 231 211 – 131	312 – 324 324 – 312 324 – 313	231 – 512 411 – 324 411 – 512	2542.91	0.81	8	0.10
Portugal	92,343.30	242 – 112 241 – 112 243 – 112	211 – 212 243 – 324 231 – 212	312 – 324 324 – 311 324 – 312	211 – 512 243 – 512 512 – 321	9821.46	10.64	14	0.76
Romania	238,405.53	211 – 112 211 – 121 243 – 131	213 – 211 231 – 211 211 – 231	324 – 311 312 – 324 311 – 324	411 – 211 512 – 411 411 – 512	3076.19	1.29	8	0.16
Serbia	77,638.18	211 – 112 242 – 112 131 – 324	211 – 231 242 – 231 211 – 242	324 – 311 311 – 324 324 – 312	231 – 512 324 – 512 211 – 512	825.14	1.06	10	0.11
Slovakia	49,022.85	133 – 512 211 – 112 133 – 511	211 – 242 231 – 324 231 – 211	312 – 324 324 – 311 324 – 313	133 – 512 133 – 511 411 – 512	1983.60	4.05	8	0.51
Slovenia	20,275.54	133 – 122 311 – 122 312 – 133	411 – 231 311 – 243 243 – 312	311 – 324 334 – 324 311 – 243	512 – 411 411 – 231 311 – 512	24.46	0.12	5	0.02
Spain	506,213.13	211 – 121 211 – 112 242 – 112	211 – 212 211 – 242 324 – 244	312 – 324 324 – 312 323 – 324	321 – 512 211 – 512 323 – 512	23,706.67	4.68	14	0.33
Sweden[a]									
Switzerland[a]									
Turkey	779,649.49	242 – 112 211 – 112 242 – 121	242 – 112 211 – 512 243 – 512	312 – 324 324 – 311 324 – 312	211 – 512 243 – 512 333 – 512	14,758.22	1.89	10	0.19
United Kingdom	248,030.56	211 – 142 211 – 112 231 – 112	211 – 142 211 – 112 231 – 112	324 – 312 322 – 312 312 – 324	412 – 312 412 – 324 211 – 512	3625.15	1.46	10	0.15

[a] Country did not participate in the CLC1990 project.

[b] Empty cells indicate no change.

Legend: CLC level 1:
1. Artificial surfaces
2. Agricultural areas
3. Forest and semi-natural areas
4.+ 5. Wetlands and water bodies

Note: The same change, for example, 211 to 112, can be contained within *Artificial surfaces* because the class 112 increased, but also within *Agricultural areas* because class 211 diminished.

FIGURE 14.2 (CONTINUED)
Three largest LC changes on the third level within the first hierarchic level CLC for 1990–2000.

hierarchic level denoted "intraclass changes" (e.g., between Level 3 classes within the *Forest* and *semi-natural areas*).

Interclass changes of *Agricultural areas* in favor of *Artificial surfaces*, for instance, 211, 231, 242, and 243 in favor of 112, 121, 122, 131, and 142 dominated in *Artificial surfaces*. Intraclass changes were not so frequent, for instance, 133 in favor of 112, 121, and 122, and 131 in favor of 132 (see Figure 14.2).

Intraclass changes prevailed in *Agricultural areas*, particularly between 211–231–242–243. Interclass changes (mostly decrease in agricultural areas) were identified to a lesser extent, for instance 211, 231, and 243 in favor of *Artificial surfaces* (112, 142, 121, and 131) (see Figure 14.2).

Intraclass changes prevailed in *Forest and semi-natural areas*, particularly those of 311, 312, and 313 in favor of 324 and vice versa. Interclass changes are scarcely represented: 211 and 231 in favor of 311 and 324 (see Figure 14.2).

Most changes identified in *Wetlands and water bodies* were interclass changes, that is, 211, 231, 242, and 243 in favor of 512, as well as 412 in favor of 324 and 312, and 133 in favor of 511 and 512. Changes of 4xx (e.g., 4xx means all classes falling under the first level "Wetlands" class; see Table 3.1) in favor of 5xx (e.g., 5xx means all classes falling under the first level "Water bodies" class; see Table 3.1) and vice versa were taken as intraclass changes (see Figure 14.2).

14.4 Conclusion

In terms of the types of changes between CLC1990 and CLC2000, the dominance of interclass changes between *Artificial surfaces* and the other hierarchic Level 1 groups should be emphasized. The combination of intraclass and interclass changes exist for the groups of *Agricultural areas* and *Wetlands and water bodies*. Intraclass changes prevailed only in *Forest and semi-natural areas*. $CLC_{1990-2000}$ data are now available at: http://land.copernicus .eu/pan-european/corine-land-cover/view.

References

Büttner, G. 2014. CORINE land cover and land cover change products. In I. Manakos and M. Braun (Eds.), *Land use and land cover mapping in Europe: Practices and trends* (pp. 55–74). Remote Sensing and Digital Image Processing 18. Dordrecht: Springer.

Feranec, J., Soukup, T., Hazeu, G. W., Jaffrain, G. 2012. Land cover and its change in Europe: 1990–2006. In G. Chandra (Ed.), *Remote sensing of land use and land cover: Principles and applications* (pp. 285–301). Boca Raton, FL: CRC Press.

15

CORINE Land Cover 2000–2006 Changes: Analysis and Assessment

Tomas Soukup, Jan Feranec, Gerard Hazeu, Gabriel Jaffrain,
Marketa Jindrova, Miroslav Kopecky, and Erika Orlitova

CONTENTS

15.1 Background

Data about $CLC_{2000-2006}$ changes are available for 39 countries (Turkey included; see Table 2.2). They characterize approximately a 6-year time interval of these changes (see Table 2.1; Feranec et al., 2012; Büttner, 2014).

The aim of this chapter is to provide an overview of the size of all changes and their percentages of the countries' total areas for the period 2000–2006. The three largest LC change types on the third level per main LC category (first hierarchic level) are shown, and the areas of all changes identified in individual countries are documented as well.

15.2 Percentages of Countries' Areas Affected by LC Changes

Figures 15.1 and 15.2 contain information about the percentages of countries' surface affected by LC changes in 2000–2006. The area of changes during the 6-year period was more than 8% in Portugal. It distinctly exceeds the size of LC change in the other 38 countries for which the $CLC_{2000-2006}$ change layer was generated. Albania, Cyprus, Estonia, Hungary, Ireland, Latvia, and Sweden represent the second group of countries in which the percentage of changes is between 2% and 3%. In the largest group of 25 countries the value of change is between 1% and 2%. The last group consists of Austria, Liechtenstein, Malta, Montenegro, Romania, Serbia, Slovenia, and Switzerland, with changes below 0.5% of the country surface.

The area of all identified LC changes in this period was largest in Sweden (12,967.4 km²), followed by Spain (8589.5 km²) and Portugal (7631.1 km²) (see Figure 15.2).

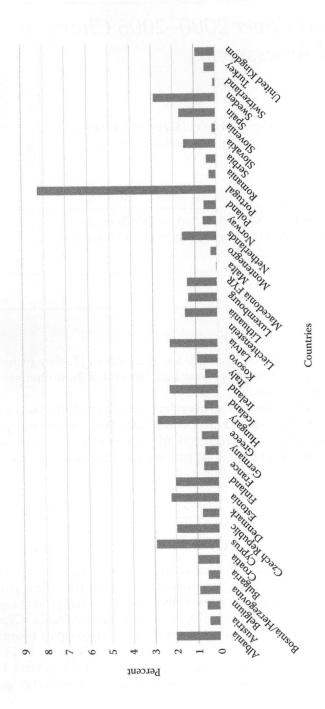

FIGURE 15.1
Percentage of a country's surface affected by LC changes in 2000–2006.

Country	Country area (km²)	Artificial surfaces	Agricultural areas	Forest and semi-natural areas	Wetlands and water bodies	Area of all changes (km²)	All changes (% of total country area)	Time difference (CLC 2000–2006), years	Yearly changes (% of total country area)
Albania	28,768.63	242 – 112 243 – 112 211 – 112	242 – 112 243 – 112 211 – 112	311 – 324 324 – 311 312 – 324	523 – 331 512 – 231 511 – 331	586.09	2.04	11	0.19
Austria	83,909.00	321 – 142 312 – 142 211 – 112	211 – 112 211 – 131 211 – 133	312 – 324 335 – 332 313 – 324	512 – 511 411 – 231 321 – 512	376.93	0.45	6	0.07
Belgium	30,671.94	211 – 121 211 – 133 322 – 133	211 – 121 211 – 133 211 – 131	312 – 324 324 – 312 311 – 324	133 – 522 211 – 512 133 – 512	181.83	0.59	6	0.10
Bosnia/Herzegovina	50,874.89	242 – 112 211 – 112 243 – 112	242 – 112 231 – 242 211 – 112	311 – 324 324 – 311 324 – 313	411 – 512 211 – 511 243 – 511	467.00	0.92	8	0.11
Bulgaria	111,013.18	211 – 131 231 – 131 243 – 131	222 – 211 211 – 131 221 – 211	311 – 324 324 – 311 312 – 324	211 – 411 523 – 123 512 – 124	572.59	0.52	6	0.09
Croatia	56,814.56	311 – 122 324 – 122 133 – 122	231 – 242 231 – 243 231 – 211	311 – 324 324 – 311 312 – 324	411 – 243 131 – 512 324 – 512	555.76	0.98	6	0.16
Cyprus[a]	9247.70	242 – 112 211 – 112 211 – 121	242 – 112 334 – 243 211 – 112	334 – 324 334 – 243 321 – 211	211 – 511 523 – 142	269.38	2.91	6	0.49
Czech Republic	78,885.99	211 – 133 132 – 231 211 – 121	211 – 231 231 – 211 211 – 221	324 – 312 312 – 324 324 – 313	131 – 512 231 – 512 211 – 512	1548.52	1.96	6	0.33
Denmark	44,076.55	211 – 112 211 – 142 211 – 133	211 – 112 211 – 142 211 – 133	312 – 324 324 – 312 324 – 313	412 – 421 512 – 521 231 – 512	329.31	0.75	6	0.12
Estonia	45,492.28	131 – 324 324 – 131 242 – 112	231 – 211 231 – 324 211 – 324	312 – 324 313 – 324 324 – 311	324 – 412 412 – 131 121 – 512	1012.15	2.22	6	0.37
Finland	338,538.75	312 – 131 312 – 112 211 – 142	412 – 211 324 – 211 312 – 211	312 – 324 324 – 312 324 – 313	412 – 211 324 – 412 312 – 412	6764.27	2.00	6	0.33
France	551,892.73	242 – 112 211 – 121 211 – 112	242 – 112 211 – 121 211 – 112	312 – 324 311 – 324 324 – 312	211 – 512 231 – 512 131 – 512	3638.63	0.66	6	0.11
Germany	361,726.57	211 – 112 211 – 131 133 – 112	231 – 211 211 – 112 211 – 131	312 – 324 324 – 313 324 – 312	333 – 512 131 – 512 324 – 312	2190.83	0.61	6	0.10
Greece	132,154.77	133 – 122 242 – 112 211 – 131	211 – 242 211 – 231 242 – 112	312 – 324 311 – 324 324 – 312	243 – 512 211 – 512 211 – 411	1001.46	0.76	6	0.13
Hungary	93,014.39	211 – 133 211 – 131 231 – 133	211 – 324 231 – 211 211 – 222	311 – 324 324 – 311 211 – 324	211 – 512 231 – 512 321 – 411	2611.24	2.81	6	0.47
Iceland	103,441.60	322 – 133 322 – 142 322 – 121	231 – 142 231 – 242 231 – 133	331 – 511 335 – 332 511 – 331	331 – 511 511 – 331 335 – 512	646.25	0.62	6	0.10
Ireland	70,757.88	231 – 112 231 – 133 211 – 112	231 – 324 231 – 112 243 – 324	312 – 324 324 – 312 412 – 324	412 – 324 412 – 131 411 – 324	1590.68	2.25	6	0.37
Italy	301,367.78	211 – 121 211 – 112 211 – 133	211 – 121 211 – 112 211 – 133	311 – 324 324 – 311 334 – 324	211 – 512 321 – 512 511 – 331	1779.04	0.59	6	0.10
Kosovo[a]	10,910.00	211 – 112 242 – 112 131 – 324	231 – 211 231 – 242 211 – 112	324 – 311 334 – 324 311 – 324		102.51	0.94	6	0.16
Latvia	64,642.13	231 – 133 231 – 131 211 – 133	231 – 211 231 – 324 211 – 231	313 – 324 312 – 324 311 – 324	412 – 324 512 – 411 131 – 512	1433.80	2.22	6	0.37

FIGURE 15.2

Three largest LC changes on the third level of the first hierarchic level CLC for 2000–2006. *(Continued)*

Country	Country area (km²)	Artificial surfaces	Agricultural areas	Forest and semi-natural areas	Wetlands and water bodies	Area of all changes (km²)	All changes (% of total country area)	Time difference (CLC 2000–2006), years	Yearly changes (% of total country area)
Liechtenstein[b]									
Lithuania	65,473.87	133 – 112	231 – 211	313 – 324	512 – 231	975.60	1.49	6	0.25
		211 – 133	211 – 231	312 – 324	412 – 324				
		242 – 133	211 – 324	324 – 311	512 – 411				
Luxembourg[a]	2594.00	132 – 324	231 – 211	324 – 311		34.69	1.34	6	0.22
		231 – 121	231 – 121	324 – 313					
		133 – 122	211 – 231	132 – 324					
Macedonia FYR	25,422.99	242 – 112	221 – 211	311 – 324	324 – 512	351.66	1.38	10	0.14
		132 – 231	211 – 221	324 – 311	243 – 512				
		131 – 231	231 – 324	324 – 312	311 – 512				
Malta[a]	315.60	323 – 132		323 – 132		0.07	0.02	6	0.00
Montenegro[a]	13,818.06	311 – 133	331 – 242	311 – 324		35.48	0.26	6	0.04
		324 – 133	243 – 133	324 – 311					
		243 – 112	243 – 112	313 – 324					
Netherlands	39,877.04	211 – 133	211 – 133	231 – 321	231 – 411	637.24	1.60	6	0.27
		231 – 133	231 – 133	211 – 321	242 – 512				
		133 – 112	242 – 133	231 – 321	231 – 512				
Norway	323,785.23	312 – 142	211 – 121	312 – 324	412 – 142	1938.22	0.60	6	0.10
		312 – 112	211 – 112	324 – 312	412 – 211				
		311 – 142	243 – 112	324 – 313	423 – 523				
Poland	312,516.11	211 – 131	211 – 324	312 – 324	131 – 512	1791.62	0.57	6	0.10
		131 – 324	211 – 222	324 – 312	231 – 512				
		211 – 133	231 – 211	211 – 324	211 – 512				
Portugal	92,345.79	133 – 112	211 – 324	312 – 324	311 – 512	7631.15	8.26	6	1.38
		133 – 121	244 – 324	311 – 324	244 – 512				
		324 – 133	243 – 324	313 – 324	211 – 512				
Romania	238,405.80	211 – 112	231 – 211	312 – 324	411 – 211	752.93	0.32	6	0.05
		211 – 121	211 – 112	311 – 324	411 – 142				
		242 – 112	211 – 121	313 – 324	211 – 512				
Serbia	77,641.00	242 – 112	231 – 211	311 – 324	242 – 512	344.72	0.44	6	0.07
		324 – 131	321 – 211	324 – 311	231 – 512				
		242 – 131	242 – 211	321 – 211	211 – 512				
Slovakia	49,023.00	211 – 133	231 – 324	312 – 324	211 – 512	732.50	1.49	6	0.25
		211 – 121	211 – 242	311 – 324	243 – 512				
		211 – 131	211 – 133	313 – 324	324 – 512				
Slovenia	20,275.65	313 – 133	242 – 133	312 – 324	231 – 512	29.63	0.15	6	0.02
		311 – 133	211 – 122	311 – 324	242 – 512				
		242 – 133	211 – 133	313 – 324	311 – 512				
Spain	506,261.83	211 – 133	211 – 223	311 – 324	211 – 512	8589.51	1.70	5	0.34
		133 – 112	211 – 212	312 – 324	244 – 512				
		242 – 133	211 – 133	334 – 324	523 – 123				
Sweden	450,023.57	312 – 133	211 – 142	312 – 324	211 – 512	12,967.37	2.88	6	0.48
		211 – 142	211 – 231	324 – 312	512 – 411				
		312 – 131	211 – 133	324 – 313	312 – 412				
Switzerland[a]	41,282.00	211 – 133	211 – 133	335 – 332		35.19	0.09	6	0.01
		211 – 131	211 – 131	312 – 324					
		211 – 142	211 – 131	313 – 324					
Turkey	779,685.91	133 – 112	211 – 212	312 – 324	242 – 512	3791.95	0.49	6	0.08
		211 – 133	321 – 211	324 – 311	512 – 331				
		133 – 121	212 – 211	311 – 324	243 – 512				
United Kingdom	248,137.67	231 – 112	231 – 112	312 – 324	412 – 324	2293.71	0.92	6	0.15
		211 – 112	211 – 112	324 – 312	231 – 412				
		231 – 142	231 – 312	321 – 312	211 – 512				

[a] Empty cells indicate no change.
[b] No changes occurred between 2000 and 2006.

Legend: CLC level 1:
1. Artificial surfaces
2. Agricultural areas
3. Forest and semi-natural areas
4. + 5. Wetlands and water bodies

FIGURE 15.2 (CONTINUED)
Three largest LC changes on the third level of the first hierarchic level CLC for 2000–2006.

Yearly change (% of total country area), which makes it possible to compare the rate of LC changes between individual countries, was largest in Portugal, with 1.38%. The dominant changes were 133 in favor of 112, 211 in favor of 324, 312 in favor of 324, and 311 in favor of 512. Other countries with high annual change rates (>0.4%) were Cyprus (0.49%, with dominant changes of 242 in favor of 112 twice [for *Artificial surfaces* and also for *Agricultural areas*], 334 in favor of 324, and 211 in favor of 511), Sweden (0.48%, with dominating changes 312 in favor of 133, 211 in favor of 142, 312 in favor of 324, and 211 in favor of 512), and Hungary (0.47%, with dominant changes of 211 in favor of 133, 211 in favor of 324, 311 in favor of 324, and 211 in favor of 512) (see Figure 15.2).

15.3 Overview of Dominant LC Changes in Individual Countries

Figure 15.2 contains an overview of the three largest change types within the *Artificial surfaces*, *Agricultural areas*, *Forest and semi-natural areas*, and *Wetlands and water bodies* (see also Section 14.3 in Chapter 14). Interclass changes of *Agricultural areas* and *Forest and semi-natural areas* in favor of *Artificial surfaces* (211, 231, 242, 243, 311, 312, and 324 in favor of 112, 121, 131, and 133) dominated in *Artificial surfaces*. These changes are characteristic for almost all countries (see Figure 15.2).

Interclass changes prevailed in *Agricultural areas*, as represented by diminishments of classes such as 211, 231, 242, and 243 in favor of *Artificial surfaces* (112, 121, 131, 133, and 142), occasionally in favor of *Forest and semi-natural areas* (211 and 231 in favor of 324). Intraclass changes (211–231–242–243) are less represented (see Figure 15.2).

Intraclass changes 311, 312, and 313 in favor of 324 and vice versa (see Figure 15.2) dominated in *Forest and semi-natural areas*.

Changes in *Wetlands and water bodies* were remarkable for their high heterogeneity. Most of the changes were interclass ones, for instance, 211, 231, and 243 in favor of 511 and 512 and 412 in favor of 324; intraclass changes were rare (see Figure 15.2).

15.4 Conclusion

Representation of LC change types between 2000 and 2006 in terms of content looks similar to that of the first time horizon. Intraclass changes prevailed in *Forest and semi-natural areas* in contrast to *Artificial surfaces* and *Wetlands and water bodies*, where interclass changes prevailed. *Agricultural areas* are characterized by a combination of inter- and intraclass changes (see Figure 15.2). $CLC_{2000-2006}$ data are now available at: http://land.copernicus.eu/pan-european /corine-land-cover/view.

References

Büttner, G. 2014. CORINE land cover and land cover change products. In I. Manakos and M. Braun (Eds.), *Land use and land cover mapping in Europe: Practices and trends* (pp. 55–74). Remote Sensing and Digital Image Processing, 18. Dordrecht: Springer.

Feranec, J., Soukup, T., Hazeu, G.W., Jaffrain, G. 2012. Land cover and its change in Europe: 1990–2006. In G. Chandra (Ed.), *Remote sensing of land use and land cover: Principles and applications* (pp. 285–301). Boca Raton, FL: CRC Press.

16

CORINE Land Cover 2006–2012 Changes: Analysis and Assessment

Tomas Soukup, György Büttner, Jan Feranec, Gerard Hazeu, Gabriel Jaffrain, Marketa Jindrova, Miroslav Kopecky, and Erika Orlitova

CONTENTS

16.1 Background

Data regarding $CLC_{2006-2012}$ changes are available for 39 countries (see Table 2.2). They characterize an approximately a 6-year interval of tracked changes (see Table 2.1; Büttner, 2014).

Like in previous chapters, the aim is to provide an overview of the size of all changes in each country and their percentages of a country's total area for the period 2006–2012, to show the three largest LC changes on the third level of the first hierarchic level and to document the area of all changes identified in individual countries.

16.2 Percentages of Countries' Areas Affected by LC Changes

In the third time horizon, that is, $CLC_{2006-2012}$, the percentage of the country's surface affected by LC changes was highest in Sweden, as the value of its change amounted to 7.4%. The LC change in Portugal between 2006 and 2012 decreased to 5.4%. The group of countries with 2% to 4% changes include the Czech Republic, Estonia, Hungary, and Latvia. The most abundant representation of countries was that within the group with 0.5% to 2% changes. Countries identified with less than 0.5% changes were Austria, Bosnia/Herzegovina, Bulgaria, Liechtenstein, Luxembourg, Malta, Montenegro, Romania, Serbia, Slovenia, and Switzerland (see Figures 16.1 and 16.2).

The summarized area of identified changes in the 6-year period was largest in Sweden ($33,400.7$ km^2), France (6650.9 km^2), Spain (6585.1 km^2), and Finland (6443.4 km^2) (see Figure 16.2).

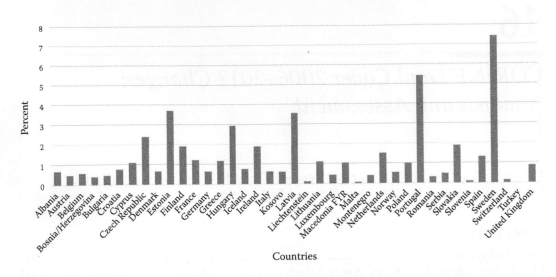

FIGURE 16.1
Percentage of a country's surface affected by LC changes in 2006–2012.

Yearly changes (% of total country area) were largest in Sweden (1.24%). The dominant changes were 312 in favor of 131, 211 in favor of 112, 324 in favor of 312, and 412 in favor of 324. Other countries with high annual change rates (>0.6%) were Portugal (0.9%, with the dominant changes of 133 in favor of 122, 211 in favor of 223, 312 in favor of 324, and 211 in favor of 512), and Estonia (0.61% with dominant changes of 324 in favor of 131, 211 in favor of 231, 324 in favor of 313, and 412 in favor of 131) (see Figure 16.2).

16.3 Overview of Dominant LC Changes in Individual Countries

Three largest area changes of LC found in individual countries are shown in Figure 16.2. The pattern of their content is similar to both interclass and intraclass changes described under Section 15.3 in Chapter 15.

16.4 Conclusion

The content of LC change types between 2006 and 2012 (third time horizon) looks similar to that in the previous two time horizons. The prevalence of intraclass changes in *Forest and semi-natural areas* remained, interclass changes prevailed in *Wetlands and water bodies* and in the combination of inter- and intraclass changes in *Agricultural areas*, as well as in *Artificial surfaces*, as Figure 16.2 demonstrates. $CLC_{2006-2012}$ data are now available at: http://land.copernicus.eu/pan-european/corine-land-cover/view.

Country	Country area (km²)	Artificial surfaces	Agricultural areas	Forest and semi-natural areas	Wetlands and water bodies	Area of all changes (km²)	All changes (% of total country area)	Time difference (CLC 2006–2012), years	Yearly changes (% of total country area)
Albania	28,768.63	242 – 121	242 – 121	323 – 334	331 – 511	188.62	0.66	6	0.11
		133 – 112	242 – 222	311 – 334	512 – 243				
		323 – 133	242 – 112	311 – 324	324 – 511				
Austria	83,909.00	312 – 142	211 – 121	312 – 324	211 – 512	389.70	0.46	6	0.08
		211 – 121	211 – 131	311 – 324	131 – 512				
		211 – 131	211 – 112	312 – 321	512 – 231				
Belgium	30,671.94	211 – 133	211 – 133	312 – 324	231 – 421	173.29	0.56	6	0.09
		242 – 133	242 – 133	313 – 324	133 – 411				
		211 – 121	211 – 121	324 – 312	512 – 133				
Bosnia/Herzegovina	50,874.89	242 – 131	243 – 222	324 – 311	512 – 211	194.45	0.38	6	0.06
		243 – 131	512 – 211	311 – 324	411 – 211				
		324 – 131	243 – 324	311 – 334	511 – 324				
Bulgaria	111,013.18	211 – 121	211 – 221	311 – 324	311 – 512	491.19	0.44	6	0.07
		211 – 131	231 – 211	324 – 311	133 – 512				
		133 – 142	211 – 222	312 – 324	512 – 211				
Croatia	56,814.56	133 – 122	231 – 211	311 – 324	512 – 411	426.20	0.75	6	0.13
		311 – 122	231 – 324	324 – 311	231 – 512				
		311 – 133	324 – 221	324 – 333	242 – 512				
Cyprus	9247.70	133 – 112	211 – 133	323 – 334	312 – 512	100.24	1.08	6	0.18
		133 – 142	242 – 112	324 – 312	131 – 512				
		211 – 133	242 – 142	312 – 324	523 – 133				
Czech Republic	78,885.99	211 – 112	211 – 231	312 – 324	131 – 512	1883.07	2.39	6	0.40
		211 – 121	231 – 211	324 – 313	211 – 512				
		211 – 131	222 – 211	324 – 312	231 – 512				
Denmark	44,076.55	211 – 112	211 – 112	312 – 324	211 – 512	288.33	0.65	6	0.11
		133 – 112	211 – 121	313 – 324	131 – 512				
		211 – 121	211 – 133	211 – 324	411 – 231				
Estonia	45,492.28	324 – 131	211 – 231	324 – 313	412 – 131	1676.71	3.69	6	0.61
		133 – 112	231 – 211	313 – 324	324 – 412				
		132 – 121	231 – 324	312 – 324	312 – 412				
Finland	338,538.75	312 – 131	324 – 211	312 – 324	312 – 412	6443.38	1.90	6	0.32
		312 – 133	211 – 412	324 – 312	211 – 412				
		324 – 121	211 – 324	324 – 313	324 – 412				
France	551,892.73	133 – 112	231 – 211	312 – 324	211 – 512	6650.86	1.21	6	0.20
		211 – 121	211 – 121	324 – 312	211 – 512				
		211 – 133	211 – 133	324 – 311	231 – 512				
Germany	361,726.57	211 – 121	231 – 211	312 – 324	131 – 512	2266.33	0.63	6	0.10
		211 – 131	211 – 231	324 – 312	211 – 512				
		231 – 121	211 – 121	324 – 311	231 – 512				
Greece	132,154.77	133 – 122	231 – 512	312 – 324	231 – 512	1522.28	1.15	6	0.19
		223 – 133	213 – 212	313 – 324	133 – 512				
		131 – 324	212 – 213	324 – 311	411 – 512				
Hungary	93,014.39	133 – 112	211 – 324	311 – 324	211 – 512	2728.73	2.93	6	0.49
		133 – 122	211 – 231	324 – 311	512 – 231				
		211 – 133	231 – 211	211 – 324	231 – 512				
Iceland	103,441.60	133 – 112	412 – 231	335 – 332	511 – 331	775.65	0.75	6	0.12
		133 – 512	321 – 231	511 – 331	331 – 511				
		133 – 142	322 – 231	331 – 511	322 – 512				
Ireland	70,757.88	133 – 112	211 – 231	312 – 324	412 – 324	1327.26	1.88	6	0.31
		133 – 122	231 – 211	324 – 312	412 – 131				
		133 – 121	231 – 324	231 – 324	411 – 324				
Italy	301,367.78	211 – 121	211 – 121	324 – 311	521 – 421	1848.77	0.61	6	0.10
		211 – 133	222 – 211	311 – 324	211 – 512				
		133 – 121	311 – 211	311 – 211	231 – 512				
Kosovo[a]	10,910.00	242 – 112	231 – 211	311 – 334		65.53	0.60	6	0.10
		211 – 121	221 – 211	311 – 324					
		211 – 124	242 – 112	324 – 131					
Latvia	64,642.13	133 – 112	231 – 211	313 – 324	324 – 412	2296.98	3.55	6	0.59
		231 – 133	211 – 231	312 – 324	411 – 133				
		324 – 131	231 – 324	311 – 324	131 – 512				

FIGURE 16.2

Three largest LC changes on the third level in the context of the first hierarchic level CLC for 2006–2012.

(*Continued*)

Country	Country area (km²)	Artificial surfaces	Agricultural areas	Forest and semi-natural areas	Wetlands and water bodies	Area of all changes (km²)	All changes (% of total country area)	Time difference (CLC 2006–2012), years	Yearly changes (% of total country area)
Liechtenstein[a]	161.00	211 – 121	211 – 121			0.16	0.10	6	0.02
Lithuania	65,473.87	133 – 112	231 – 211	312 – 324	411 – 324	713.30	1.09	6	0.18
		211 – 133	243 – 324	313 – 324	131 – 512				
		131 – 324	211 – 324	311 – 324	324 – 412				
Luxembourg[a]	2594.00	231 – 133	231 – 133	312 – 324		10.78	0.42	6	0.07
		121 – 133	242 – 133	324 – 142					
		242 – 133	211 – 133	313 – 324					
Macedonia FYR	25,422.99	132 – 231	211 – 213	311 – 324	324 – 512	261.75	1.03	6	0.17
		211 – 121	132 – 231	324 – 311	411 – 512				
		231 – 132	211 – 221	313 – 324	512 – 411				
Malta[a]	315.60	323 – 132		323 – 132		0.19	0.06	6	0.01
Montenegro[a]	13,818.06	324 – 131	231 – 221	311 – 334		53.27	0.39	6	0.06
		142 – 124	211 – 231	324 – 334					
		311 – 142	243 – 131	334 – 324					
Netherlands	39,877.04	133 – 112	211 – 133	231 – 321	231 – 411	603.16	1.51	6	0.25
		211 – 133	231 – 133	211 – 321	523 – 133				
		133 – 121	242 – 133	133 – 321	211 – 411				
Norway	323,785.23	312 – 142	312 – 211	312 – 324	412 – 142	1759.94	0.54	6	0.09
		311 – 142	211 – 122	324 – 312	412 – 133				
		312 – 112	211 – 112	312 – 142	412 – 324				
Poland	312,516.11	211 – 133	231 – 324	312 – 324	211 – 512	3057.08	0.98	6	0.16
		211 – 131	211 – 133	324 – 312	231 – 512				
		211 – 112	211 – 324	324 – 313	131 – 512				
Portugal	92,345.79	133 – 122	211 – 223	324 – 311	211 – 512	4999.97	5.41	6	0.90
		324 – 133	212 – 223	311 – 324	243 – 512				
		133 – 121	211 – 212	312 – 324	133 – 512				
Romania	238,405.80	211 – 133	211 – 411	312 – 324	211 – 411	708.78	0.30	6	0.05
		211 – 131	211 – 121	311 – 324	511 – 512				
		211 – 122	211 – 222	324 – 311	231 – 512				
Serbia	77,641.00	131 – 324	231 – 211	311 – 324	231 – 512	367.27	0.47	6	0.08
		324 – 131	222 – 211	324 – 311	211 – 512				
		211 – 131	211 – 222	131 – 324	512 – 324				
Slovakia	49,023.00	211 – 133	211 – 133	312 – 324	211 – 512	916.63	1.87	6	0.31
		211 – 121	231 – 324	311 – 324	313 – 512				
		133 – 122	211 – 121	324 – 312	243 – 512				
Slovenia[a]	20,275.65	133 – 122	242 – 121	312 – 324	242 – 512	16.84	0.08	6	0.01
		313 – 133	243 – 131	324 – 311	211 – 512				
		312 – 133	311 – 211	324 – 313					
Spain	506,261.83	211 – 133	221 – 242	311 – 324	321 – 512	6585.10	1.30	6	0.22
		133 – 112	221 – 211	312 – 324	211 – 512				
		211 – 121	211 – 133	324 – 311	323 – 512				
Sweden	450,023.57	312 – 131	211 – 112	324 – 312	412 – 324	33,400.70	7.42	6	1.24
		133 – 312	211 – 133	312 – 324	312 – 412				
		211 – 112	211 – 231	324 – 313	412 – 131				
Switzerland[a]	41,282.00	211 – 121	211 – 121	335 – 332		50.99	0.12	6	0.02
		211 – 131	211 – 131	324 – 313					
		133 – 122	211 – 112	324 – 311					
Turkey[b]									
United Kingdom	248,137.7	211 – 131	211 – 131	312 – 324	131 – 512	2112.20	0.85	6	0.14
		312 – 121	211 – 133	324 – 312	412 – 121				
		133 – 112	231 – 131	313 – 324	412 – 131				

[a] Empty cells indicate no change.
[b] Country participated in CLC2012 project, results will be available in the future.

Legend: CLC level 1:
1. Artificial surfaces
2. Agricultural areas
3. Forest and semi-natural areas
4. + 5. Wetlands and water bodies

FIGURE 16.2 (CONTINUED)
Three largest LC changes on the third level in the context of the first hierarchic level CLC for 2006–2012.

Reference

Büttner, G. 2014. CORINE land cover and land cover change products. In I. Manakos and M. Braun (Eds.), *Land use and land cover mapping in Europe: Practices and trends*. Remote Sensing and Digital Image Processing 18 (pp. 55–74). Dordrecht: Springer.

Section IV

Case Studies: Solution of the European Environmental Problems Using the CORINE Land Cover Data

17

Land Cover of Europe

Tomas Soukup, Jan Feranec, Gerard Hazeu, Gabriel Jaffrain, Marketa Jindrova, Miroslav Kopecky, Erika Orlitova, and Katerina Jupova

CONTENTS

17.1 Introduction

CoORdination of Information on the Environment (CORINE) Land Cover (CLC) data are an excellent carrier and mediator of the variety of European landscapes and, of course, an original information source about natural but also modified (cultivated) and artificial (created by humans) landscape structures and objects. Information potential of CLC data is enhanced by the compatibility of their content and the fact that they represent an almost 25-year period (Feranec et al., 2012). The aim of this chapter is to offer an overview of the geographical distribution of 44 CLC classes, their dominant representation in the European countries, as well as the summarized areas in four time horizons: 1990, 2000, 2006, and 2012.

17.2 Geographical Distribution of CLC Classes

The most recent CLC2012 data layer (see Figure 17.1) is a seamless mosaic of LC in 38 countries rendering the most topical view of the European LC at a scale of 1:100,000. Its information potential contributes to the knowledge of the present structure of the European landscapes.

The *Forest landscape* (CLC classes 311–324) covers 43.9% of the surface according to the most recent CLC2012 statistics, making it the most extensive landscape in Europe. Forest landscape is linked mostly to mountain ranges but also to parts of lowlands and hilly landscape (see Figure 17.1).

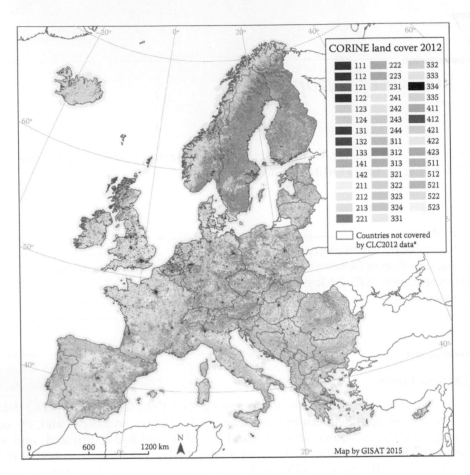

FIGURE 17.1
Spatial distribution of 44 CLC classes of Europe for the year 2012 (*CLC2012 data update for Turkey is still ongoing).

Classes of *Agricultural landscape* (211–244) cover approximately 41.7% of Europe's surface (according to CLC2012), following *Forest landscape*. Areas of these classes were identified in lowland and hilly areas. Figure 17.1 provides insight into their spatial distribution.

Artificial surfaces (classes 111–142) cover (CLC2012 statistics) 4.4% of Europe's surface. They are newly created and changed parts of the landscape. Although they represent a small area, they can be encountered everywhere in Europe, starting with coasts and lowlands over hills and basins and ending in mountain valleys (see Figure 17.1). Their spatial distribution over Europe is very heterogeneous.

Classes of *Wetlands* (4xx) and *Water bodies* (5xx) covering 2.8% and 2.7%, respectively (CLC2012) complement the mosaic of the European landscape. They are to be found principally on the coastal and flat parts of Europe (see Figure 17.1).

However, the complete mosaic of the European landscapes also includes *Open space with little or no vegetation* (33x). These classes cover 4.5% of the European surface. All CLC data layers contain information about their spatial distribution. In Figure 17.1 the most recent of them, that is, CLC2012, is visualized.

17.3 Representation of Dominant CLC2012 Classes in European Countries

Figure 17.2 shows maps of Europe in which countries are classified according to the percentages of CLC2012 classes on the first hierarchic level (*Artificial surfaces, Agricultural areas, Forest and semi-natural areas,* and *Wetlands and water bodies*).

A high share (6.1%–30.0%) of *Artificial surfaces* (all classes 1xx; their codes are explained in Table 3.1) is found in Belgium, Cyprus, Czech Republic, Denmark, Germany, Great Britain, Hungary, Malta, the Netherlands, and Switzerland. A medium share (2.1%–6.0%) is typical for a substantial part of the European countries (e.g., France, Poland, Portugal, Romania,

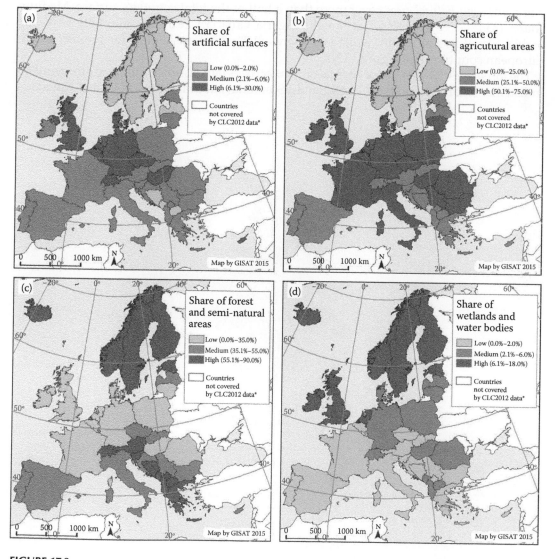

FIGURE 17.2
Areas as percentages of the total country area in 2012 (*CLC2012 data update for Turkey is still ongoing).
(a) *Artificial surfaces.* (b) *Agricultural areas.* (c) *Forest and semi-natural areas.* (d) *Wetlands and water bodies.*

Spain, and other). Low shares (0.1%–2.0%) exist in Bosnia/Herzegovina, Latvia, Macedonia FYR, Montenegro, and the Scandinavian countries (see Figure 17.2a).

Low shares (0.1%–25.0%) of *Agricultural areas* (all classes of 2xx) are characteristic for the Montenegro and Scandinavian countries. Medium share values (25.1%–50.0%) are found in Albania, Austria, Bosnia/Herzegovina, Croatia, Greece, Kosovo, Macedonia FYR, Portugal, Slovakia, Slovenia, Spain, and Switzerland. High shares (50.1%–75.0%), that is, more than one-half of the country's territory is covered by *Agricultural landscape,* is encountered in the rest of Europe (see Figure 17.2b).

The top shares (55.1%–90.0%) of *Forest and semi-natural areas* (all classes 3xx) are characteristic for the Albania, Austria, Bosnia/Herzegovina, Estonia, Greece, Iceland, Kosovo, Macedonia FYR, Malta, Montenegro, Scandinavian countries, Slovenia, and Switzerland. *Forest and semi-natural areas* cover more than a half of the total country's surface. Medium shares (35.1%–55.0%) are associated with Bulgaria, Croatia, Czech Republic, Italy, Portugal, Serbia, Slovakia, and Spain, while low shares (0.1%–35%) are typical for the rest of the European countries (see Figure 17.2c).

The top shares (6.1%–18.0%) of *Wetlands and water bodies* (all classes 4xx and 5xx) are identified in Estonia, Iceland, Ireland, the Netherlands, Scandinavian countries, and United Kingdom.

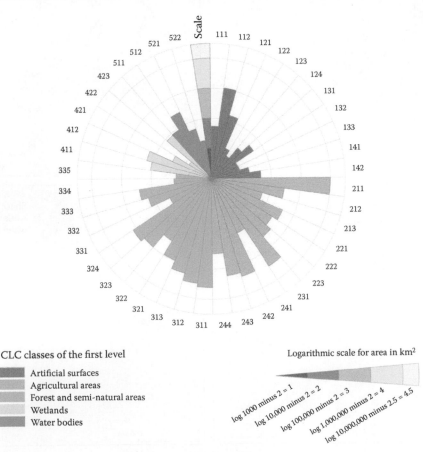

FIGURE 17.3
Representation of CLC classes in Europe in 1990. The logarithmic scale was used because of the large variation in extent of represented CLC class areas ranging from hundreds to millions of square kilometers. The third level CLC nomenclature three-digit codes are explained in Table 3.1.

Medium share values (2.1%–6.0%) represent Albania, Bulgaria, Denmark, Germany, Hungary, Latvia, Lithuania, Macedonia FYR, Montenegro, Poland and Switzerland. Low shares (0.1%–2.0%) belong to the remaining European countries (see Figure 17.2d).

17.4 Representation of CLC Classes in Four Time Horizons

Figures 10.2, 11.2, 12.2, and 13.2 represent areas (in km²) of CLC classes on the first hierarchic level in the European countries in 1990, 2000, 2006, and 2012. Individual CLC classes on the third hierarchic level disappear in this aggregated form. As they are important components of the European landscape pattern, a more detailed quantitative presentation is desirable. Figures 17.3 through 17.6 provide information about the representation of 43 CLC classes for the entire European territory (class 523, *Sea and ocean*, was not included in this survey) contribute to the fulfillment of this request.

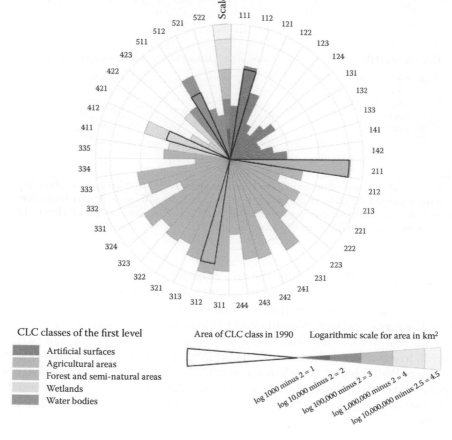

FIGURE 17.4

Representation of CLC classes in Europe in 2000. The logarithmic scale was used because of the large variation in extent of represented CLC class areas ranging from hundreds to millions of square kilometers. Reference areas of 1990, that is, the red triangles, are presented only for largest CLC classes of the first level. The third level CLC nomenclature three-digit codes are explained in Table 3.1.

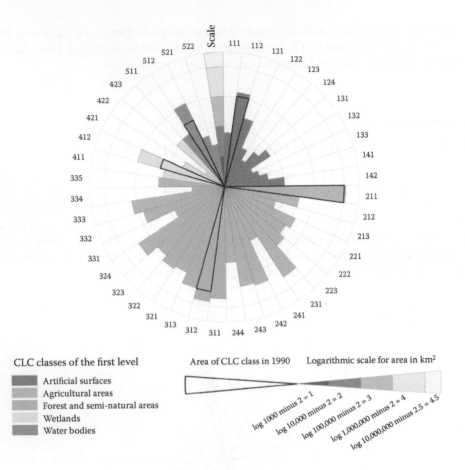

FIGURE 17.5

Representation of CLC classes in Europe in 2006. The logarithmic scale was used because of the large variation in extent of represented CLC class areas ranging from hundreds to millions of square kilometers. Reference areas of 1990, that is, the red triangles, are presented only for largest CLC classes of the first level. The third level CLC nomenclature three-digit codes are explained in Table 3.1.

Figures 17.3 through 17.6 show that the area of class 211 dominates in all four time horizons (although only 26 European countries participated in the CLC1990 Project, the missing data of 12 countries with small areas of class 211 do not distort the final statistics of this class). The *Forest landscape* is dominated by class 312. The exception is the year 1990, as Scandinavian countries, where this class is abundantly represented, did not participate in CLC Project. In *Artificial surfaces* class 112 is the largest while classes 412 and 512 cover the largest areas of *Wetlands* and *water bodies*, respectively. The dynamics of the most dominant class for each class at the first hierarchic level is visualized by highlighting in red the reference year 1990 in the figures, reflecting the situation for 2000, 2006, and 2012 (Figures 17.4 through 17.6).

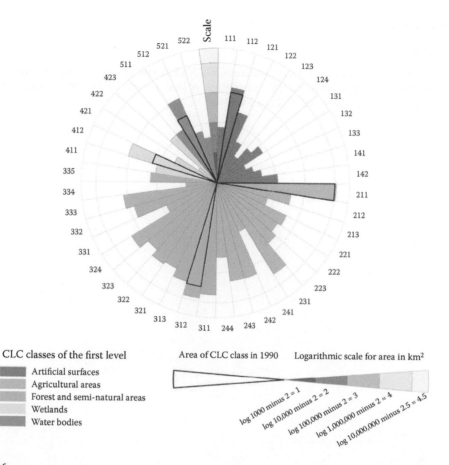

FIGURE 17.6
Representation of CLC classes in Europe in 2012. The logarithmic scale was used because of the large variation in extent of represented CLC class areas ranging from hundreds to millions of square kilometers. Reference areas of 1990, that is, the red triangles, are presented only for largest CLC classes of the first level. The third level CLC nomenclature three-digit codes are explained in Table 3.1.

17.5 Conclusion

CLC data are a valuable information source that contribute to the discernment of the European landscapes such as *Artificial surfaces, Agricultural areas, Forest and semi-natural areas,* and *Wetlands and water bodies.* The main benefits of CLC data is the harmonized way that land cover is mapped for the European territory, which makes it possible to compare countries with each other. The compatibility of content represented by 44 LC classes and the temporal sequence of four time horizons makes it possible to identify, analyze, and assess the landscape changes.

Reference

Feranec, J., Soukup, T., Hazeu, G. W., Jaffrain, G. 2012. Land cover and its change in Europe: 1990–2006. In G. Chandra (Ed.), *Remote sensing of land use and land cover: Principles and applications* (pp. 285–301). Boca Raton, FL: CRC Press.

18

Trend of Land Cover Changes in Europe in 1990–2012

Tomas Soukup, Jan Feranec, Gerard Hazeu, Gabriel Jaffrain, Marketa Jindrova, Miroslav Kopecky, Erika Orlitova, and Katerina Jupova

CONTENTS

18.1 Background

Changes of land cover (LC), caused by various socioeconomic and natural processes, are presented for the individual periods of 1990–2000, 2000–2006, and 2006–2012 in Chapters 14 through 16. However, it is important to recognize and point to the main developments in the European landscape during the entire period 1990–2012. Therefore, the aim of this chapter is to provide an overview of seven important processes expressed through LC changes for nearly a quarter of a century.

As LC is indivisible from landscape, it reflects its state in different stages of development. This is why LC changes can be regarded as a relevant information source about processes (flows) in landscape. The methodology applied in Haines-Young and Weber's study (2006) categorizes LC changes into LC flows (LCFs) on the basis of the second CoORdination of Information on the Environment (CORINE) Land Cover (CLC) data level and presents the spatial aspects of LC changes through LCF intensity maps. The changes, grouped into LCFs, represent seven major LU processes (Feranec et al., 2010, 2012):

- (LCF1) Urbanization is defined as a flow, which represents the change of agricultural land (classes 21, 22, and 23), forest land (classes 31, 32, and 33), wetlands (classes 41 and 42), and water bodies (51 and 52) into urbanized as well as industrialized land. The construction of buildings designed for living, education, healthcare, recreation, and sport as well as the construction of facilities for production, all forms of transport, and electric power generation are grouped under the process of urbanization.

- (LCF2) Intensification of agriculture is a flow that represents the transition of LC types associated with lower use intensity into higher use intensity. For example, changes from natural areas (classes 32, 33, 41, and 42) into agricultural land (classes 21, 22, 23, and 24) or changes in the frame of agricultural land from a lower use intensity (e.g., classes 23 and 24) into higher use intensity (e.g., classes 21 and 22).

- (LCF3) Extensification of agriculture is defined as a flow that represents the transition of the LC type associated with a higher use intensity (classes 21 and 22) into a lower use intensity (classes 23, 24, 32, and 33).

- (LCF4) Afforestation is defined as a flow that represents forest regeneration, that is, the establishment of forests by planting and/or natural regeneration, for example, change of classes 21, 22, 23, 24, 33, 41, and 42 into classes 31 and 32.

- (LCF5) Deforestation is defined as a flow involving forest land (class 31) changes into another LC or damaged forest (classes 21, 22, 23, 24, 32, 33, and 41). It means that the tree canopy falls below a minimum percentage threshold of 30%.

- (LCF6) Construction and management of water bodies is a flow involving the change of mainly agricultural land (classes 21, 22, 23, and 24) and forest land (classes 31 and 32) into water bodies (classes 51 and 52).

- (LCF7) Other changes are changes caused by various anthropic activities such as "recultivation" of former mining areas, dump sites, unclassified changes, and so forth. More detailed characteristics of LCFs are quoted in Feranec et al. (2010).

Figures 18.1 through 18.7 are representations of the seven LCFs showing the trend for each of them in relation to the reference period 1990–2000. The LC changes grouped into seven LCFs are calculated per 3 × 3 km square (in ha; recalculated for 1 year) for the three time horizons 1990–2000, 2000–2006, and 2006–2012. Any small change (e.g., around 5–10 ha) is represented through the grid 3 × 3 km, which may lead to overestimation but only in the map. The real change is only, for example, 10 ha within a square of 9 km^2, although color covers the full square. The LCFs$_{1990-2000}$ or grouped changes for the period 1990–2000 are taken as reference values. The figures show for each 3 × 3 km square within Europe three possible trends in intensity (ha of change/year) per LCF:

- Increasing trend, that is, reaching greater value as the reference value (red).
- Decreasing trend, that is, reaching smaller value than the reference value (blue).
- Mixed trend (magenta), that is, the trend for period 2000–2006 is increasing and for the period 2006–2012 (or the other way around) is decreasing with reference to the period 1990–2000. Also, if the intensity in one period did not change in relation to the intensity for the reference period 1990–2000 and also if no changes are reported for a specific period the squares are classified as mixed. Table 2.2 presents the countries where for one of the time horizons no changes are reported.

18.2 Urbanization (LCF1)

Figure 18.1 shows the trend in occurrence of changes, particularly those of agricultural and forest landscapes in favor of urbanization in relation to the same type of changes in

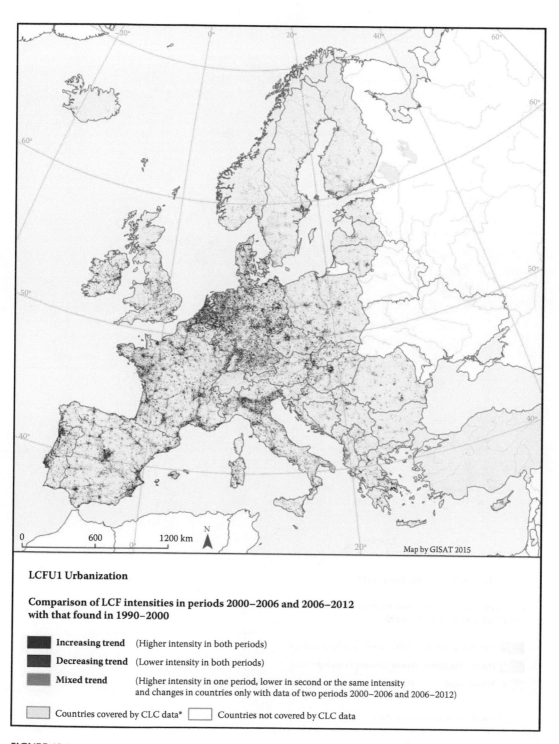

LCFU1 Urbanization

**Comparison of LCF intensities in periods 2000–2006 and 2006–2012
with that found in 1990–2000**

■ **Increasing trend**	(Higher intensity in both periods)
■ **Decreasing trend**	(Lower intensity in both periods)
■ **Mixed trend**	(Higher intensity in one period, lower in second or the same intensity and changes in countries only with data of two periods 2000–2006 and 2006–2012)

☐ Countries covered by CLC data*   ☐ Countries not covered by CLC data

FIGURE 18.1

Spatial distribution of urbanization in European countries in 1990–2000–2006–2012 (*LCF unprocessed for Turkey as CLC2012 data update is still ongoing).

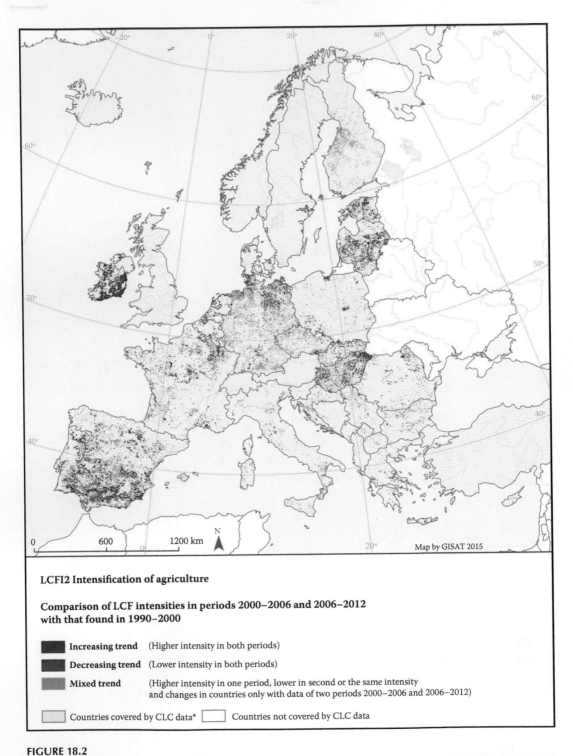

LCFI2 Intensification of agriculture

Comparison of LCF intensities in periods 2000–2006 and 2006–2012
with that found in 1990–2000

Increasing trend (Higher intensity in both periods)

Decreasing trend (Lower intensity in both periods)

Mixed trend (Higher intensity in one period, lower in second or the same intensity
and changes in countries only with data of two periods 2000–2006 and 2006–2012)

Countries covered by CLC data*   Countries not covered by CLC data

FIGURE 18.2

Spatial distribution of intensification of agriculture in European countries in 1990–2000–2006–2012 (*LCF
unprocessed for Turkey as CLC2012 data update is still ongoing).

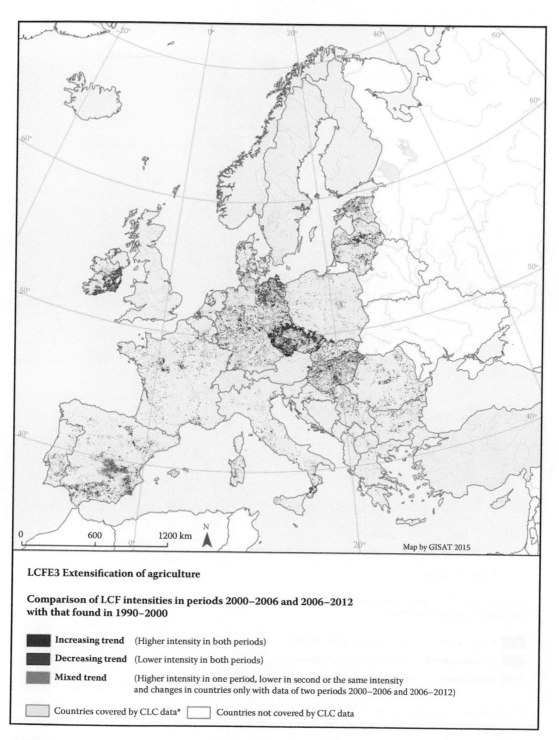

FIGURE 18.3
Spatial distribution of extensification of agriculture in European countries in 1990–2000–2006–2012 (*LCF unprocessed for Turkey as CLC2012 data update is still ongoing).

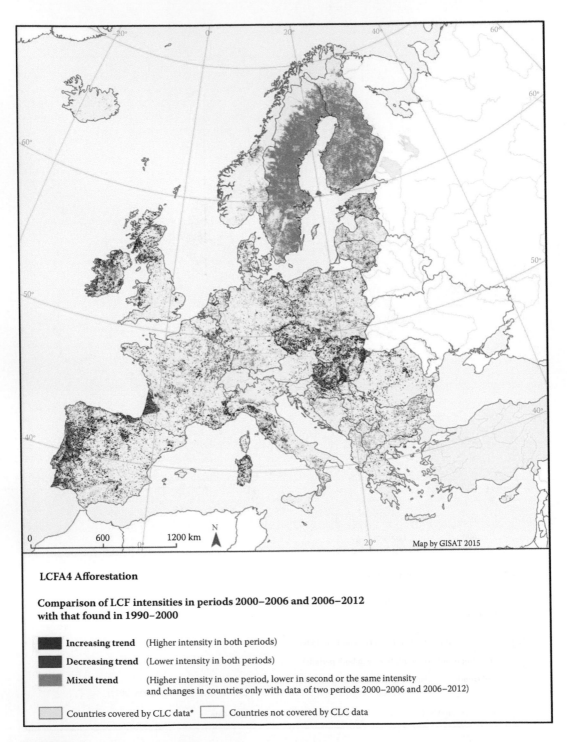

LCFA4 Afforestation

**Comparison of LCF intensities in periods 2000–2006 and 2006–2012
with that found in 1990–2000**

Increasing trend   (Higher intensity in both periods)

Decreasing trend   (Lower intensity in both periods)

Mixed trend   (Higher intensity in one period, lower in second or the same intensity
and changes in countries only with data of two periods 2000–2006 and 2006–2012)

Countries covered by CLC data*   Countries not covered by CLC data

FIGURE 18.4
Spatial distribution of afforestation in European countries in 1990–2000–2006–2012 (*LCF unprocessed for
Turkey as CLC2012 data update is still ongoing).

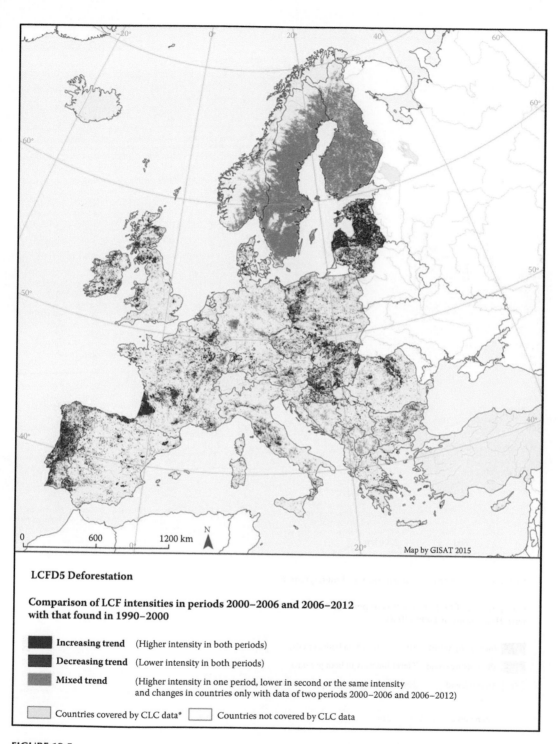

LCFD5 Deforestation

**Comparison of LCF intensities in periods 2000–2006 and 2006–2012
with that found in 1990–2000**

Increasing trend (Higher intensity in both periods)

Decreasing trend (Lower intensity in both periods)

Mixed trend (Higher intensity in one period, lower in second or the same intensity
and changes in countries only with data of two periods 2000–2006 and 2006–2012)

Countries covered by CLC data*   Countries not covered by CLC data

FIGURE 18.5
Spatial distribution of deforestation in European countries in 1990–2000–2006–2012 (*LCF unprocessed for Turkey as CLC2012 data update is still ongoing).

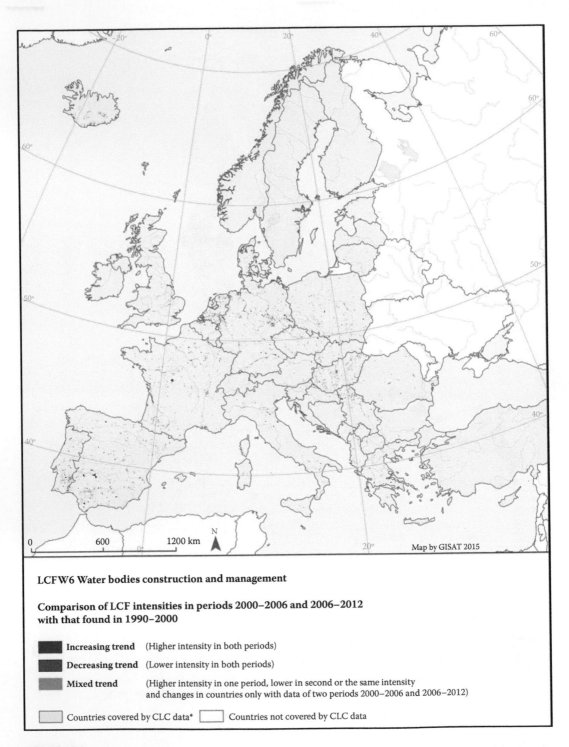

LCFW6 Water bodies construction and management

Comparison of LCF intensities in periods 2000–2006 and 2006–2012
with that found in 1990–2000

■ Increasing trend   (Higher intensity in both periods)

■ Decreasing trend   (Lower intensity in both periods)

■ Mixed trend   (Higher intensity in one period, lower in second or the same intensity
                  and changes in countries only with data of two periods 2000–2006 and 2006–2012)

☐ Countries covered by CLC data*   ☐ Countries not covered by CLC data

FIGURE 18.6
Spatial distribution of construction and management of water bodies in European countries in 1990–2000–
2006–2012 (*LCF unprocessed for Turkey as CLC2012 data update is still ongoing).

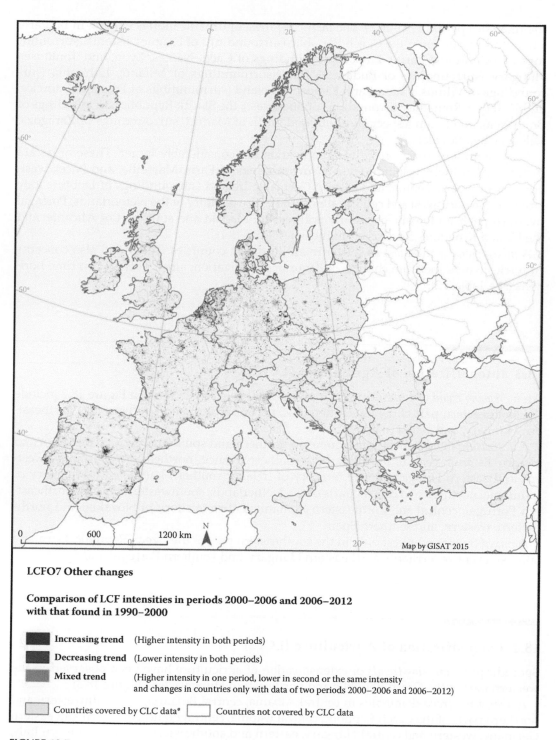

LCFO7 Other changes

Comparison of LCF intensities in periods 2000–2006 and 2006–2012 with that found in 1990–2000

Increasing trend (Higher intensity in both periods)

Decreasing trend (Lower intensity in both periods)

Mixed trend (Higher intensity in one period, lower in second or the same intensity and changes in countries only with data of two periods 2000–2006 and 2006–2012)

Countries covered by CLC data* Countries not covered by CLC data

FIGURE 18.7

Spatial distribution of other changes in European countries in 1990–2000–2006–2012 (*LCF unprocessed for Turkey as CLC2012 data update is still ongoing).

the reference period 1990–2000. The *increasing trend* of urbanization is evident in Bulgaria (surroundings of Sofia), the Czech Republic (surroundings of Prague), Estonia (surroundings of Tallinn and Tartu), France (surroundings of Caen, Nantes, Lyon, and Toulouse), Hungary (surroundings of Budapest), Italy (surroundings of Milano), Lithuania (surroundings of Vilnius, Kaunas, and Klaipeda), Poland (surroundings of Poznan, Katowice, and Krakow), Romania (surroundings of Bucharest), the Slovak Republic (surroundings of Bratislava), and Spain (especially south and north of Madrid, surroundings of Zaragoza, Valencia, and Seville).

Areas with a *decreasing trend* of urbanization are considerably larger. These areas are distinct in northern Belgium; France (surroundings of Paris, Marseille, and Nice); western, southwestern, central, and eastern Germany; Ireland (surroundings of Dublin); Italy (areas in the northwest and center); almost the entire territory of the Netherlands; Portugal (western part); and Spain, especially northeast of Madrid and southeast of Alicante; and the United Kingdom (center and southeast) (Figure 18.1).

A *mixed trend* is manifest as scatter in all European countries. This trend also concerns countries where there are changes in favor of urbanization, albeit only in two time horizons (see Table 2.2).

18.3 Intensification of Agriculture (LCF2)

An *increasing trend* of agriculture is almost negligible. Areas evident in Figure 18.2 include the southeastern part of the Czech Republic, northern Germany, southern and northeastern Hungary, and southern Spain.

A *decreasing trend* is most conspicuous in northern and southern Belgium; northern and eastern Estonia; central, northern, and southern France; northeastern Germany; western and eastern Hungary; entire territory of Ireland; southern Latvia; entire territory of Lithuania; northern and central parts of the Netherlands; southwestern Poland; northeastern Portugal; central and northeastern Romania; entire territory of Slovakia; and northeastern, western, and southern Spain.

A *mixed trend* is manifest only in the southeastern part of the Czech Republic, Germany (almost its entire territory), northeastern Hungary, and southern Portugal and Spain.

18.4 Extensification of Agriculture (LCF3)

Sporadically *increasing trends* of extensification of agriculture have been observed in the western part of the Czech Republic and northeastern Hungary (see Figure 18.3).

A *decreasing trend* dominates in central Albania, central Bulgaria, the southwestern and northern parts of the Czech Republic, central Estonia, central France, northern and eastern Germany, western and central Hungary, eastern and southeastern Ireland, southern Italy, southern Latvia, eastern Lithuania, central and eastern Romania, and southern Spain.

A *mixed trend* stands out only in the central part of the Czech Republic, northwestern Estonia, and southeastern Spain.

18.5 Afforestation (LCF4)

An *increasing trend* of LC changes in favor of forest regeneration prevails in central and eastern Estonia, northern and southern Hungary, the south and southeast of Ireland, central Italy, southeastern Lithuania, northwestern and southern Poland, western Portugal, northern Spain, and the northern part of the United Kingdom.

A *decreasing trend* is observable in the southern part of Belgium; central and eastern Bulgaria; northeastern Croatia; central and southern parts of the Czech Republic; central, eastern, and southern France; northeastern Germany; northern and western Hungary; western and northern Ireland; northwestern Italy; Sardinia; western Poland; northeastern Portugal; central and northeastern Romania; central Serbia; western and northern parts of Spain; and the northwestern part of the United Kingdom.

A *mixed trend* is dominant in Finland and Sweden. These countries have the CLC data only for the 2000–2006 and 2006–2012 periods. Owing to the large number of small changes spread all over the total territory of both countries, and as no changes are reported for the period 1990–2000, both countries appear in magenta. The mixed trend is also distinct in central Bosnia/Herzegovina, entire territory of Estonia, southern Germany, central Ireland, Latvia and Lithuania, the entire territory of Macedonia FYR, northern and southern Poland, and the entire territory of Portugal (see Figure 18.4).

18.6 Deforestation (LCF5)

An *increasing trend* is observable in central Austria, southern Belgium, northeastern Bulgaria, Estonia, southwestern France, southwestern and northeastern Hungary, central Italy, Latvia, western Poland, northeastern Portugal, central Romania, northern Slovakia, northern Spain, and the northern part of the United Kingdom (Figure 18.5).

A *decreasing trend* is evident in three Baltic countries, northeastern Croatia, northeastern part of the Czech Republic, southern Denmark, central and northeastern France, southern Germany, southern Poland, northeastern Portugal, northeastern Romania, northern and central Slovakia, and in central and northern Spain.

Areas of the *mixed trend* occur in almost all European countries but mainly in Finland, Sweden, and southeastern Norway (see explanation in Section 18.4).

18.7 Construction and Management of Water Bodies (LCF6)

The *increasing trend* of these changes is very sporadic in the eastern part of Germany and central Hungary.

The *decreasing trend* is patchy in central France, western Germany, southern part of the Netherlands, and southwestern Spain.

Likewise, the *mixed trend* scarcely appears in eastern Germany, northern Italy, southern Poland, and southern Portugal and Spain (see Figure 18.6).

TABLE 18.1

Summarized Overview of LCFs in Europe 1990–2000, 2000–2006, and 2006–2012

Type of Change	Changes 1990–2000		Changes 2000–2006		Changes 2006–2012		Tendency +/−		
	(in km²)	(in km²/year*)	(in km²)	(in km²/year)	(in km²)	(in km²/year)	1990–2000	2000–2006	2006–2012
Urbanization	9706.68	888.13	5801.27	1014.10	4832.95	805.49	*	+	−
Intensification of agriculture	14,911.54	1520.23	3890.03	715.24	3993.01	665.50	*	−	−
Extensification of agriculture	12,969.30	1453.65	1520.10	259.91	3583.30	597.22	*	−	−
Afforestation	26,397.32	2488.37	9600.60	1645.95	11,902.16	1983.69	*	−	−
Deforestation	20,676.38	2146.48	17,428.10	2967.29	16,195.62	2699.27	*	+	+
Water body construction	1204.41	112.00	624.76	105.95	516.12	86.02	*	−	−
Other changes	2041.81	197.81	1784.83	309.90	2091.08	348.51	*	+	+
Total	87,907.44	8806.67	40,649.69	7018.42	43,114.24	7185.70			

Note: + +, increasing trend; − −, decreasing trend; + −, mixed trend; *, yearly reference value.

18.8 Other Changes (LCF7)

An *increasing trend* of other changes is obvious in southern France, central Germany, southern Lithuania, entire territory of the Netherlands, southern Poland, and central and southern Spain.

A *decreasing trend* is almost negligible. It appears sporadically in the northwestern part of the Czech Republic and eastern Germany.

Although the *mixed trend* areas are scarce, they are dissipated in several European countries (see Figure 18.7).

18.9 European Figures

Table 18.1 displays an overview of trends in landscape changes for Europe in general for the periods 1990–2000, 2000–2006, and 2006–2012. For the sake of correctness of the comparison, the rate of LC changes has been recalculated to km^2/year, as the assessed changes do not represent equally long periods.

Urbanization is characterised by a *mixed trend* (+ –, see Table 18.1). After an increased rate in construction to 1014.18 km^2/year in 2000–2006 (for the period 1990–2000 the rate of urbanization was 888.13 km^2/year), a distinct drop to 805.49 km^2/year occurred in the 2006–2012 period (see Table 18.1).

Intensification of agriculture progressively dropped (– –, see Table 18.1) from 1520.23 km^2/year in the reference period 1990–2000 to 665.50 km^2/year in the period 2006–2012.

Extensification of agriculture decreased (– –, see Table 18.1) from 1453.65 km^2/year (1990–2000) to 597.22 km^2/year (2006–2012).

Afforestation also experienced a decreasing trend (– –, see Table 18.1) from 2488.37 km^2/year (1990–2000) to 1983.69 km^2/year (2006–2012).

Deforestation experienced an increasing trend (+ +, see Table 18.1) compared to the reference period 1990–2000. It increased from 2146.48 km^2/year to 2967.29 km^2/year and to 2699.27 km^2/year for the periods 2000–2006, respectively 2006–2012.

A decreasing trend (– –, see Table 18.1) was also observed for *construction and management of water bodies*, gradually from 112.00 km^2/year to 105.95 km^2/year and even to 86.02 km^2/year.

An increasing trend (+ +, see Table 18.1) was observed in *other changes*. The rate of change gradually increased from 197.81 km^2/year to 309.90 km^2/year and to 348.51 km^2/year for the 1990–2000, 2000–2006, and 2006–2012 periods, respectively.

References

Feranec, J., Jaffrain, G., Soukup, T., Hazeu, G. 2010. Determining changes and flows in European landscapes 1990–2000 using CORINE land cover data. *Applied Geography* 30:19–35.

Feranec, J., Jaffrain, G., Soukup, T., Hazeu, G. 2012. Land cover and its change in Europe: 1990–2006. In C. Giri (Ed.), *Remote sensing of land use and land cover: Principles and applications* (pp. 285–301). Boca Raton, FL: CRC Press.

Haines-Young, R., Weber, J.-L. 2006. Land accounts for Europe 1990–2000. Towards integrated land and ecosystem accounting. EEA Report 11. Copenhagen: European Environment Agency.

19

Monitoring of Urban Fabric Classes and Their Validation in Selected European Cities (Urban Atlas)

Gabriel Jaffrain, Christophe Sannier, and Jan Feranec

CONTENTS

19.1 Introduction

According to information contained in *The European Environment: State and Outlook 2010: Land use* (European Environment Agency [EEA], 2010) publication, areas of artificial surfaces in Europe have been increasing, (by 0.6% in 2000–2006), which is one of the reasons why ever greater attention is given to the issue. Occupation of land by construction is the phenomenon referred to by the term soil sealing (imperviousness as a synonym means the covering of the soil surface with impervious materials as a result of urban development and infrastructure construction*). Soil sealing in the EEA glossary (2006) refers to the soil that becomes an impermeable medium as it is covered by impermeable materials (concrete, metal, glass, asphalt, or plastics). The basic aspects to identify and measure soil-sealed areas are described in the study of Kampouraki et al. (2006). Projects dealing with soil sealing include, for example, Monitoring Urban Dynamics (MURBANDY) and Monitoring Land Use-Cover Change Dynamics (MOLAND) (Burghardt et al., 2004). An overview of current approaches to the definition, phenomenology, and conceptual and empiric modeling related to soil sealing—with the emphasis on urban areas (zones) in Europe—is presented in the Scalenghe and Marsan (2009) study. These authors also define

---

* http://eusoils.jrc.ec.europa.eu/library/themes/Sealing/.

the sealed soil as surface covered by impermeable material, which occupies about 9% of Europe's surface.

The soil sealing layer represents a new type of information in the European environmental assessment, being the first example of the planned series of European high-resolution land monitoring layers (Maucha et al., 2011). Molini and Salgado (2012) discuss the situation of urban sprawl in Spain and Madrid and place it in a pan-European context. They define the residential sprawl as urban growth with single-family houses. A commercial or other type of sprawl is considered when the growth of the artificial surface is referred to. Artificial surfaces allude to an overall urban growth and include residential areas, industrial and commercial areas, transport infrastructure, mining areas, dumps, areas under construction, sports and leisure facilities, and green urban areas (Molini and Salgado, 2012). Different levels of soil sealing are integrated in the land use/land cover (LU/LC) artificial classes of the Urban Atlas (UA)—a joint project of the European Commission Directorate General (EC DG) for Regional Policy and EC DG for Enterprise and Industry with the support of the European Space Agency (ESA) and the EEA in the context of Global Monitoring for Environment and Security (GMES)/Copernicus Land (GIO Land) activities (see Chapters 1 and 28).

19.2 Objectives of Urban Atlas 2006 (UA2006) and Urban Atlas 2012 (UA2012)

CoORdination of Information on the Environment (CORINE) Land Cover (CLC) provides a resolution of 1:100,000 (25 ha) (see Chapter 2), which is unsatisfactory for several services needed by the EC. As in Europe urban areas accommodate more than three-quarter of the population and these areas have grown rapidly in recent decades, there was an urgent need for pan-European, reliable, and intercomparable urban planning data (Montero et al., 2014). The work on UA project consists of the production of a series of LU/LC maps of larger urban zone (LUZ) for a selected number of European cities. Recently the acronym LUZ has been changed into functional urban area (FUA).

UA started in the year 2008 with the production of 305 urban LC/LU datasets spread over the EU27, and addressing the LUZs as defined in the Urban Audit. This UA provides the first European wide urban LC/LU database that allows for comparison of more than 300 LUZs as it is based on one harmonized methodology and common LU/LC nomenclature. This first edition of the UA covered all EU capitals and a large sample of large and medium-sized cities with more than 100,000 inhabitants. The UA is meant to complement the statistical information with a geospatial component. From the beginning of UA2006, EC has been planned to produce one update step to build a time series toward the reference year 2012 with the idea to monitor the urban extension over time of major urban agglomerations in Europe. This new project called UA2012 consists of an update of the previous 305 LUZs as defined in the UA2006 and is extended with a list of 392 additional LUZ (see Figure 19.1). The main input datasets for UA2012 are the UA2006 datasets and/or the Very High Resolution (VHR SPOT 5 with 2.5 m resolution) satellite images of the reference year 2012.

The aim of this chapter is to focus on the UA2006 production and validation process by using some illustrations in terms of small, medium, large, and very large LUZs.

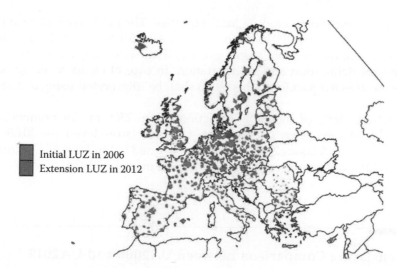

FIGURE 19.1
Distribution of LUZs in Europe in 2006 and the extension in 2012.

19.3 Methodology of the Urban Atlas

The implementation of the UA project was awarded to the French company Systems Inform Refer Spatial (SIRS) in December 2008. The methodology description is based on UA2006 Project; the complete organization and the different production steps of UA2006 project are described in the report SIRS, Institut Géographique National France International (IGN FI), Romanian Geospatial Engineering Company (EUROTOPO) (2011).

The 305 cities (>100,000 inhabitants) have been mapped using very high resolution (2.5 m) Earth observation (EO) data (SPOT 5, Formosat 2, Kompsat 2, Quick Bird, Rapideye, and ALOS data) for the reference year 2006 ± 1 year. The UA cities are mapped at a scale of approximately 1:10,000, using about 20 classes of which the 17 urban classes have a minimum mapping unit (MMU) of 0.25 ha and the non-urban classes a MMU of 1 ha. For UA2006, urban fabric (CLC classes 111 and 112) is differentiated by their degree of imperviousness, which is integrated from the Land Monitoring Core Service (LMCS) high-resolution soil sealing layer (Montero et al., 2014). The same MMUs are applied for UA2012, i.e., 0.25 ha for urban fabric and the nonurban classes a MMU of 1 ha.

The production of UA2006 is based on a mix of computer-assisted visual interpretation (CAVI) (Nunes de Lima, 2005) and automatic classification. Automated segmentation and classification is used to achieve a first differentiation into basic LC classes (urban vs. forest vs. water vs. other LC). An internal quality control has been carried out by the production team (SIRS) and concerned the consistency of the attribute table, the geometrical precision, and a specific cross-control between photointerpreters. A visual assessment between photointerpreters have been implemented consisting of a "cross control." Indeed, at regular intervals and after a significant LUZ produced, each operator submitted his work to another operator for control.

The delineation is carried out on the satellite images that are considered as primary (guiding) data source. The interpretation of the object is done using the EO data, the topographic maps, commercial off the shelf (COTS) navigation data (for the road network), and other auxiliary information including local expertise. The LUZ area has been interpreted with at least 100 m extension (100 m buffer) to ensure accuracy and continuity of polygons. In some areas where two or more scenes overlap, it must be ensured that the most recent data are used for delineation and interpretation. In case of cloud coverage on the most recent scene the affected part (only this part!) shall be interpreted using a cloudless older scene.

The contractual MMU of thematic extractions were 2500 m^2 for nomenclature posts relating to artificial areas (class 1 at highest UA nomenclature level) and 10,000 m^2 for the others main UA classes (classes 2, 3, 4, and 5) as explained in the Table 19.1. Minimum mapping width (MMW) between two objects for distinct mapping: 10 m.

19.4 Nomenclature Comparison between UA2006 and UA2012

The UA nomenclature respects strictly the code structuration of European CLC nomenclature and is illustrated in the Table 19.1.

The main difference between nomenclatures used for UA2006 and UA2012 is the description of the nonurban classes (see Table 19.1) corresponding to the second level of CLC nomenclature.

19.5 External Quality Control Assessment

The external quality control is independent from the production phase, and is performed by European high level experts appointed by IGN FI on the datasets received from the production team. The main goal of the external quality assessment is to guarantee the quality of the UA mapping product (mainly the quality of the interpretation process) and the final product. It is also important to produce reproducible and traceable results for the quality control as mentioned in the GMES Service Element (GSE) Land Information Services (2008).

The independent accuracy assessment is applied to the following products:

- UA2006 (initial 305 LUZ)
- UA2012 (initial 305 LUZ [with UA2006] and 350 new LUZ)
- Changes between UA2006 and UA2012 (initial 305 LUZ)

The main goal of an accuracy assessment independent from the production is to guarantee the quality of the UA2006 and UA2012 mapping products with reference to the accuracy thresholds:

- ≥85% for the urban classes
- ≥80% for the rural classes

TABLE 19.1

UA Nomenclatures 2006 and 2012

Nomenclature for UA2006	Nomenclature for UA2012
1 Artificial surfaces	**1 Artificial surfaces**
11 Urban fabric	11 Urban fabric
11100 Continuous urban fabric (sealing degree [S.D.]: > 80%)	11100 Continuous urban fabric (S.D.: > 80%)
11210 Discontinuous dense urban fabric (S.D.: 50%–80%)	11210 Discontinuous dense urban fabric (S.D.: 50%–80%)
11220 Discontinuous medium density urban fabric (S.D.: 30%–50%)	11220 Discontinuous medium density urban fabric (S.D.: 30%–50%)
11230 Discontinuous low density urban fabric (S.D.: 10%–30%)	11230 Discontinuous low density urban fabric (S.D.: 10%–30%)
11240 Discontinuous very low density urban fabric (S.L. < 10%)	11240 Discontinuous very low density urban fabric (S.L. < 10%)
11300 Isolated structures	11300 Isolated structures
12 Industrial, commercial, public, military, and private units	12 Industrial, commercial, public, military, and private units
12100 Industrial, commercial, public, military, and private units	12100 Industrial, commercial, public, military, and private units
12210 Fast transit roads and associated land	12210 Fast transit roads and associated land
12220 Other roads and associated land	12220 Other roads and associated land
12230 Railways and associated land	12230 Railways and associated land
12300 Port areas	12300 Port areas
12400 Airports	12400 Airports
13 Mine, dump, and constructions sites	13 Mine, dump, and constructions sites
13100 Mineral extraction and dump sites	13100 Mineral extraction and dump sites
13300 Construction sites	13300 Construction sites
13400 Land without current use	13400 Land without current use
14 Artificial non-agricultural vegetated areas	14 Artificial non-agricultural vegetated areas
14100 Green urban areas	14100 Green urban areas
14200 Sports and leisure facilities	14200 Sports and leisure facilities

(Continued)

TABLE 19.1 (CONTINUED)

UA Nomenclatures 2006 and 2012

Nomenclature for UA2006	Nomenclature for UA2012
2 Agricultural + Semi-natural + Wetlands areas	**2 Agricultural + Semi-natural + Wetlands areas**
	21 Arable land (annual crops)
	22 Permanent crops/orchards
	23 Pastures
	24 Complex and mixed cultivation patterns
	25 Orchards
	32 Herbaceous vegetation associations
	33 Open spaces with little or no vegetation
	4 Wetlands
	31 Forests transitional woodland
	5 Water
3 Forest/transitional woodland	
5 Water	

Source: Meirich, S. 2008. Mapping guide for a European Urban Atlas. GSE Land Consortium. Report ITD-0421-GSELand-TN-01. Available at: http://www.eea.europa .eu/data-and-maps/data/urban-atlas/mapping-guide. With permission.

19.5.1 Methodology Used for External Quality Control

The protocol used for the quality control of UA2006 is described here. The sampling design for LUZs covered in the UA2006 was reused for the UA2012 update. The rationale is that changes should occupy a small portion of the LUZs' area (1%–2% for Level 1 classes excluding changes within the urban classes) and that most of the points should be valid. For the complementary 350 LUZs from UA2012, the sampling design will be kept and applied for the new LUZs.

19.5.2 Determining the Primary Sampling Unit

The selected sampling method is a stratified random sampling of EEA grid cells with a variable sampling rate, depending on the stratum (urban/non-urban) and the size of the LUZ (number of points proportional to the LUZ size with a minimum number of points for very small LUZ).

The two additional datasets are needed for the sampling for the following reasons:

- The EEA European grid is used for the assessment of landscapes and ecosystems. This grid is made up of regular 1 × 1 km squares on the Lambert Azimuthal Equal Area (LAEA) projection.
- A stratification aimed at better targeting of urban areas is established from CLC2006 Level 1—Class 1 for Urban, and the grouping of the other Level 1 class for Non-Urban (see Figure 19.2).

As the focus of the UA is the urban theme (with the majority of LC/LU classes in that theme), the purpose of the stratification is to make it possible to apply a higher sampling rate for predominantly urban areas. It is proposed that a cell be considered belonging to the urban stratum if more than 33% of its area falls within the artificial class as defined by CLC2006. The remaining cells will be considered as rural cells.

The samples are set by the experts responsible for evaluating the quality and are not communicated to the team in charge of production. As such the independence of the validation is guaranteed.

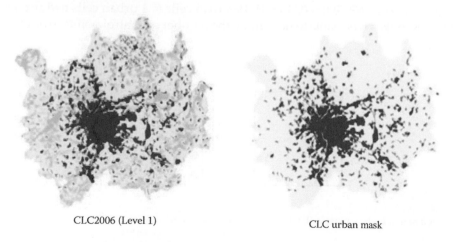

CLC2006 (Level 1)   CLC urban mask

FIGURE 19.2
CORINE Land Cover urban mask creation: example LUZ of Lyon, France.

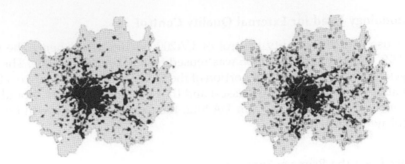

Classification urban/non-urban + European grid   CLC urban mask + European grid + sample (blue)

FIGURE 19.3
Urban/non-urban sample; example LUZ of Lyon, France.

The selection of Primary Sampling Units (PSUs) is determined randomly, as illustrated in Figure 19.3. Initially, a sampling fraction of 7% is drawn for the rural stratum and 10% for the urban stratum to fit with an average case, but these values can be modified to adapt the number of PSUs depending on the LUZ size and the proportion of artificial areas as detected from CLC2006.

19.5.3 Determining the Secondary Sampling Unit: Number of Control Points

Within each PSU, a 200-m grid is applied. Secondary Sampling Units (SSUs) are point observations corresponding to the centroid of each 200 × 200 m cell as illustrated in Figure 19.4. Each of the 25 potential SSU locations is called a replicate. Replicates are selected sequentially starting from 1 until 25. Care was taken that sequential replicates are never adjacent to one another to maximize the distance between them.

A different sampling rate is applied depending on whether the PSU is located in the rural or the urban stratum. With a view to guarantee a sufficient number of observations per class and taking into account the overall balance between rural and urban themes. Based on CLC2006, we know that there is a ratio, that is, number of urban cells 72,267/ number of rural cells 639,290 = 0.113 ≈ 11.3% urban cells (≈ 1 urban cells to 8 rural cells on average). This makes it possible to determine the number of control points to be considered

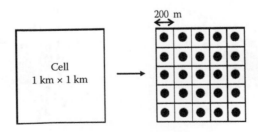

FIGURE 19.4
Grid 200 × 200 m inside the square cell. Basic grid for SSU selection.

TABLE 19.2

Total of Control Points to Be Checked according to the Categories of LUZ

LUZ	Number of Urban Cells	Number of Control Points in Urban Cells	Number of Rural Cells	Number of Control Points in Rural Cells
Weakly urbanized	7084	70,840	205,306	20,531
Moderately urbanized	33,148	99,444	321,462	32,146
Highly urbanized	32,035	64,070	112,522	11,252
Total	72,267	234,354	639,290	63,929

(see Table 19.2). Considering the focus of the UA on urban classes, the following ratio was selected:

	Urban	Rural
UA2006 nomenclature	80% (17 classes)	20% (3 classes)

This implies that one point in rural areas should be selected for four points in urban areas. Considering that there is on average one cell identified as urban for every eight cells, this means that to keep the number of points to a reasonable level, not all rural cells will be sampled.

The aforementioned thresholds could be more or less reached by selecting one in ten rural PSUs randomly and selecting all urban PSUs with one replicate in rural and 10 replicates in urban areas. Replicate 1 would be selected for rural areas and replicates 1–8 for urban areas, as illustrated in Figure 19.5.

However, this is valid only for an average and the number of replicate selected needs to be adapted according to the relative proportion of urban cell within each LUZ. As shown in Figure 19.6, the distribution of urban cells can be quite different.

However, to have a balanced sampling, the LUZs have been classified into three categories according to the percentage of urban cells (see Table 19.3).

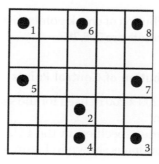

FIGURE 19.5
Example of SSUs replicate selection.

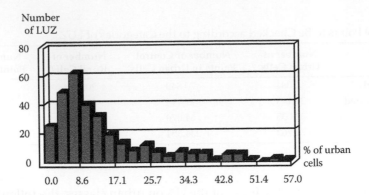

FIGURE 19.6

Distribution of LUZ according to the % of urban cells.

TABLE 19.3

LUZ Categories Distribution according to Urban Cell Density

LUZ	Percentage of "Urban Cells"/LU8	Number of LU8 (UA2006) Concerned
Weakly urbanized	Between 0 and 6%	81
Moderately urbanized	Between 6% and 17%	146
Highly urbanized	Above 17%	78

The final selection of (sample) points (SSUs) for the validation is based on the following rules applied for the different LUZ categories:

- LUZ weakly urbanized: 1 PSU for every 10 "rural cells" and 1 SSU selected, all PSUs selected, and 10 SSUs selected for "urban cells"
- LUZ moderately urbanized: 1 PSU for every 10 "rural cells" and 1 SSU selected, all PSUs selected, and 3 SSUs selected for "urban cells"
- LUZ highly urbanized: 1 PSU for every 10 "rural cells" and 1 SSU selected, all PSUs selected, and 2 SSUs selected for "urban cells"

The Table 19.2 shows the distribution of check points distributed according to the categories of LUZ. The total of check points are 298,283.

19.5.4 Determining the Distribution of Control Points

Figure 19.4 shows the basic grid 200 × 200 m used for the second SSU and Figure 19.5 shows an example of replicate selection.

For each of the urban categories specified in the LUZ (weakly, moderately, or highly urbanized) the proposed distribution is shown in Figures 19.7 through 19.9.

LUZ weakly urbanized

LUZ moderately urbanized

LUZ highly urbanized

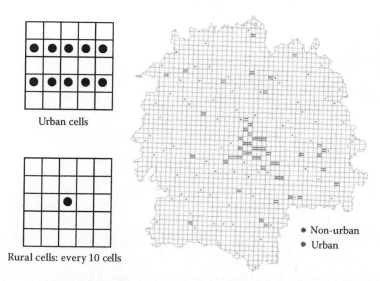

FIGURE 19.7
Example of LUZ weakly urbanized: distribution of control point—Amiens, France.

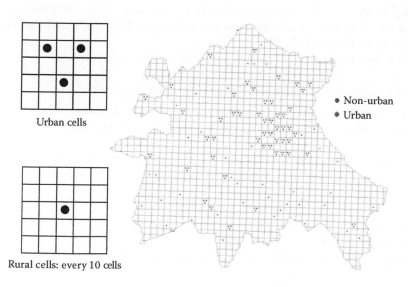

FIGURE 19.8
Example of LUZ moderately urbanized: distribution of control point—Cambridge, UK.

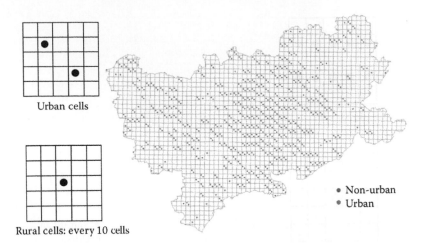

FIGURE 19.9
Example of LUZ highly urbanized: distribution of control point—Cologne, Germany.

19.6 Final Results and Accuracy Assessment

The quality assessment is not subject to any interference from production teams. It is thus an external assessment that approves/rejects the data produced. This method used is easily applicable and highly automated by using Geographical Information System (GIS) software in terms of sample generation and the quality assessment.

According to the specifications, the statistical accuracy of the database was assessed through the calculation of a confusion matrix relating to the quality control points created by European experts from IGN FI.

Each cell of the matrix gives a number of points sorted out with

- *Line*: the point attributed by the interpreter when mapping
- *Column*: the same observation as analyzed with outside data (aerial photographs, field data, other reference databases) during the control process

The marginal and diagonal totals of this matrix are used to assess two types of accuracy, commonly called "user accuracy" and "producer accuracy." Measurements are given by attribution categories.

For each of the LUZ produced, three confusion matrixes have been calculated. The first one focuses on urbanized areas, the second on non-urbanized areas, and the third matrix gives an overall accuracy.

As an example, the urban and rural confusion matrices for the Cologne LUZ are presented in Tables 19.4 and 19.5.

The combined confusion matrix (Table 19.6) including the rural and urban matrix shows an overall accuracy close to 85%. But the most important is that the overall accuracy of each matrix (rural and urban) is not less than 85% (see Tables 19.4 and 19.5). The quality controller in charge of the LUZ specified the thematic problems to the operator, who then had to correct the thematic discrepancies for the entire LUZ. In the end, the matrices were recalculated until the LUZ became compliant.

TABLE 19.4

Urban Confusion Matrix between Results from Classification (Line) and Control Process (Column) for the Cologne LUZ

Classification		110	113	121	122	123	124	131	133	134	141	142	600	Total	Producer Accuracy (%)	Commission Error (%)
						Control Point										
	110	129		2				1		1			1	134	96.3	3.7
	113			1										1		100.0
	121	4		56		5								65	86.2	13.8
	122	1		1	13									15	86.7	13.3
	123			1		11				2				14	78.6	21.4
	124						8							8	100.0	
	131			1	1			29						31	93.5	6.5
	133									1				1		100.0
	134									2			1	3	66.7	33.3
	141			4						1	30	1	3	39	76.9	23.1
	142	1										9		10	90.0	10.0
	600*	2		1	5			8		1	2	2	165	186	88.7	11.3
	Total	137		67	19	16	8	38		8	32	12	170	507		
	Producer accuracy (%)	94.2		83.6	68.4	68.8	100.0	76.3		25.0	93.8	75.0	97.1			
	Omission error (%)	5.8		16.4	31.6	31.3		23.7		75.0	6.3	25.0	2.9			

TABLE 19.5

Rural Confusion Matrix between Results from Classification (Line) and Control Process (Column) for the Cologne LUZ

| | Control Point | | | | | | | | | | | | Producer | Commission |
	100	210	220	230	240	250	310	320	330	400	500	Total	Accuracy (%)	Error (%)
Classification 100	316			1			4					321	98.4	1.6
210	14	66		6			10					96	68.8	31.3
220														
230	4	3		10		1						18	55.6	44.4
240														
250														
310	2						50					52	96.2	3.8
320		1		1			2					4		100.0
330														
400														
500	1										15	16	93.8	6.3
Total	337	70		18		1	66				15	507		
Producer accuracy (%)	93.8	94.3		55.6		100.0	75.8				100.0			
Omission error (%)	6.2	5.7		44.4		100.0	24.2							

TABLE 19.6

Combined Confusion Matrix between Results from Classification (Line) and Control Process (Column)

Classification		110	113	121	122	123	124	131	133	134	141	142	210	220	230	240	250	310	320	330	400	500	Total	Producer Accuracy (%)	Commission Error (%)
															Control Point										
	110	129		2				1		1								1					134	96.3	3.7
	113			1																			1		100.0
	121	4		56		5																	65	86.2	13.8
	122	1		1	13																		15	86.7	13.3
	123			1		11				2													14	78.6	21.4
	124						8																8	100.0	
	131			1				29		1													31	93.5	6.5
	133							1															1		100.0
	134									2	1												3	66.7	33.3
	141	1		4						1	30	1						2					39	76.9	23.1
	142										1	9											10	90.0	10.0
	210	2			5			5		1		1	66		6			10					96	68.8	31.3
	220																								
	230			1				2				1	3		10		1	1					18	55.6	44.4
	240																								
	250																								
	310														1			50					52	96.2	3.8
	320												1		1			2					4		100.0
	330																								
	400																								
	500																					15	16	93.8	6.3
	Total	137		67	19	16	8	38		8	32	12	70		18		1	66				15	507		
	Producer accuracy (%)	94.2		83.6	68.4	68.8	100.0	76.3		25.0	93.8	75.0	94.3		55.6		100.0	75.8				100.0			
	Omission error (%)	5.8		16.4	31.6	31.3		23.7		75.0	6.3	25.0	5.7		44.4			24.2							

19.7 Conclusions

The internal validation procedure developed as part of the production of the UA2006 demonstrated its efficiency and contributed to strengthening the uptake of the UA among the user community. The UA2006 is now the second most downloaded dataset after CLC from the EEA portal. In addition, the accuracy measurement results were confirmed by the external validation process conducted by the EC. A similar internal validation approach is now applied as part of the UA2012 update and extension. The UA2012 production is now ongoing and should be completed in 2016. It now covers nearly 1 million km², that is, almost 25% of the EU28 areas and includes close to 700 functional urban areas representing all cities/towns with a population of more than 50,000 inhabitants.

References

Burghardt, W., Banko, G., Hoeke, S., Hursthouse, A., de L'Escaille, T., Ledin, S., Marsan, F. A., Sauer, D., Stahr, K. 2004. TG 5—Soil sealing, soils in urban areas, land use and land use planning. In L. Van-Camp et al. (Eds.), Reports of the Technical Working Groups Established under Thematic Strategy for Soil Protection 6. Luxembourg: Office for Official Publications of the European Communities.

EEA (European Enivironment Agency). 2006. EEA glossary website. Available at: http://glossary .eea.europa.eu/EEAGlossary/S/soil_sealing.

EEA (European Environment Agency). 2010. *The European environment: State and outlook 2010: Land use.* Luxembourg: Office for Official Publications of the European Union.

EEA (European Environment Agency). 2013. GIO land High Resolution Layers. Available at: http:// land.copernicus.eu/user-corner/publications/gio-land-high-resolution-layers/view.

GSE Land Information Services–Service Validation protocol. 2008. ITDO-421-RP-0003-C5. ESRIN/ Contract No.10407/05/I-LG.

Maucha, G., Buttner, G., Kosztra, B. 2011. European validation of GMES FTS soil sealing enhancement data. In L. Halounova (Ed.), *Remote sensing and geoinformation not only for scientific cooperation* (pp. 223–238). Prague: EARSeL and Czech Technical University in Prague. Available at: http://www.earsel.org/symposia//2011-Prague/Proceedings/index.htm.

Meirich, S. 2008. Mapping guide for a European Urban Atlas. GSE Land Consortium. Report ITD-0421-GSELand-TN-01. Available at: http://www.eea.europa.eu/data-and-maps/data/urban -atlas/mapping-guide.

Molini, F., Salgado, M. 2012. Sprawl in Spain and Madrid: A low starting point growing fast. *European Planning Studies* 20: 1075–1092.

Montero, E., Van Wolvelaer, J., Garzon, A. 2014. The European Urban Atlas. In I. Manakos and M. Braun (Eds.), *Land use and land cover mapping in Europe: Practices and trends* (pp. 115–124). Remote Sensing and Digital Image Processing 18 Dordrecht: Springer.

Nunes de Lima, M. V. (Ed.). 2005. IMAGE2000 and CLC2000. Products and methods. Ispra: European Commission and Joint Research Centre.

Scalenghe, R., Marsan, A. F. 2009. The anthropogenic sealing of soil in urban areas. *Landscape and Urban Planning* 90:1–10.

SIRS, IGN FI, EUROTOPO. 2011. Urban Atlas final delivery of land use/cover maps of major European urban agglomerations—Final report. European Commission.

20

Landscape Fragmentation in Europe

Jochen A. G. Jaeger, Tomas Soukup, Christian Schwick,
Luis F. Madriñán, and Felix Kienast

CONTENTS

20.1 What Is Landscape Fragmentation and Why Is It a Problem?

20.1.1 Definition of Landscape Fragmentation and Its Effects on the Environment

Landscape fragmentation is the result of transforming large habitat patches into smaller, more isolated fragments of habitat. This process is most obvious in intensively used regions, where fragmentation is the product of the linkage of built-up areas via linear infrastructure, such as roads and railroads (Saunders et al., 1991; Forman, 1995). Despite many improvements in legislation to better protect biodiversity, urban sprawl is still increasing in Europe and new transport infrastructure is being constructed at a rapid pace. Fragmentation has significant effects on various ecosystem services and wildlife populations (Table 20.1).

There is a growing volume of evidence of negative ecological impacts of roads (Forman et al., 2003). For example, Fahrig and Rytwinski (2009) reviewed 79 studies that provide data on population-level effects (abundance and density) and found that, overwhelmingly, roads and traffic have a negative effect on animal abundance, with negative effects outnumbering positive effects by a factor of five. The four main effects of roads and traffic that affect animal populations detrimentally are that they decrease habitat amount and quality; enhance mortality due to collisions with vehicles; prevent access to resources on the other side of the road; and subdivide animal populations into smaller and more vulnerable fractions (Figure 20.1). Roads also enhance human access to wildlife habitats and facilitate the spread of invasive species, and the subdivision and isolation of subpopulations interrupts metapopulation dynamics (Hanski, 1999) and reduces genetic variability (Forman and Alexander, 1998; IUCN, 2001). Landscape fragmentation increases the risk of populations of becoming extinct, as isolated populations are more vulnerable to natural stress factors such as natural disturbances (e.g., weather conditions, fires, diseases), that is, lower resilience.

The possibility for two animals of the same species to find each other in the landscape is a prerequisite for the persistence of animal populations (e.g., because of the need for genetic exchange between populations and for the recolonization of empty habitats). There may be various additional effects about which our knowledge is still very limited, such as cumulative effects (combination with other human impacts), response times of wildlife populations, and effects on ecological communities (e.g., cascading effects). Therefore, the precautionary principle should be employed in road planning.

The extent of the negative impacts of habitat fragmentation on animal and plant populations is difficult to quantify because the full extent of the ecological effects of landscape alterations will become evident only decades afterwards (Figure 20.2). Even if all further habitat fragmentation were stopped, some wildlife populations would still disappear over the coming decades owing to their long response times to the alterations that have already occurred. This effect is called the "extinction debt" of altered landscapes (Tilman et al., 1994). Therefore, indicators are needed that measure various pressures or threats to biodiversity. For example, the threat to biodiversity due to landscape fragmentation can be quantified by the "mesh size" of the network created by the fragmenting elements present in the landscape.

The purpose of the study *Landscape Fragmentation in Europe* (European Environment Agency [EEA] and Swiss Federal Office for the Environment [FOEN], 2011) was to quantify landscape fragmentation of land areas in all countries in Europe for which the necessary data were available. The study considered fragmentation caused by transportation infrastructure and built-up areas. However, other anthropogenic alterations of the landscapes in Europe also contribute to fragmentation (e.g., intensive agriculture, fences, shale gas exploitation).

TABLE 20.1

Effects of Landscape Fragmentation on the Environment and Various Ecosystem Services (Grouped into Seven Themes): Examples of the Consequences of Linear Infrastructure Facilities Such as Roads, Railways, and Power Lines (Not Including the Effects of Construction Sites Such as Excavation and Deposition of Soils, Vibrations, and Acoustic and Visual Disturbances)

Theme	Consequences of Linear Infrastructure Facilities
Land Cover	• Land occupation for road surface and shoulders • Soil compaction, sealing of soil surface • Alterations to geomorphology (e.g., cuts, embankments, dams, stabilization of slopes) • Removal of vegetation, alteration of vegetation
Local Climate	• Modification of temperature conditions (e.g., heating up of roads, increased variability in temperature) • Accumulation of cold air at embankments of roads (cold air build-ups) • Modification of humidity conditions (e.g., lower moisture content in the air due to higher solar radiation, stagnant moisture on road shoulders due to soil compaction) • Modification of light conditions • Modification of wind conditions (e.g., due to aisles in forests) • Climatic thresholds
Emissions	• Vehicle exhaust, pollutants, fertilizing substances leading to eutrophication • Dust, particles (abrasion from tires and brake linings) • Oil, fuel, etc. (e.g., in case of traffic accidents) • Road salt • Noise • Visual stimuli, lighting
Water	• Drainage, faster removal of water • Modification of surface water courses • Lifting or lowering of groundwater table • Water pollution
Flora and Fauna	• Death of animals caused by road mortality (partially due to attraction of animals by roads or railways: "trap effect") • Higher levels of disturbance and stress, loss of refuges • Reduction or loss of habitat; sometimes creation of new habitat • Modifications of food availability and diet composition (e.g., reduced food availability for bats due to cold air build-ups along road embankments at night) • Barrier effect, filter effect to animal movement (reduced connectivity) • Disruption of seasonal migration pathways, impediment of dispersal, restriction of recolonization • Subdivision and isolation of habitats and resources, breaking up of populations • Disruption of meta-population dynamics, genetic isolation, inbreeding effects and increased genetic drift, interruption of the processes of evolutionary development • Reduction of habitat below required minimal areas, loss of species, reduction of biodiversity • Increased intrusion and distribution of invasive species, pathways facilitating infection with diseases • Reduced effectiveness of natural predators of pests in agriculture and forestry (i.e., biological control of pests more difficult)
Landscape Scenery	• Visual stimuli, noise • Increasing penetration of the landscape by roads, posts, and wires • Visual breaks, contrasts between nature and technology; occasionally vivification of landscapes (e.g., by avenues with trees) • Change of landscape character and identity

(Continued)

TABLE 20.1 (CONTINUED)

Effects of Landscape Fragmentation on the Environment and Various Ecosystem Services (Grouped into Seven Themes): Examples of the Consequences of Linear Infrastructure Facilities Such as Roads, Railways, and Power Lines (Not Including the Effects of Construction Sites Such as Excavation and Deposition of Soils, Vibrations, and Acoustic and Visual Disturbances)

Theme	Consequences of Linear Infrastructure Facilities
Land Use	• Consequences of increased accessibility for humans due to roads, increase in traffic volumes, increased pressure for urban development and mobility • Farm consolidation (mostly in relation with construction of new transport infrastructure) • Reduced quality of agricultural products harvested along roads • Reduced quality of recreational areas due to shrinkage, dissection, and noise

Source: Jaeger, J. 2003. II-5.3 Landschaftszerschneidung. In W. Konold, R. Böcker, and U. Hampicke (Eds.), *Handbuch Naturschutz und Landschaftspflege* (1999 ff.), 11. Erg.-Lieferung. Landsberg, Germany: Ecomed-Verlag; based on various sources.

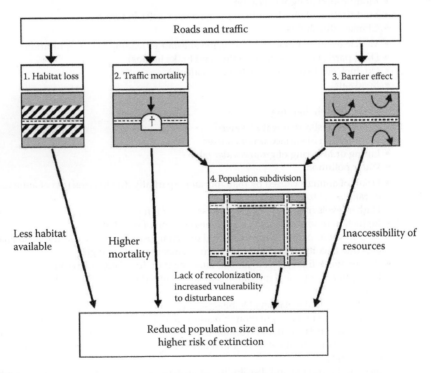

FIGURE 20.1

The four main effects of transportation infrastructure on wildlife populations. Both traffic mortality and barrier effect contribute to population subdivision and isolation. (From Jaeger, J. A. G., Bowman, J., Brennan, J., Fahrig, L., Bert, D., Bouchard, J., Charbonneau, N., Frank, K., Gruber, B., Tluk von Toschanowitz, K. 2005. Predicting when animal populations are at risk from roads: An interactive model of road avoidance behavior. *Ecological Modelling* 185:329–348.)

20.1.2 Socioeconomic Drivers of Landscape Fragmentation

Even though there is a general agreement about the negative effects of landscape fragmentation, the interactions among geophysical, ecological, and anthropogenic drivers are still poorly understood. However, such information is essential for management and restoration efforts (Laurance, 1999; Geist and Lambin, 2001; Bayne and Hobson, 2002).

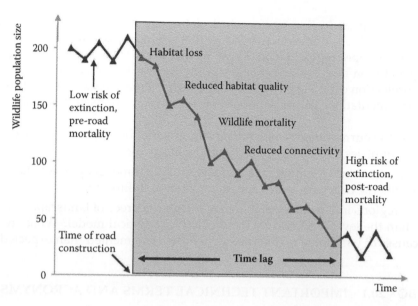

FIGURE 20.2

Four ecological impacts of roads on wildlife populations and their time lags for their combined effect. After the time lag (often in the order of decades), population size is smaller, exhibits greater relative fluctuations over time, and is more vulnerable. (Modified after Forman, R. T. T. et al. 2003. *Road ecology—Science and solutions.* Washington, DC: Island Press.)

Therefore, studies are urgently needed that address the driving forces of landscape fragmentation.

20.1.3 Research Questions

Most countries in Europe are now emphasizing the need to preserve biodiversity and ensure connectivity between the remaining natural areas for the movement of animals, including migration and dispersal, for access to different types of habitats and other resources, for recolonization of empty habitats, and for genetic exchange between populations. The UN Convention on Biological Diversity considers fragmentation by infrastructure and other land use (LU) a major threat to the populations of many species. This is reflected in the pan-European biological and landscape diversity strategy (PEBLDS) (http://www.strategy guide.org), the European Community biodiversity strategy (European Commission [EC], 1998), and the habitats directive (Council of the European Communities, 1992). In May 2011, the European Union (EU) adopted a Biodiversity Strategy to halt biodiversity loss in Europe by 2020. Target 2 of the strategy states that "by 2020, ecosystems and their services are maintained and enhanced by establishing Green Infrastructure and restoring at least 15% of degraded ecosystems." Accordingly, the EU also adopted a Green Infrastructure Strategy in May 2013 (EC, 2013).

However, the discrepancy between the political objectives and the real development has grown. The next important task is to integrate habitat fragmentation into transportation planning and monitoring studies and to develop agreements about environmental standards such as limits and targets to curtail landscape fragmentation.

Our study (EEA and FOEN, 2011) quantified the degree of landscape fragmentation caused by transportation infrastructure and built-up areas in more than 500 NUTS-X regions in 28 countries in Europe (for which data were available) for the first time (Box 20.1). We also identified socioeconomic factors that are most likely to explain the observed patterns of landscape fragmentation. These results provide a baseline to measure landscape fragmentation and to track changes in the future. The effectiveness of policies of landscape protection can then be evaluated through comparison with these time series of landscape fragmentation. The most important research questions of this study were

1. What is the current degree of landscape fragmentation in Europe at three spatial scales (countries, NUTS-X regions, 1-km^2 LEAC grid)?
2. What is the relative importance of socioeconomic and geophysical factors that determine the degree of landscape fragmentation in Europe?
3. Which regions in Europe exhibit higher or lower degrees of landscape fragmentation than the level predicted by the predictive statistical model? What are potential causes of why some regions are more or less fragmented than expected?

BOX 20.1   IMPORTANT TECHNICAL TERMS AND ACRONYMS

FG	A **fragmentation geometry** (FG) is a set of various types of barriers that are considered relevant in a landscape. These barriers are combined to determine the degree of fragmentation by calculating the effective mesh density. Three fragmentation geometries were used: FG-A1 = Major anthropogenic fragmentation FG-A2 = Major and medium anthropogenic fragmentation FG-B2 = Fragmentation of non-mountainous land areas For further information see Section 20.2.3.
m_{eff}	The **effective mesh size** (m_{eff}) serves to measure landscape connectivity, that is, the degree to which movement between different parts of the landscape is possible. It expresses the probability that any two points chosen randomly in a region are connected. The more barriers fragmenting the landscape, the lower the effective mesh size (measured in square kilometers).
s_{eff}	The **effective mesh density** (s_{eff}) is a measure of landscape fragmentation, that is, the degree to which movement between different parts of the landscape is interrupted by barriers. It gives the effective number of meshes per 1000 km^2, or the *density* of the meshes, by determining how many times the effective mesh size fits into an area of 1000 km^2. The more barriers fragmenting the landscape, the higher is the effective mesh density (further information in Sections 20.2.1 and 20.2.2).
NUTS-X	The **Nomenclature of Statistical Territorial Units** (NUTS) is a hierarchical system for dividing up the territory of the EU for the purpose of the collection of regional statistics, socioeconomic analyses of the regions and framing of policies. NUTS-X denotes a combination of NUTS-2 (basic regions for the application of regional policies) and NUTS-3 (small regions for specific diagnoses).
LEAC grid	This grid is the European reference grid used for activities in the frame of **Land and Ecosystem Accounting** (LEAC) and other tasks. It has a resolution of 1 km^2.
CLC	**Coordination of Information on the Environment (CORINE) Land Cover** (CLC) is a digital map of land cover types in Europe that is consistent across the continent and is useful for environmental analysis and assessment and for policy making. CLC2006 is the third dataset in a series (after 1990 and 2000).
TeleAtlas	**TeleAtlas** is a provider of digital maps for navigation services, covering 200 counties around the world. The data of roads and railways used in this report were taken from the TeleAtlas 2009 dataset.

4. What are the implications of these results and how can they be applied to traffic planning and regional planning in Europe?

The results demonstrate that there is an urgent need for action. Therefore, we discuss fragmentation analysis as a tool in transportation planning and regional planning and for performance review, and we recommend a set of measures to control landscape fragmentation, such as more effective protection of the remaining unfragmented areas and the setting of targets and limits.

20.2 How Can We Measure Landscape Fragmentation?

20.2.1 Methods for Measuring Landscape Fragmentation

Researchers have used various metrics to quantify landscape fragmentation (Gustafson, 1998; Leitão et al., 2006). Each metric has specific strengths and weaknesses. To select a reliable measure for monitoring landscape change over time, the behavior of the metrics needs to be studied and compared carefully (Li and Wu, 2004). Mostly, the average size of remaining nonfragmented land parcels (or average patch size [*APS*]) and the density of transportation lines (*DTL*, in relation to total area of the landscape) are examples of proposed measures of fragmentation. More recently, Jaeger (2000) introduced the method of the effective mesh size and effective mesh density. Our study of landscape fragmentation in Europe used this method because it has several advantages over most other metrics:

- It takes into account all patches remaining in the "network" of transportation infrastructure and built-up area.
- It can be used for comparing the fragmentation of regions with differing total area and with differing proportions of land taken up by housing, industry, and transportation infrastructure.
- Its reliability has been confirmed on the basis of nine suitability criteria through a systematic comparison with other quantitative measures (Jaeger, 2000; Girvetz et al., 2007) and 17 criteria for indicator selection for monitoring sustainable development (Jaeger et al., 2008). The suitability of other metrics was limited, as they met only a subset of the criteria.
- It can be extended to include the permeability of roads and railways for wildlife moving in the landscape (i.e., filter effect; Jaeger, 2002, 2007).

One reason why the average patch size is not a suitable metric of landscape fragmentation is that it does not react consistently to different fragmentation phases (Forman, 1995), for example, it *decreases* when habitat patches are dissected, when the size of a patch decreases, or when large patches are lost, but it *increases* when small patches are lost (Jaeger, 2000, 2002). The average patch size and the density of transportation lines do not behave in a suitable fashion in the phases of shrinkage and attrition of habitat patches (EEA and FOEN, 2011), which contribute to landscape fragmentation. Therefore, the suitability of these metrics for quantifying landscape fragmentation is limited.

20.2.2 Effective Mesh Size and Effective Mesh Density

The method of the effective mesh size (m_{eff}) and effective mesh density (s_{eff}), which is applied to measure the degree of landscape fragmentation, is based on the probability that two locations chosen randomly in a region are connected, that is, are located in the same patch (Jaeger, 2000). This can be understood as the probability that two animals, placed in different locations somewhere in a region, can find each other and meet without having to cross a barrier such as a road, built-up area, or major river. It indicates the ability of animals to move around freely in the landscape without encountering such barriers. If one of the locations (or both) is placed within a fragmenting landscape element, for example, in a built-up area, it is considered isolated from all other locations. The smaller the effective mesh size, the more fragmented the landscape (Box 20.2). The maximum value of the effective mesh size is reached with a completely unfragmented area: m_{eff} then equals the size of the whole area. If an area is divided up into patches of equal size, then m_{eff} equals the size of these patches.

Landscape fragmentation can be understood as a reduction in landscape connectivity, which is defined as "the degree to which the landscape facilitates or impedes movement among resource patches" (Taylor et al., 1993: 571; see also Tischendorf and Fahrig, 2000). The effective mesh size is a straightforward quantitative expression of landscape connectivity: it corresponds directly with the suggestion by Taylor et al. (1993: 572) that "landscape connectivity can be measured for a given organism using the probability of movement between all points or resource patches in a landscape." As a consequence, this method has substantial advantages, for example, it meets all scientific, functional, and pragmatic requirements of environmental indicators (Jaeger et al., 2008).

The effective mesh size has been widely implemented as an indicator for environmental monitoring by various countries, for example, Switzerland (Bertiller et al., 2007; Jaeger et al., 2007, 2008), Germany (as one out of 24 core indicators in the National Sustainability Report and in the National Strategy on Biological Diversity; Schupp, 2005; Federal Ministry for the Environment, Nature Conservation and Nuclear Safety, 2007), Baden-Württemberg (State Institute for Environment, Measurements and Nature Conservation Baden-Württemberg, 2006), and South Tyrol, Italy (Tasser et al., 2008).

Alternatively, the degree of fragmentation can be expressed as the *effective mesh density* s_{eff}, that is, the effective number of patches per 1000 km^2 (Jaeger et al., 2007, 2008) which is related to effective mesh size according to $s_{eff} = 1/m_{eff}$ (Box 20.2). For reading trends off graphs, it is easier to use effective mesh density as increases in s_{eff} indicate increasing landscape fragmentation (Figure 20.4). Therefore, we mostly present the results using s_{eff}.

We used the Cross-Boundary Connections procedure (CBC procedure) for calculating the effective mesh size and effective mesh density (Moser et al., 2007). This procedure removes any bias due to the boundaries of the reporting units when quantifying landscape structure. It accounts for the connections within unfragmented patches that extend beyond the boundaries of the reporting units.

20.2.3 Fragmentation Geometries, Base Data, and Reporting Units

To analyze landscape fragmentation, it is necessary to identify the landscape elements that are relevant to fragmentation. The choice of a specific set of fragmenting elements defines a so-called "fragmentation geometry." To identify the relative contributions of various types of barriers to the overall fragmentation of the landscape, three different fragmentation geometries (FGs) were used, where each FG handles man-made and natural barriers in a different way (Table 20.2).

BOX 20.2   DEFINITION OF EFFECTIVE MESH SIZE m_{eff}
AND EFFECTIVE MESH DENSITY s_{eff}

The definition of the *effective mesh size* m_{eff} is based on the probability that two points chosen randomly in an area are connected and are not separated by any barriers. This leads to the formula:

$$m_{eff} = \left(\left(\frac{A_1}{A_{total}} \right)^2 + \left(\frac{A_2}{A_{total}} \right)^2 + \left(\frac{A_3}{A_{total}} \right)^2 + \ldots + \left(\frac{A_n}{A_{total}} \right)^2 \right) \cdot A_{total} = \frac{1}{A_{total}} \sum_{i=1}^{n} A_i^2,$$

where n is the number of patches, A_1 to A_n represent the patch sizes from patch 1 to patch n, and A_{total} is the total area of the region investigated.

The first part of the formula gives the probability that two randomly chosen points are in the same patch. The second part (multiplication by the size of the region) converts this probability into a measure of area. This area is the "mesh size" of a regular grid pattern showing an equal degree of fragmentation (Figure 20.3) and can be directly compared with other regions. The smaller the effective mesh size, the more fragmented is the landscape.

FIGURE 20.3
Barriers in the landscape (left) and the corresponding effective mesh size represented in the form of a regular grid (right). (From Bertiller, R. et al. 2007. *Degree of landscape fragmentation in Switzerland: Quantitative analysis 1885–2002 and implications for traffic planning and regional planning.* Project report [in German]. FEDRO Report No. 1175. Berne: Swiss Federal Roads Authority.)

(Continued)

**BOX 20.2 (CONTINUED)   DEFINITION OF EFFECTIVE
MESH SIZE m_{eff} AND EFFECTIVE MESH DENSITY s_{eff}**

The *effective mesh density* s_{eff} gives the effective number of meshes per square kilometer, that is, the density of the meshes. It is often more convenient to count the effective number of meshes per 1000 km² rather than per 1 km² (the difference is visible in the unit following the number). This number is very easy to calculate from the effective mesh size: it is simply a question of how many times the effective mesh size fits into an area of 1000 km².

For example, for $m_{\text{eff}} = 25$ km², the corresponding effective mesh density is $s_{\text{eff}} = 1$ mesh/(25 km²) = 0.04 meshes per km² = 40 meshes per 1000 km².

This relationship is therefore expressed as

$$s_{\text{eff}} = \frac{1000 \text{ km}^2}{m_{\text{eff}}} \frac{1}{1000 \text{ km}^2} = \frac{1}{m_{\text{eff}}}.$$

The effective mesh density value rises when fragmentation increases (Figure 20.4). The two measures contain the same information about the landscape, but the effective mesh density is more suitable for detecting trends and changes in trends (see Jaeger [2000, 2002] for a detailed description).

As a simple example of calculating m_{eff} and s_{eff}, consider a landscape that is fragmented by highways into three patches:

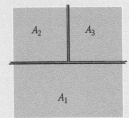

$$A_{\text{total}} = 2 \text{ km} \cdot 2 \text{ km} = 4 \text{ km}^2$$

The probability that two randomly chosen points will be in patch 1 (and therefore connected) is

$$\left(\frac{A_1}{A_{\text{total}}}\right)^2 = 0.5 \cdot 0.5 = 0.25.$$

The corresponding probability is $0.25^2 = 0.0625$ for both patches 2 and 3. The probability that the two points will be in patch 1 or 2 or 3 is the sum of the three probabilities, resulting in 0.375. Multiplying this probability by the total area of the region investigated gives the value of the effective mesh size:

$$m_{\text{eff}} = 0.375 \cdot 4 \text{ km}^2 = \mathbf{1.5 \text{ km}^2}.$$

The effective mesh density s_{eff} is then $s_{\text{eff}} = 666.7$ meshes per 1000 km².

The relationship between mesh density and mesh size is such that a percentile increase in mesh density is different from a percentile decrease in mesh size (Figure 20.4). For example, an increase in effective mesh density by 100% (e.g., from 20 to 40 meshes per 1000 km²) corresponds to a decrease in effective mesh size by 50% (e.g., from 50 to 25 km²).

(Continued)

BOX 20.2 (CONTINUED)   DEFINITION OF EFFECTIVE MESH SIZE m_{eff} AND EFFECTIVE MESH DENSITY s_{eff}

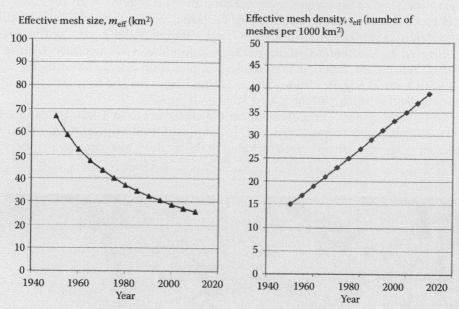

FIGURE 20.4

Example illustrating the relationship between effective mesh size and effective mesh density (= effective number of meshes per 1000 km²). In this hypothetical example, the trend remains constant. A linear rise in effective mesh density (right) corresponds to a $\frac{1}{x}$ curve in the graph of the effective mesh size (left). A slower increase in fragmentation results in a flatter curve for effective mesh size, and a more rapid increase produces a steeper curve. It is therefore easier to read trends off the graph of effective mesh density (right). (From Jaeger, J., Bertiller, R., Schwick, C. 2007. *Degree of landscape fragmentation in Switzerland—Quantitative analysis 1885–2002 and implications for traffic planning and regional planning—Condensed version*. Neuchâtel: Bundesamt für Statistik.)

TABLE 20.2

Definition of the Three Fragmentation Geometries Used for Analyzing Landscape Fragmentation in Europe

		Natural Barriers	
		None	Mountains, lakes, and major rivers are considered as barriers; the fragmentation of the remaining land area is reported.
Man-made barriers (motorways, other roads, railway lines, built-up areas)	Roads up to class 2 (major roads)	**FG-A1:** "Major anthropogenic fragmentation"	–
	Roads up to class 4 (connecting roads)	**FG-A2:** "Major and medium anthropogenic fragmentation"	**FG-B2:** "Fragmentation of non-mountainous land areas"

Fragmentation geometries A1 and A2 include only man-made barriers: roads, railways, and built-up areas. Lakes, rivers, and high mountains also play a major role in acting as natural barriers. In some regions, their impact is so important that it is not meaningful to compare the level of landscape fragmentation in such regions with regions without lakes and mountains. Therefore, in geometry B2, lakes, major rivers, and high mountains were considered as barriers, and the level of fragmentation of those parts of these regions that are covered by land areas that are not high mountains was calculated (Figure 20.5). Thus, FG-B2 reflects the fact that man-made fragmentation affects biodiversity in combination with natural fragmentation. The resulting values for the fragmentation of non-mountainous land areas can be compared among all regions.

If a road or railway goes through a tunnel that is longer than 1 km, the landscape in this area was considered as connected and almost not affected by disturbance from traffic noise. However, shorter tunnels were included in the analysis as normal transport routes.

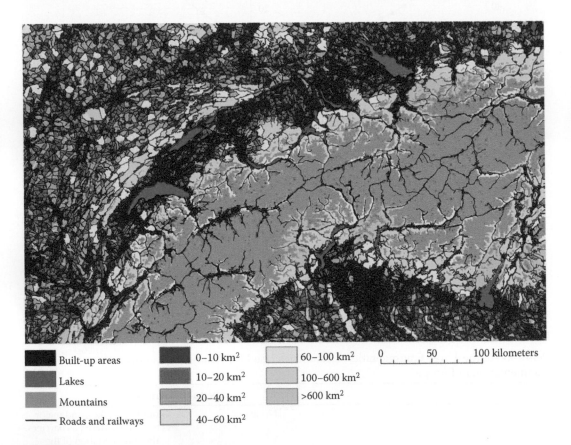

■ Built-up areas	■ 0–10 km²	▨ 60–100 km²
▦ Lakes	▨ 10–20 km²	□ 100–600 km²
▨ Mountains	▨ 20–40 km²	□ >600 km²
— Roads and railways	□ 40–60 km²	

0   50   100 kilometers

FIGURE 20.5

Illustration of fragmentation geometry B2 using an example from Switzerland and adjacent countries. The barriers are shown in black (built-up areas, roads, railways), and the colors indicate the sizes of the remaining patches in the landscape. Lake Constance and Lake Geneva are visible at the top and on the left of the map, respectively. In FG-B2, the lakes (blue) and mountains (gray) were considered as barriers, but were not included in the reporting units. (From EEA [European Environment Agency] and FOEN [Swiss Federal Office for the Environment]. 2011. *Landscape fragmentation in Europe. Joint EEA-FOEN report*. EEA Report No. 2/2011 (J. A. G. Jaeger, T. Soukup, L. F. Madriñán, C. Schwick, and F. Kienast, authors). Luxembourg: Publications Office of the European Union.)

Every fragmentation geometry has its strengths and wem aknesses. The choice of the most appropriate FG depends on a study's objectives and context. This condensed book chapter mostly shows the results for FG-B2 because it is the most suitable one for comparing different regions. For a more detailed analysis of landscape fragmentation in the context of environmental impact assessment, cumulative impact assessment, and strategic environmental assessment, a combination of all three (or even more) FGs may be more appropriate than any single FG.

Our analysis used the CLC2006 for the built-up areas at a scale of 1:100,000 (minimum mapping unit [MMU] size of 25 ha). For the linear features, we used the 2009 TeleAtlas dataset (Table 20.3). The classification of roads was based on the road classes used in the TeleAtlas Multinet® platform (scale 1:100,000). The TeleAtlas dataset is currently the only available dataset of road classes that is consistent across Europe at this resolution. These maps are highly accurate and provide standardized cover for 28 countries in Europe, and therefore, the results also have a similar level of accuracy. The roads were buffered (on either side) to reflect the loss of habitat due to their surface. We had also intended to use the TeleAtlas of 2002 in combination with the CLC2000 to investigate landscape fragmentation for an earlier point in time and determine the direction and rate of change in landscape fragmentation. However, it turned out that the TeleAtlas data for 2002 and 2009 were not comparable because of various changes in road classes between the two points in time. In addition, there was a concern about the quality of the data for Romania in the year 2009: Not all roads of classes 3 and 4 in Romania seemed to be represented in the TeleAtlas dataset. Therefore, our calculations of the degree of landscape fragmentation for Romania for FG-A2 and -B2 underestimated the true level of fragmentation.

TABLE 20.3

Datasets and Fragmenting Elements Used to Create the Fragmentation Geometries

Dataset	Year	Fragmenting Elements
CORINE Land Cover (CLC)	2006	1.1: Continuous urban fabric, discontinuous urban fabric
		1.2: Industrial and commercial units, road and rail networks and associated land, port areas, and airports
		1.3: Mineral extraction sites, dump sites, and construction sites
		1.4.1: Green urban areas
		1.4.2: Sport and leisure facilities (only included as a barrier if they were completely surrounded by the previous classes)
		4.2.2: Salines
		5.1.2: Water bodies
TeleAtlas Multinet®	2009	Class 00 "Motorways" (buffer $2 \cdot 15$ m)
		Class 01 "Major roads" (buffer $2 \cdot 10$ m)
		Class 02 "Other major roads" (buffer $2 \cdot 7.5$ m)
		Class 03 "Secondary roads" (buffer $2 \cdot 5$ m)
		Class 04 "Local connecting roads" (buffer $2 \cdot 2.5$ m)
		Railroads (buffer $2 \cdot 2$ m)
Nordregio	2004	Criterion 1: Elevation is higher than 2500 m
		Criterion 2: Elevation is higher than 1500 m and the slope is steeper than 2°
WorldClim	2009	Mean July temperature < 9.5°C (mean 1950–2000, 30″)
CCM2: Catchment Characterization and Modeling version 2.1	2007	Catchment areas greater than 3000 km^2

We considered as high mountains all elevations above 2500 m and also those elevations above 1500 m that had a slope of more than 2 degrees. In addition, we used the 9.5°C isoline for identifying mountainous areas, which applies mostly in Scandinavia following the rationale that the growing season is very short in areas with temperatures lower than this isoline, primary production is very limited (no trees and the presence of glaciers), and that these regions have limited accessibility for humans (virtually no construction of buildings, towns, or roads). An additional rationale is that these mountainous areas are not suitable for settlement and usually have no or very few roads. Therefore, they should be removed from the reporting units before comparing regions with differing amounts of such areas for the comparison to be meaningful.

Only rivers with catchment areas greater than 3000 km^2 were included as fragmenting elements because such water bodies tend to be navigable and thus become a relevant obstacle for the free movement of many terrestrial animals. We used the CCM2 (Catchment Characterisation and Modelling version 2.1) river and catchment data for Europe (Vogt et al., 2007), as data about river width were not available. For the lakes, we used the CLC nomenclature and selected class 5.1.2 (*Water bodies*).

As a consequence of the resolution of the CLC data with an MMU of 25 ha, the evaluation of the results in terms of fragmentation effects for species that are sensitive to habitats smaller than 25 ha is limited. Such species that are still important in food chains and for certain ecosystem services may also be affected by fragmentation processes at smaller scales. For example, to capture low-density sprawl in units <10 ha, data at a higher resolution would be required. The 25-ha resolution is appropriate for quantifying the degree of fragmentation at the scale of a 1-km^2 grid and for larger reporting units. However, for traffic planning at the regional scale, additional, more detailed information will be needed along with information about intensive agriculture, intensive silviculture, mining activities, and so forth.

This study used three scales or types of reporting units for which fragmentation were calculated and reported. The available data resulted in the following set of regions for the fragmentation analysis:

1. 28 countries in Europe
2. 580 NUTS-X regions
3. 1 km × 1 km grid units within the 28 countries

NUTS is a hierarchical classification that subdivides each member state into a whole number of NUTS-1 regions, each of which is in turn subdivided into a whole number of NUTS-2 regions, and so on for NUTS-3, -4, and -5 levels. The NUTS-X layer is a combination of the NUTS-2 and NUTS-3 layers to create reference regions that are more homogeneous in size than the other two. NUTS-X regions are a synthesis of NUTS 2 and 3 regions that provide region-specific information on rurality, urban structure, socioeconomic profiles, and landscape character. The 1 km ×1 km grid was used to describe fine detail fragmentation patterns in our study. This grid is the reference grid used for EEA activities in the frame of LEAC.

20.2.4 Predictive Models of Landscape Fragmentation Based on Geophysical and Socioeconomic Variables

To examine how strongly socioeconomic and geophysical parameters are related to fragmentation levels, a set of variables was selected for their potential importance as

TABLE 20.4

Physical and Socioeconomic Variables Used in the 2009 Analysis (Name of Variable, Source of Data: Country Level or NUTS for NUTS-2 or NUTS-3 Level, Number of Complete Cases in NUTS-X Regions, and Unit of Each Variable)

Variable[a]	Source	Complete Cases	Unit
Population density (*PD*)	NUTS	498	Number of inhabitants per km^2
GDP per capita (*GDPc*)	NUTS	481	PPS per person in 2007[b]
Quantity of goods loaded and unloaded per capita (*QGLUc*)	NUTS	418	1000 tons loaded and unloaded per person
Volume of passenger transport density (*VPD*)	Country	487	1000 pkm per km^2
Environmental expenditure (*EEc*)	Country	494	PPS per person[b]
Unemployment rate (*UR*)	NUTS	504	% of active population in 2007
Education, per capita (*EDc*)	Country	500	PPS[b] per person in 2007
Island size Index (*IsI*)	NUTS	530	No unit (0 < IsI < 1)
Hills—% (*Hills*)	NUTS	530	%
Mountain-and-slope—% (*MtSl*)	NUTS	530	%

[a] The three variables *IsI*, *Hills*, and *MtSl* are available for all of Europe. However, we did not include some countries in the socioeconomic analysis (Sweden, Bulgaria, San Marino, and Vatican City) because of lack of data. Without them, the highest number of cases possible for the socioeconomic models was 530.

[b] The unit PPS (purchasing power standards) indicates that the variable was adjusted for differences in price levels between the countries.

drivers of landscape fragmentation and for their availability at the European level. The variables used in this study can help identify the main driving forces of landscape fragmentation in Europe, as driving forces and the resulting levels of landscape fragmentation would be expected to covary. An increase or decrease of the value of some predictor variables (e.g., in population density or gross domestic product [*GDP*]) by a given amount may suggest an increase in fragmentation in the near future (Table 20.4). The source of all socioeconomic data was Eurostat (http://nui.epp.eurostat.ec.europa.eu/).

To examine how the socioeconomic variables are related to fragmentation levels, we used generalized linear models (GLMs). A global linear regression model contained all the geophysical and socioeconomic variables. Model selection was done by successively adding variables for which we had a hypothesis about its potential effect on fragmentation. All possible combinations of the explanatory variables were examined (for further information, see EEA and FOEN, 2011).

20.3 Results

20.3.1 What Is the Degree of Landscape Fragmentation in Europe?

This section presents the results for FG-B2 "Fragmentation of Non-Mountainous Land Areas" for 2009 on three scales: countries (Section 20.3.1.1), 1-km^2 grid (Section 20.3.1.2), and NUTS-X regions (Section 20.3.1.3). For a more detailed presentation see EEA and FOEN (2011).

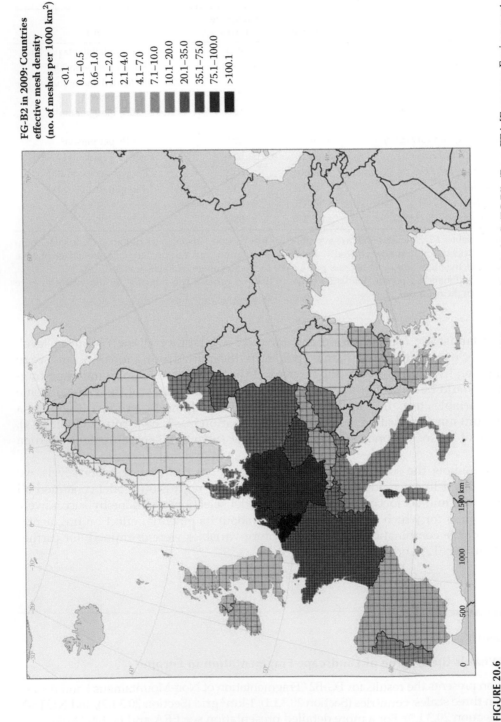

FG-B2 in 2009: Countries effective mesh density (no. of meshes per 1000 km²)

<0.1
0.1–0.5
0.6–1.0
1.1–2.0
2.1–4.0
4.1–7.0
7.1–10.0
10.1–20.0
20.1–35.0
35.1–75.0
75.1–100.0
>100.1

FIGURE 20.6

Level of landscape fragmentation in the countries, represented as regular grid (cell size of m_{eff} in km²) (FG-B2). (From EEA [European Environment Agency] and FOEN [Swiss Federal Office for the Environment]. 2011. *Landscape fragmentation in Europe. Joint EEA-FOEN report.* EEA Report No. 2/2011 (J. A. G. Jaeger, T. Soukup, L. F. Madriñán, C. Schwick, and F. Kienast, authors). Luxembourg: Publications Office of the European Union.)

20.3.1.1 Landscape Fragmentation in the Countries

The effective mesh density values cover a large range, from low values in the Iberian and Scandinavian peninsulas to very high values in the Benelux countries and Germany (Figure 20.6). Large parts of Europe are highly fragmented by transportation infrastructure and urban development. High fragmentation values are often found in the vicinity of large urban centers and along major transportation corridors. The value of the effective mesh density for all 28 investigated countries together is s_{eff} = 1.749 meshes per 1000 km².

The Benelux countries are clearly the most fragmented part in Europe (s_{eff} > 60 meshes per 1000 km²), with one of the highest population densities (*PD*) in the world and a very dense road network (Figure 20.7). Further infrastructure development is expected and will probably be the most important cause of destruction of the remaining natural habitats in these countries. In part, this development is due to this area's role as a crossroad region in the European context (Froment and Wildman, 1987). Belgium and the Netherlands have followed two different types of urban development, even though they both have high *PD*s. Belgium has a high urbanization level of 97.3% and an average population density of 330 inhabitants per square kilometer (Antrop, 2004). The Netherlands, which has an urbanization level of 83% and a *PD* of 399 inhabitants per square kilometer (*CIA Factbook*, 2010) most noticeably has a polycentric urban structure in the Randstad region that exhibits some concentration of the population in urban centers at a national level. This is in clear contrast with Belgium, which exhibits a disperse urbanization pattern due to continuous increase of smaller urban centers in the countryside (Nijkamp and Goede, 2002).

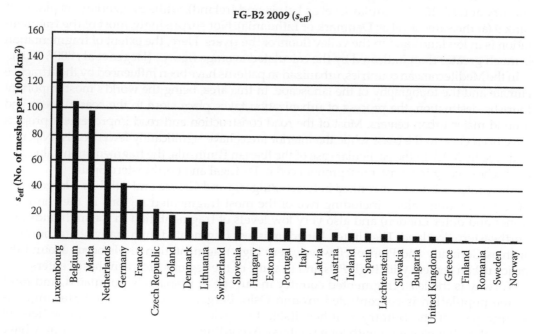

FIGURE 20.7
Effective mesh density values by country for FG-B2 in 2009. (From EEA [European Environment Agency] and FOEN [Swiss Federal Office for the Environment]. 2011. *Landscape fragmentation in Europe. Joint EEA-FOEN report.* EEA Report No. 2/2011 [J. A. G. Jaeger, T. Soukup, L. F. Madriñán, C. Schwick, and F. Kienast, authors]. Luxembourg: Publications Office of the European Union.)

After the Benelux countries, Germany is one of the most heavily fragmented countries in Europe in all three FGs. This country has a long history of road and motorway construction. Its motorway network is one of the most closely meshed in the world and exhibits one of the highest volumes of passenger and freight transport in Europe. Its central location in Europe, high levels of industrialization, and the lack of major topographical obstacles against the construction of transportation infrastructure explain this high level of landscape fragmentation, among other factors.

The next most heavily fragmented country is France. France exhibits a wide range of fragmentation patterns. The country includes some of the most heavily fragmented regions in Europe. The northern part of the country (e.g., around the metropolitan area of Paris) has fragmentation levels similar to those of the neighboring countries of Belgium, the Netherlands, and Germany, and also similar to them, has large areas of intensive agriculture and no mountains. In contrast, the fragmentation levels in the southern, more mountainous part of the country are more similar to those of Spain and Italy.

The Czech Republic and Poland rank sixth and seventh among the 28 countries investigated. The topographical conditions in these countries are similar to those of Germany. However, major differences exist between their economic development since the middle of the last century. In addition, Poland's PD is lower. Nevertheless, quite similar to Germany, high fragmentation values are observed in both countries.

The countries that are strongly influenced by the Alps, such as Switzerland and Austria, exhibit some of the lowest effective mesh density values in Europe for both FG-A1 and FG-A2, even lower than in Norway. However, once the Alps are considered as barriers and removed from the reporting units in FG-B2, the fragmentation value for Austria places the country at rank 16 (at a similar level as Latvia and Ireland), while Switzerland is placed at rank 9 (at the same level as Denmark or Lithuania). Not surprisingly, most of the fragmentation is in lowlands and in the valley floors of the rivers. Here, the extent of fragmentation is much greater than indicated by the calculated average figures (Jaeger et al., 2007, 2008).

In the Mediterranean countries, urbanization patterns have been influenced by the attractive climate and the topography of the landscape. In this area, being the world's most important touristic destination, the process of urbanization takes place close to the coastal areas and around major urban centers. Most of the road construction and road improvement projects have been close to the coast while the interior areas have significantly fewer roads (Nijkamp and Goede, 2002). In the particular case of the Iberian Peninsula, the fragmentation values are the highest along the coast, most pronounced in Portugal and northeastern Spain.

The United Kingdom is ranked between Bulgaria and Greece and exhibits a large range of fragmentation values, including two of the most fragmented regions in the continent (inner and outer London) and also very low levels of fragmentation in the highlands and northeastern Scotland. The United Kingdom is less fragmented than many other countries such as Germany and France. However, very high fragmentation levels occur around the major urban centers with high PDs and in some of the regions between these centers.

Norway is the least fragmented country in Europe. It is sparsely populated, and most of the population is concentrated around Oslo, Bergen, Trondheim, and Stavanger as a large part of the country is inhospitable for agriculture and permanent settlement (areas at higher elevation with long winters). Accordingly, most parts of the country have low fragmentation values, even though its population has a very high GDP per capita. Finland is much less mountainous than Norway but still has low levels of fragmentation (ranking fourth to fifth last among the 28 countries in all FGs). The degree of fragmentation decreases from south to north, related to less favorable climatic conditions and lower population densities.

Many fragmentation patterns and explanations as in the countries discussed in the preceding text also apply to the other countries.

20.3.1.2 Landscape Fragmentation at 1 km² Resolution

At the outer boundaries of Europe, fragmentation levels are lower than in the central part (Figure 20.8). These areas include Scandinavia, Eastern European countries, Mediterranean countries, Ireland, and Scotland. An exception to this general pattern is the western part of Portugal which is more fragmented. Southern Europe is more fragmented than northern Europe, but not as much as the central area of Europe. In the central part of Europe, the major transportation corridors and the neighboring areas are highly fragmented, including areas of high urban sprawl.

Most of Western and Central Europe is heavily fragmented, with values ranging between 35 and 100 meshes per 1000 km². The coastal areas of the Iberian and Italic Peninsulas and most of England are also in this category of high fragmentation. The Alps, the Pyrenees, and the Scandinavian mountain ranges are clearly visible as the least fragmented parts in Europe. In contrast, the Apennines are visible in FG-A1 as they have lower levels of fragmentation, but they differ less strongly from the rest of the landscape in FG-A2 and FG-B2. A similar observation is also made for the mountains in Greece, the Carpathians, and the Balkan mountain range.

We present one region in higher resolution to distinguish more detail. This region shows the Alps (Figure 20.9). The mountains and lakes were considered as barriers and excluded from the reporting units (i.e., the cells of the 1-km² grid). Areas of low fragmentation surround the Alps or are associated with other regions of higher elevation, for example, in the Black Forest (Germany) and in the Apennines (Italy).

20.3.1.3 Landscape Fragmentation in the NUTS-X Regions

The analysis of the NUTS-X regions confirmed that the most fragmented NUTS-X regions (>50 meshes per 1000 km²) are located in Belgium, the Netherlands, Luxembourg, France, Germany, Denmark, the Czech Republic, Poland, the United Kingdom, and Slovenia (Figure 20.10). Some of them have effective mesh densities even above 100 meshes per 1000 km². The NUTS-X regions in Norway, Sweden, Finland, and Romania are among the least fragmented.

Figures 20.10 and 20.11 show the distribution of the fragmentation values within the 28 countries investigated. In some countries, only a small proportion of NUTS-X regions is highly fragmented, for example, in Ireland and Greece, whereas in other countries a much larger proportion is highly fragmented, for example, in Germany. The most fragmented NUTS-X regions in Europe are metropolitan Paris (FR101, FR105, FR106), inner London (UKI1), Brussels (BE10), and Vatican City (VC) with more than 1000 meshes per 1000 km²; Copenhagen (DK001), Val-de-Marne (FR107), Berlin (DE30), West Midlands (UKG3), and Vlaams-Brabant (BE24) with more than 300 meshes per 1000 km²; and the regions of outer London (UKI2), Bucharest (RO081), and Budapest (HU101) with more than 275 meshes per 1000 km². The least fragmented regions are Finnmark (NO073), Lappi (FI1A3), Vrancea (RO026), Troms (NO072), Jämtlands län (SE072), Norrbottens län (SE082), Nord-Trøndelag (NO062), Covasna (RO073), Nordland (NO071), and Buzau (RO022), all with less than 0.34 meshes per 1000 km².

There is a clear difference in the levels of fragmentation between eastern and western Germany. This indicates opportunities for protecting biodiversity by conserving the

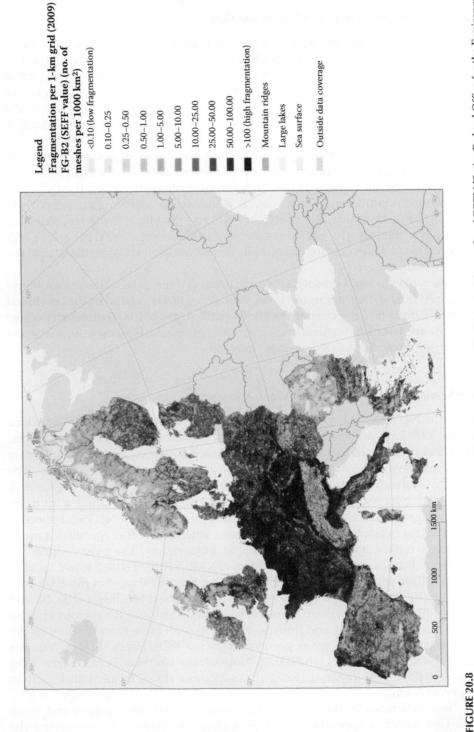

Legend

Fragmentation per 1-km grid (2009)
FG-B2 (SEFF value) (no. of
meshes per 1000 km²)

<0.10 (low fragmentation)
0.10–0.25
0.25–0.50
0.50–1.00
1.00–5.00
5.00–10.00
10.00–25.00
25.00–50.00
50.00–100.00
>100 (high fragmentation)

Mountain ridges
Large lakes
Sea surface
Outside data coverage

FIGURE 20.8
Landscape fragmentation per 1-km² grid in 2009 (FG-B2). (From EEA [European Environment Agency] and FOEN [Swiss Federal Office for the Environment]. 2011. *Landscape fragmentation in Europe. Joint EEA-FOEN report.* EEA Report No. 2/2011 [J. A. G. Jaeger, T. Soukup, L. F. Madriñán, C. Schwick, and F. Kienast, authors]. Luxembourg: Publications Office of the European Union.)

Legend

Fragmentation per 1-km grid (2009) FG-B2
FG-B2 (SEFF value)(no. of meshes per 1000 km²)

- <0.10 (low fragmentation)
- 0.10–0.25
- 0.25–0.50
- 0.50–1.00
- 1.00–5.00
- 5.00–10.00
- 10.00–25.00
- 25.00–50.00
- 50.00–100.00
- >100 (high fragmentation)
- Mountain ridges
- Water bodies
- Sea surface
- Outside data coverage

FIGURE 20.9

Landscape fragmentation per 1-km² grid in the region around the Alps in 2009 (FG-B2). (From EEA [European Environment Agency] and FOEN [Swiss Federal Office for the Environment]. 2011. *Landscape fragmentation in Europe. Joint EEA-FOEN report.* EEA Report No. 2/2011 [J. A. G. Jaeger, T. Soukup, L. F. Madriñán, C. Schwick, and F. Kienast, authors]. Luxembourg: Publications Office of the European Union.)

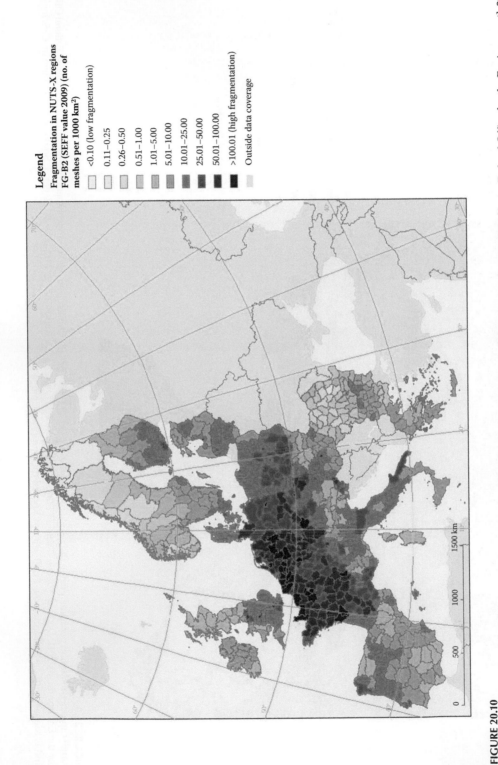

FIGURE 20.10

Landscape fragmentation in the NUTS-X regions in 2009. (From EEA [European Environment Agency] and FOEN [Swiss Federal Office for the Environment]. 2011. *Landscape fragmentation in Europe. Joint EEA-FOEN report.* EEA Report No. 2/2011 [J. A. G. Jaeger, T. Soukup, L. F. Madriñán, C. Schwick, and F. Kienast, authors]. Luxembourg: Publications Office of the European Union.)

Legend

Fragmentation in NUTS-X regions FG-B2 (SEFF value 2009) (no. of meshes per 1000 km²)

- <0.10 (low fragmentation)
- 0.11–0.25
- 0.26–0.50
- 0.51–1.00
- 1.01–5.00
- 5.01–10.00
- 10.01–25.00
- 25.01–50.00
- 50.01–100.00
- >100.01 (high fragmentation)
- Outside data coverage

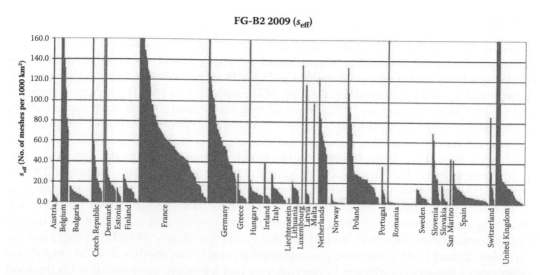

FIGURE 20.11

Effective mesh density values in the NUTS-X regions for FG-B2 in 2009, grouped by 28 countries (plus Liechtenstein and San Marino). (From EEA [European Environment Agency] and FOEN [Swiss Federal Office for the Environment]. 2011. *Landscape fragmentation in Europe. Joint EEA-FOEN report.* EEA Report No. 2/2011 [J. A. G. Jaeger, T. Soukup, L. F. Madriñán, C. Schwick, and F. Kienast, authors]. Luxembourg: Publications Office of the European Union.)

remaining relatively large unfragmented areas. Switzerland has been considered one of the leading countries in the world in promoting a more sustainable use of the landscape, with a strong legislation that limits the amount of roads to be constructed. By the year 2010, 93% of the motorways that are included in the national transportation plan of the country had already been built (Galliker, 2009). In a referendum held in 1994, Swiss voters rejected a plan to increase the road capacities in sensitive areas of the Alps (Bundesverfassung der Schweizerischen Eidgenossenschaft 1999, Art. 84 Alpenquerender Transitverkehr). According to international agreements with the EU, no more roads will be constructed to cross the Alps and most freight transport crossing the Alps is limited to using the railway connections through the tunnels of Gotthard and Lötschberg. The strict protection of the forest area since 1876 is noteworthy in this context as well. The agricultural areas are much less protected. More than two thirds of the Swiss population lives in cities and large agglomerations. However, urban sprawl is progressing rapidly in the Swiss Lowlands and in the valley floors of the Alpine rivers (Schwick et al., 2012), causing the destruction of valuable agricultural soils, a reduction in the diversity of landscape elements, affecting outdoor recreation, the beauty of the landscape, and the overall quality of life (Ewald and Klaus, 2009).

20.3.2 Predictive Socioeconomic Models

This section compares two approaches for predicting levels of fragmentation based on socioeconomic variables. The comparison shows that the pan-European model presented first is limited in its predictive value. Predicting landscape fragmentation based on a combination of six models (one for each of six large regions in Europe) is more appropriate and more relevant.

Fragmentation geometry B2 is the most important fragmentation geometry, as it is possible to compare regions with differing geophysical conditions, that is, percentage of area covered by mountains or lakes as these areas were excluded from the reporting units. In addition, this fragmentation geometry includes the most relevant barriers (Table 20.2). Therefore, we used this FG-B2 to conduct our socioeconomic analysis.

The pan-European model that incorporated all physical and socioeconomic variables exhibited a fairly strong relationship (R^2 = 46.1%) with effective mesh density. The most relevant variables at the continental scale with 443 complete cases (NUTS-X regions) were *PD* and volume passenger density (*VPD*), followed by gross domestic product per capita (*GDPc*), education per capita (*EDc*), hills in % (*Hills*), high mountain areas in % (*MtSl*), environmental expenditure per capita (*EEc*), and the quantity of goods loaded and unloaded per capita (*QGLUc*), with an R^2 of 45.9% for these eight variables alone which is fairly high. As *GDP* is highly correlated with *PD* (regions with higher *PD* have a higher *GDP*), we used *GDP* per capita (*GDPc*), which varies independently of the *PD* of a region (and of the size of a region).

No other model among the competing models was more parsimonious than the one including all 10 variables (i.e., the global model). This implies that all variables are important to some degree. However, a clear tendency was found: any model that included the variables *PD* and *GDPc* were more parsimonious than all other models that did not include either one of these two variables. Seven variables had a positive relationship with the level of fragmentation and three variables (*MtSl*, *UR*, and *QGLUc*) showed a negative relationship. The "second best model" besides the global model included seven variables (*VPD*, *GDPc*, *QGLUc*, *EEc*, *EDc*, *MtSl*, *Hills*).

According to this pan-European predictive model, many regions in France and Germany are more fragmented than expected, while most of the United Kingdom and the Mediterranean, Scandinavian, and large parts of the Eastern European countries are less fragmented than expected. This is due primarily to the fact that different variables are most strongly related to the observed levels of fragmentation in different groups of regions, and one single model cannot capture all of these differences. The resulting overall model is a compromise between all regions. The clustering of regions for which the differences between the predicted and observed values are positive (or negative, respectively) indicates that the relationship with the predictor variables differs in different sets of countries. Therefore, different predictive models would be more appropriate to reveal these different relationships and the relevant variables.

When the global European model was observed more closely, we found several clusters of regions that can be distinguished from each other. For Germany and the Benelux countries, the new models resulted in an R^2 higher than 80%. France did not follow this trend; the R^2 was just 50%. The northern part of the country behaved in a similar way as the other western countries, but the s_{eff} values in the southern part were more similar to values found in the Mediterranean regions. Therefore, we studied the cloud of residuals from the pan-European model and identified six main clusters (Figure 20.12):

1. Group A: Belgium, Denmark, Germany, northern France, Luxembourg, the Netherlands (countries with access to the sea, but excluding the southern part of France, which is more similar to the Mediterranean countries)

2. Group B: Austria, the Czech Republic, and Switzerland (countries close to the Alps and with "continental" characteristics)

3. Group C: southern France, Greece, Italy, Malta, Portugal, Spain (all Mediterranean countries, including the southern part of France as it shows patterns that are more similar to this group of regions than to Group A)

4. Group D: Finland, Norway, Sweden (Scandinavian countries)

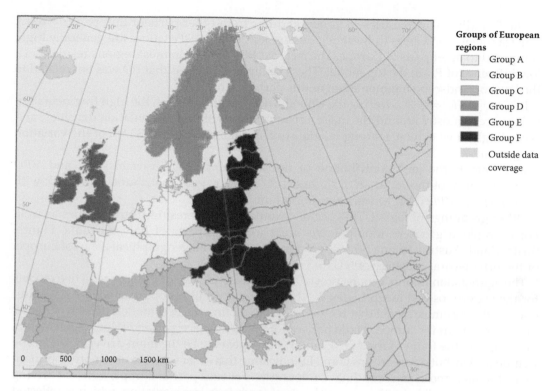

FIGURE 20.12

Six groups of regions used for separate analysis. (From EEA [European Environment Agency] and FOEN [Swiss Federal Office for the Environment]. 2011. *Landscape fragmentation in Europe. Joint EEA-FOEN report.* EEA Report No. 2/2011 [J. A. G. Jaeger, T. Soukup, L. F. Madriñán, C. Schwick, and F. Kienast, authors]. Luxembourg: Publications Office of the European Union.)

5. Group E: Ireland and the United Kingdom

6. Group F: Bulgaria, Estonia, Hungary, Latvia, Lithuania, Poland, Romania, Slovakia, and Slovenia

These six clusters match well with the classification of the four European regions recognized by the UN (United Nations, 2010): western Europe comprises Groups A and B, southern Europe = Group C, northern Europe comprises Groups D and E, Eastern Europe = Group F.

The six models provided a much better fit than the pan-European model. The regions of Group A cover the most densely populated part of Europe and also the most fragmented (R^2 = 76%). Accordingly, population density has an important role in explaining landscape fragmentation in this part of Europe: it appears to be the main driver of landscape fragmentation in this group. The second most important variable is *GDP* per capita.

In Group B (R^2 = 86%), most of the variation in s_{eff} values was explained by *PD* and volume of passenger density. This group includes very heterogeneous fragmentation values, and regions with relatively high and low fragmentation levels are often adjacent to each other. In this group of regions seven independent variables behaved as expected in our original hypothesis.

In Group C, the highest R^2 (49%) resulted for the model that included all 10 variables. This was the lowest R^2 among all six groups, indicating that other drivers of fragmentation are also relevant in this group that are not covered by the 10 predictor variables.

The Scandinavian countries in Group D differ from other parts of Europe (R^2 = 88%): The fragmentation values in these regions are low, while $GDPc$ values are generally high, compared to other parts of Europe. The fact that PDs are low in most regions resulted in a low influence of PD in the top model. These findings suggested that PD was not related to the level of landscape fragmentation in this part of Europe.

In Group E (R^2 = 91%), covering the United Kingdom and Ireland, the high fragmentation values occur mostly around the major urban centers with high population densities, and accordingly, they exhibit a strong relationship: PD alone explained 82.5% of the variation in s_{eff} values.

In Group F, the joint contribution of the two independent variables PD and VPD explained most of the variation in s_{eff} values (47.3%). The joint contributions of $GDPc$ with PD (7.7%) and VPD (14.47%) were also important.

Other groupings of the NUTS-X regions would be of interest in future studies as well. For example, a group of mountainous areas (covering parts of Italy, Slovenia, France, Switzerland, Austria, among others), groups according to biogeographical zones of Europe, or the urban–rural typology of NUTS-3 regions (EC, 2007).

The aggregation into six groups proved to be successful. Five of the six predictive models for the six groups had much higher R^2 than for the overall European model. As a consequence, the ensemble model that is created by assembling the best models of each of the six groups of regions for Europe should be used for predicting levels of landscape fragmentation in Europe. In the four groups A, B, E, and F, PD appears to be the main driver of landscape fragmentation, but not in groups C and D, indicating that other variables are more important in the Mediterranean and Scandinavian countries. For each part in Europe, different driving forces are responsible for the current levels of landscape fragmentation, which is reflected in the large differences between the predictive models that ranked best. This information can be used to identify regions that have performed better than others in terms of avoiding landscape fragmentation while serving the LU needs of their population and their economic development, as reflected in the seven socioeconomic variables included in the analysis.

Which regions are more or less fragmented than expected? The ensemble model created by assembling the best model of the six groups of regions, predicted fragmentation levels in most regions in Europe well (often within +10 and −10 meshes per 1000 km^2). Many predicted values for the regions belonging to Western and Central Europe are almost identical to the observed values for s_{eff}, while the regions of Eastern Europe are often a little less fragmented than expected. The regions in Scandinavia and the United Kingdom and Ireland are very often somewhat more fragmented than expected. In the Mediterranean regions, the predicted values are often close to the observed values, but we also find several regions that are much more or much less fragmented than expected.

According to our results, the most relevant driving force is human PD. Average GDP per capita is relevant in some cases, but not always, for example, it is relevant when the settlement structure is disperse such as in Belgium. However, contrary to our initial hypothesis, if human populations with a high average GDP per capita are concentrated in cities, their level of fragmentation tends to be lower with increasing $GDPc$, as is the case in many countries in Western Europe (i.e., groups A, B, C). In the first three groups (Western Europe and the Mediterranean countries), we found that $GDPc$ is often high in regions where PD is already high, but fragmentation is not as high as predicted by PD alone (even when we used the square root of PD in the model). The population here is often concentrated, and higher $GDPc$ and VPD do not necessarily contribute to higher levels of fragmentation. In this situation, higher values of $GDPc$ reduce the predicted value of fragmentation in the NUTS-X regions to provide a better fit of the models, that is, $GDPc$ shows a negative relationship with landscape fragmentation in models that include *PD*. This was not the case in Scandinavia, the United Kingdom and Ireland, and the eastern countries, where higher *GDPc* contributed to higher levels of fragmentation.

To answer the question why certain regions are more (or less) fragmented than predicted by the statistical model, more detailed research about the history and the political and economic conditions of different parts of Europe would be required that are not captured by the 10 predictive variables. As an example, the laws about regional planning are not well adopted and put into action in northwestern Switzerland, contributing to the higher levels in landscape fragmentation than in other parts of Switzerland.

Some relationships between the level of fragmentation and the 10 independent variables indicate that these variables are drivers of landscape fragmentation, but this may not always be true. For example, large transport volumes of passengers (*VPD*) and goods (*QGLUc*) may not only *cause* higher levels of fragmentation (reflecting a high demand for transportation infrastructure), but may also be a *response* to the availability of roads and railways to some degree. Thus, there are *feedback loops*, which can be studied using a systems-theoretic approach. In addition, roads are often built to stimulate economic development and increase *GDPc* rather than as a response to economic growth. These questions would be interesting subjects for future studies.

20.4 Policy Relevance and Implications

20.4.1 Need for Monitoring the Degree of Landscape Fragmentation

The aims of environmental monitoring are to discover and better understand changes in the environment. Our results are applicable in several types of monitoring: biodiversity monitoring, environmental monitoring, sustainability monitoring, and landscape quality monitoring. Accordingly, the findings of this study should be adopted in the European monitoring systems, and in the land accounting and ecosystem accounting efforts of the European Environment Agency (EEA, 2006b; Romanowicz et al., 2007; Weber, 2007; see also Chapter 23 in this book). They have already been included as a pressure indicator in the *2009 Environment Policy Review* of the European Commission (EC, 2010) and in the *European State-of-Environment Report* (SOER) 2010 (EEA, 2010a). Ideally, the updates would be done every 3–5 years, but this will depend on data availability. Potential driving forces that are not yet monitored should also be observed and included in future statistical analysis.

The issues related to data inconsistency in the TeleAtlas dataset between the years 2002 and 2009 highlight the need for a rigorous and consistent definition of the fragmentation geometries in space and time to provide reliable monitoring data on the European level. This implies that future monitoring should be based on exactly the same FGs and fragmenting elements as listed in Table 20.3. The use of the same FGs is a necessary precondition for being able to compare the results between countries.

The FGs presented in this report are considered appropriate for most countries in Europe, most importantly FG-B2, "Fragmentation of Non-Mountainous Land Areas." Therefore, the results of this study should also be included in the national monitoring systems of the 28 countries investigated, unless better data about the level of fragmentation that contain earlier points in time are already available, as in Switzerland (Figure 20.13a).

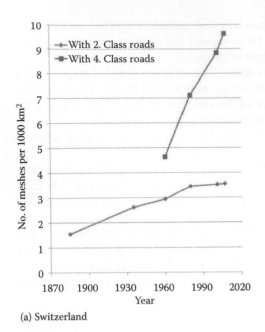

(a) Switzerland

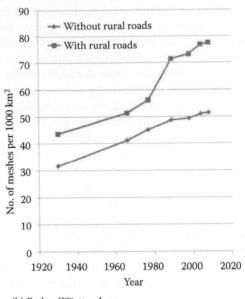

(b) Baden-Württemberg

FIGURE 20.13

Examples of the use of effective mesh density in monitoring systems of sustainable development, biodiversity, and landscape quality. (a) **Switzerland**: the data are used in the Swiss Landscape Monitoring System (LABES), the Biodiversity Monitoring Switzerland, and the Swiss Monitoring System of Sustainable Development (MONET). Two fragmentation geometries are shown: "CH-1: Degree of Fragmentation class 2" (shown by diamond symbols) includes land areas below 2100 m with roads up to class 2, and "CH-2: Degree of Fragmentation class 4" (shown by quadrant symbols) with roads up to class 4 for 1960–2008. (From Roth, U. et al. 2010. *L'état du paysage en Suisse—Rapport intermédiaire du programme Observation du paysage suisse (OPS). État de l'environnement* 2010. Berne: Office Fédéral de l'Environnement.) (Note that "Degree of Fragmentation class 2" also includes values for 1885 and 1935, which are based on a different dataset; Bertiller, R. et al. 2007. *Degree of landscape fragmentation in Switzerland: Quantitative analysis 1885–2002 and implications for traffic planning and regional planning.* Project report [in German]. FEDRO Report No. 1175. Berne: Swiss Federal Roads Authority.) (b) **Baden-Württemberg**: times series since 1930 for two fragmentation geometries: with and without municipal roads. (From Esswein, H., Schwarz-von Raumer, H.-G. 2008. *Landschaftszerschneidung in Baden-Württemberg: Neuberechnung des Landschaftszerschneidungsgrades 2008. Verbindungsräume geringer Zerschneidung.* Universität Stuttgart, im Auftrag der Landesanstalt für Umwelt, Messungen und Naturschutz Baden-Württemberg. Karlsruhe.) The values in Switzerland are for the entire country, including the Jura Mountains and parts of the Alps up to 2100 m. Therefore, they are much lower than the values in Baden-Württemberg. However, the level of fragmentation in the Swiss Lowlands is much higher than the average and similar to the values in Baden-Württemberg.

The effective mesh size and effective mesh density method has already been implemented in various monitoring systems. For example, the Swiss Federal Statistical Office (SFSO), the Swiss Federal Office for the Environment (FOEN), and the Swiss Federal Office for Spatial Development (ARE) launched the Monitoring Sustainable Development project (MONET) in 2000 to establish a system of indicators for sustainable development in Switzerland. The effective mesh size is also applied as a pressure indicator in the Swiss Biodiversity Monitoring system (http://www.biodiversitymonitoring.ch/) and in the novel Swiss Landscape Monitoring system (LABES; Roth et al., 2010; Kienast et al., 2015).

Another instructive example is given by the Environmental Report from Baden-Württemberg, Germany, where m_{eff} was implemented in 2003 (Figure 20.13b; State Institute for Environment, Measurements and Nature Conservation Baden-Württemberg, 2006).

In Germany, effective mesh size is applied as one of 24 core indicators for environmental monitoring (Schupp, 2005). It is also used in the German National Strategy on Biological Diversity (Federal Ministry for the Environment, Nature Conservation and Nuclear Safety, 2007; EEA, 2010b).

To measure progress toward existing objectives and targets on the European level, EEA developed the Transport and Environment Reporting Mechanism (TERM) and has published TERM reports since 2000 (EEA, 2010c). Including landscape fragmentation indicators in TERM would contribute to a better understanding of spatial effects of transportation infrastructure.

Many other fragmenting elements also have an influence on the landscapes such as power lines, ski lifts, pressure lines, and infrastructure of shale gas development (access roads and pipelines; Racicot et al., 2014) and can be included in additional fragmentation geometries. It is expected that recreation activities in Europe will grow annually with rates of about 5% over the next several years (European Travel Commission, 2010). This will lead to further growth of the built-up areas and new transport infrastructure, and will result in higher landscape fragmentation in many recreational areas. These landscapes are in high danger of being more and more fragmented and of losing much of their remaining recreational quality and beauty. Thus, there is an urgent need for action.

The reassuring experiences and results about the monitoring of landscape quality in Switzerland (Kienast et al., 2015) should be applied for implementing a monitoring system of landscape quality at the European level as well in the near future. Landscape fragmentation is an indicator of high importance for monitoring landscape quality, and the results from this report can be directly included for this purpose at the European scale.

20.4.2 Implications for Nature Conservation, Traffic, and Urban Planning

20.4.2.1 Application as a Tool for Performance Review

Once quantitative targets or limits for the future degree of landscape fragmentation will be available, the degree of fragmentation can be recalculated after new roads have been built or existing roads have been removed, and compared to the target or limit (Box 20.3). This is already possible in the planning stages of the construction or removal of transportation infrastructure. Maps of planned transportation infrastructure can be combined with models for predicting future LU changes, and the resulting degree of fragmentation can be compared to the targets.

The effective mesh density is an important criterion for consideration in transportation planning and regional planning. However, habitat amount and habitat quality in the entire landscape are at least as important for all relevant species in the study area, both current and potential habitats. If this is ignored, there is a danger that road construction may be considered unproblematic by decision makers if the new roads are combined with the construction of wildlife passages and fences. This is deceptive when habitat amount and quality in these landscapes continue to decline (Fahrig, 2001, 2002). Therefore, the conservation and restoration of wildlife habitats must be the first priority. Wildlife passages will be useless if there is not enough habitat left to connect or if the matrix surrounding the habitat islands is too hostile to support wildlife.

20.4.2.2 Relevance for Biodiversity

In many parts of Europe, populations of large terrestrial mammals are endangered or the animals live in small numbers, and many of these species have large habitat requirements

BOX 20.3   APPLYING THE METHOD

The method of effective mesh density and effective mesh size can be used at any level (e.g., regions, districts, or local scale) as an instrument of analysis for the following purposes:

1. Data on planned future development reveal the extent to which planned transport routes will increase fragmentation and can be compared to the targets and to previous trends. This approach will take into account the cumulative effects of several projects combined (including the predicted expansion of urban areas) on the effective mesh density.

2. Various planning alternatives can be assessed and compared with respect to their impacts on the effective mesh density. Consideration should be given to the cumulative effects of all planned future developments and their interactions. The method can be broadened to encompass the issue of landscape quality, for example, through the inclusion of weights for landscape character, recreational quality, or the ecological quality of affected habitats; the inclusion of noise bands; and the inclusion of wildlife passages and probabilities of crossing success of transport routes.

3. The extent to which each category of transport route contributes to the total degree of fragmentation can be determined. Such values can, for example, indicate threat levels for the remaining ecological networks, as smaller transport routes may serve as an indication where expansion might be envisaged by planners in the future if traffic levels rise.

4. Specific suggestions can be put forward for the removal of transport routes, which would have a particularly positive effect on effective mesh size.

5. It would be informative to study the extent to which regions are successful in decoupling their economic welfare from their level of landscape fragmentation, that is, where economic development is accompanied by increase in landscape fragmentation and where it is occurring without additional fragmentation.

and require long distance migrations and dispersal (Boitani, 2000; Mysterud et al., 2007). In areas where the effective mesh size is smaller than the typical size of the home range of a species, the animals encounter roads and other barriers on a daily basis. The long response times of many species to changes in landscape structure present a particular challenge. Given that the negative effects of habitat fragmentation and isolation often become apparent only after several decades, it is likely that further population losses will be incurred in the coming decades as a result of the landscape changes that have already taken place (Findlay and Bourdages, 2000). If a decline in wildlife populations is documented, it may already be too late to take measures to stabilize the populations, as in the case of brown hare in Aargau Canton, Switzerland (Roedenbeck and Voser, 2008).

We recommend drawing up guiding concepts for the landscapes in Europe (together with the member states) that include the identification of regionally and nationally important unfragmented areas (Selva et al., 2011) and priority areas for defragmentation. To make these guiding concepts more tangible, it is desirable to adopt appropriate benchmarks or targets for the degree of landscape fragmentation. For example, Waterstraat et al. (1996) recommended the protection of large unfragmented low-traffic areas in Germany. More recently, the German Federal Environment Agency suggested that region-specific limits to control landscape fragmentation should be introduced (Box 20.4 in Section 20.4.3.2).

Objectives and measures should be elaborated that are made binding for European and national offices and should state what measures should be taken and where and how they should be implemented, in connection with ongoing EU initiatives for a green infrastructure such as the Green Infrastructure Strategy of May 2013 (Green Infrastructure, 2007; EC, 2013). This work could build on the achievements of the previous EU COST 341 Action (Iuell et al., 2003) and the Infra Eco Network Europe (IENE) (http://www.iene.info).

Continued increase in landscape fragmentation will also increase the future costs for re-connecting separated habitats, for the restoration of wildlife corridors and the rescue of endangered wildlife populations. The map of the wildlife corridors of national importance

---

BOX 20.4   EXAMPLE OF A RECOMMENDATION FOR LIMITING LANDSCAPE FRAGMENTATION

The German Federal Environment Agency has issued recommendations for limiting landscape fragmentation based on the effective mesh size (Umweltbundesamt, 2003; Penn-Bressel, 2005). Considering predicted increases in landscape fragmentation, the agency recommended to curtail the rate of increase in Germany. The remaining large unfragmented areas are to be preserved and enlarged where possible; and in areas that are already highly fragmented, the trend is to be slowed down (Table 20.5). By 2015, the reduction in effective mesh size in highly fragmented areas should be at least half the rate had the situation been left unchecked. The following specific aims were set:

a. The number and total area of each, as yet, unfragmented, low-traffic areas above 140, 120, 100, 80 and 60 square kilometres shall not decrease further and instead will be increased through defragmentation measures so that their current proportion of 20.6% of Germany's territory will be raised to 23% by 2015.

b. The degree of fragmentation of highly fragmented regions shall be limited by additional criteria (as listed in Table 20.5) (Penn-Bressel, 2005).

The authors of this book chapter support this recommendation.

TABLE 20.5

Values for Limiting the Rate of Increase of Landscape Fragmentation in Highly Fragmented Regions in Germany, as Put Forward by the German Federal Environment Agency

Starting Point at the End of 2002: Value of the Effective Mesh Size (m_{eff})	Goal to Be Reached by 2015: Further Decrease in Effective Mesh Size (m_{eff}) to Be Less Than (%)
<10 km²	1.9
10–20 km²	2.4
20–35 km²	2.8
>35 km²	3.8

Note: The maximum size of reporting units should not be larger than 7000 km² (for details, see Umweltbundesamt, 2003 and Penn-Bressel, 2005).

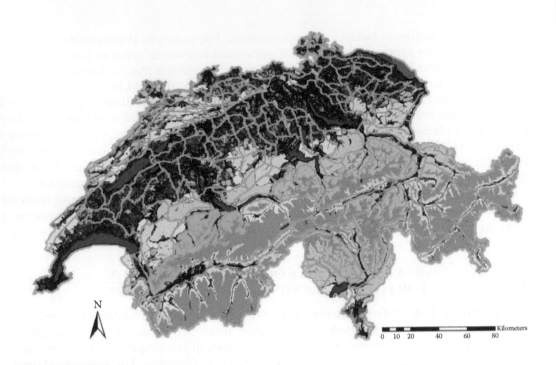

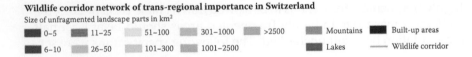

Wildlife corridor network of trans-regional importance in Switzerland

Size of unfragmented landscape parts in km²

■ 0–5	■ 11–25	51–100	301–1000	>2500	■ Mountains	■ Built-up areas
■ 6–10	26–50	101–300	1001–2500		■ Lakes	—— Wildlife corridor

FIGURE 20.14
Overlay of the wildlife corridor network of trans-regional importance in Switzerland (shown in blue) with the Swiss fragmentation geometry FG4 "Land areas below 2100 m." The trans-regional corridor network in Switzerland for terrestrial fauna includes the wildlife corridors and the trans-regional movement axes. Red, yellow, and green indicate the sizes of the remaining patches. (From Bertiller, R. et al., 2007, after Holzgang, O. et al. 2001. *Korridore für Wildtiere in der Schweiz—Grundlagen zur überregionalen Vernetzung von Lebensräumen*. BUWAL [Bundesamt für Umwelt, Wald und Landschaft], SGW [Schweizerische Gesellschaft für Wildtierbiologie] and Vogelwarte Sempach, Schriftenreihe Umwelt Nr. 326. Bern, Switzerland.)

in Switzerland (Figure 20.14) provides an idea of the size of the task of restoring wildlife corridors, which will be huge, once the landscape has become heavily fragmented. Therefore, it is wise policy to implement effective measures to avoid an increase of the level of fragmentation from the beginning as much as possible.

Previous research has demonstrated that there are thresholds in the effects of landscape fragmentation on the viability of wildlife populations (Jaeger and Holderegger, 2005; Jaeger et al., 2006). If wildlife populations have so far survived all new road construction in a landscape, this does *not* imply that the populations will also be able to withstand further densification of the transportation network. When the threshold has been reached, it is highly likely that the next new road will bring about the populations' extinction. The exact thresholds for a population or a species are largely unknown, and it is unlikely that they will be known any time soon. Therefore, any hopes for a general hard number for the maximum acceptable level of fragmentation will be disappointed. Rather, the precautionary principle should be applied in the assessment of landscape fragmentation (Kriebel et al., 2001), and the implementation of limits requires a consultation process, just as it has been the case with other limits that are in use for water quality and air quality (Streffer et al., 2003).

The important role of large roadless areas for biodiversity conservation has been emphasized in the scientific literature (e.g., DeVelice and Martin, 2001; Selva et al., 2011), and it is equally important for preserving landscape quality. The protection of the remaining large unfragmented areas is a measure of high priority and should be implemented immediately, based on the existing maps and existing knowledge about habitat types, habitat amount, and habitat quality.

The synergistic effects of roads and other factors that operate simultaneously (e.g., agricultural intensification, increased urbanization) have rarely been investigated. This lack of knowledge is often used as a justification for not preventing the construction of new roads or for not including more substantive mitigation measures, by arguing that not enough is known and more research is needed before road construction might slow down. Therefore, there is a danger to think that the addition of wildlife passages to new roads will make it possible to construct new roads without negative consequences to wildlife populations. This attitude would ignore all other negative effects of roads and the critical importance of habitat amount and habitat quality (Fahrig, 2001, 2002).

20.4.2.3 Future Research Needs

There is an urgent need for further research about landscape fragmentation in Europe. The analysis of the relationship between the level of landscape fragmentation and biodiversity is one of the most important areas for future research.

What will the effective mesh density be when the entire transportation infrastructure that is currently in the planning stages in Europe—at least that on the European and national levels—will be constructed? Its cumulative effects should be considered in the decision-making process because all these projects will interact and their *combined impact* is relevant for wildlife populations. It is desirable to compare different scenarios of how many and where the new roads and railway lines will be located. Future research studies should also consider other types of reporting units, such as watersheds, ecosystem types (e.g., forests, grasslands, wetlands) and corridor–habitat networks. Species-specific studies can apply the effective mesh density to the habitat of a particular species rather than to the landscape in general, that is, all nonhabitat would be considered as barriers of varying degrees of permeability.

Earlier points in time should be investigated to determine the rate of increase in landscape fragmentation. More in-depth research about the driving forces and improvements to the predictive model of landscape fragmentation are also desirable to gain a better understanding of regions that are clearly more fragmented than expected or of those that are less fragmented than expected.

The young research field of road ecology is confronted with many urgent unanswered questions (Roedenbeck et al., 2007; van der Ree et al., 2011). Most importantly, research in road ecology needs to move toward larger scales. There is a paucity of studies that explicitly examine the population, community, ecosystem, or landscape-level effects of roads and mitigation measures. The future of road ecology research will be best enhanced when multiple road projects in different states or countries are combined and studied as part of integrated, well-replicated research projects (van der Ree et al., 2011, 2015; Fraser et al., 2013).

In addition, a research approach is required that will address the remaining uncertainties which to a large degree are irreducible, for example, through building on the precautionary principle (EEA, 2001; Kriebel et al., 2001) and the concept of environmental threat (Jaeger,

2002; Scheringer, 2002). Further research should also address the question of how the spatial configuration of transportation systems can be improved to keep landscapes unfragmented.

20.4.3 Recommendations for Controlling Landscape Fragmentation

Generally, four types of measures to address the problem of landscape fragmentation can be distinguished: (1) to minimize negative impacts during the planning and construction stages of new transportation infrastructure, (2) to restore connectivity across existing transportation infrastructure, (3) to prevent further increase of the density of the transportation network, and (4) to remove existing transportation infrastructure. Without better methods and higher awareness and consideration of the remaining uncertainties it will be impossible to resolve the increasing conflicts about LU and landscape fragmentation in a responsible manner. This implies that different measures may be needed in regions with different current levels of fragmentation, with different departures of the observed from the predicted levels of fragmentation, and with different prevailing driving forces.

20.4.3.1 Measures in Traffic Planning and Regional Planning

- *Tunnels and wildlife passages*

  Existing roads and railways can be made more permeable for wildlife through tunnels, crossing structures (wildlife overpasses and underpasses), or by raising roads up on pillars so that wildlife can cross underneath.

- *Priority of upgrading of existing roads over construction of new roads*

  The widening of existing highways and railways will increase their barrier effect, and higher traffic volumes will contribute to the stronger barrier effect. However, the upgrading of existing highways is still less detrimental than the construction of new highways at another location in most cases.

- *Bundling of transport routes*

  The tighter that transport routes are bundled together, the larger the remaining unfragmented areas of land. In addition, wildlife passages could then be placed so that all the transport routes could be crossed over or under in one go.

- *Keep bypass routes close to settlement areas*

  If bypasses (and other roads) are sited close to developed areas, their fragmentation effect is lower compared to the construction of bypasses away from settlements.

- *Dismantling of transportation routes*

  Transport infrastructure that is not urgently needed any more (e.g., owing to the construction of new routes or changing requirements) should be removed. The potential for removal of roads is probably higher than the current practice suggests.

- *Reduction of the width of roads with decreasing traffic volume*

  Roads, on which traffic volumes have decreased as a result of the construction of other transportation infrastructure or because of changing conditions, should be downgraded and physically reduced in width.

- *Limiting urban areas, and internal urban development based on densification*

  To preserve open space in the countryside, it is necessary to limit the size of urban areas (EEA, 2006a). Regional planning legislation should more effectively require local authorities to treat land sparingly in their LU plans.

- *Oasis concept*

  The oasis concept means that small communities and areas suitable for pre-serving biodiversity or for recreational use (refuges or "oases") will be kept free from trans-regional traffic (Arbeitskreis Strassen im VCD-Kreisverband Ludwigsburg, 1996; Jaeger, 2002). Road traffic will be concentrated onto a small number of roads located at a clear distance from such oases. Small communities will be connected by access roads. Current roads that route traffic directly from community to community will then be dismantled. The major advantages of the concept are that communities are freed from through-traffic, that areas for preserving biodiversity or for recreational use are protected from through-traffic, and that it halts the trend of continually building new bypasses around communities.

20.4.3.2 Measures at a Strategic Level

- *Preserving and restoring wildlife movement corridors*

  The restoration of damaged or severed wildlife corridors is a significant step in recreating the opportunities for species to migrate and disperse. Ongoing efforts for implementing a system of green infrastructure (EC, 2013) aim at addressing this issue on the European level. Perhaps the most prominent country-wide plan is the German Habitat Corridor Network (Böttcher et al., 2005; Reck et al., 2008). An example from the Netherlands is the Dutch Long-Term Defragmentation Programme (van der Grift, 2005). However, these wildlife corridor plans often do not match well along the boundaries between different countries. Therefore, better coordination of these efforts at the European level is needed.

- *A European defragmentation strategy*

  Landscape fragmentation must no longer continue to increase within trans-regionally important wildlife corridors. Rather, transport infrastructure that is not absolutely necessary should be removed or tunneled under or bridged over. Likewise, built-up areas should be strictly prevented from expanding in these areas.

  Various ecological network initiatives exist on the European level (e.g., Tillmann, 2005). Three important examples are

1. The Pan-European Ecological Network (PEEN) under the aegis of the Council of Europe (CE), the United Nations Environmental Programme (UNEP), and the European Centre of Nature Conservation (ECNC) (Jongman et al., 2011)

2. The Emerald Network, also known as Network of Areas of Special Conservation Interest, launched in 1989 by the CE (Council of Europe, 2009)

3. The Trans-European Wildlife Networks Project (TEWN) (EuroNatur, 2010)

  Ideally, the PEEN should integrate all network initiatives. The PEEN will consist of core areas, corridors, and buffer zones and will identify restoration areas where they are considered necessary. It aims to conserve the full range of ecosystems, habitats, species, and landscapes of European importance and to counteract the main causes for decline by creating the right spatial and environmental conditions (Council of Europe, 1996).

- *Effective protection of remaining large unfragmented areas*

  The protection of the remaining large unfragmented areas is a measure of high priority and we recommend it to be implemented immediately, based on the existing maps and existing knowledge about habitat types, habitat amount, and habitat quality (Selva et al., 2011). These areas should cover habitats of a range of species. The maps can help identify areas where further fragmentation is an imminent threat and their rapid preservation is critical.

  In regions with a rapid pace of development, where there are still significant amounts of large unfragmented areas and important pockets of biodiversity left, the mistakes that many regions in the Western European countries have committed should not be repeated. The countries in Eastern, Central, and Northern Europe hold most of the remaining megafauna in the continent along with the highest levels of endemisms (EuroNatur, 2010). The importance of this measure in the Eastern European countries is clear when the many new roads and railways that are planned in these countries are included in the fragmentation analysis.

- *Targets and limits*

  There is very little knowledge available about the question where the exact threshold for a particular population is, and by how much the threshold will shift due to diminishing resources, reduced genetic exchange, or changes in climate. A particular challenge is given by the long response times of long-lived animals to changes in landscape structure. This situation makes it all the more essential that a precautionary approach is adopted that guides landscape fragmentation in the desired direction. Rather, targets and limits for the future degree of landscape fragmentation should be broadly discussed and implemented. Such targets and limits are urgently needed by government offices and administrations for being able to act and justify their decisions and actions toward better protection of the environment (see example in Box 20.4). The socioeconomic models can support the definition of these limits. The values decided upon can be revised at a later point in time when knowledge about the effects of landscape fragmentation will have increased. As long as the knowledge about the thresholds of landscape fragmentation is insufficient, the precautionary principle should be applied (Kriebel et al., 2001).

20.4.4 Most Immediate Priorities

The current trend of continued increase of landscape fragmentation is clearly in contradiction to the principles of sustainability, and there is an urgent need for action. The authors of this chapter recommend putting into practice the following three measures with the highest priority:

1. Protection of large unfragmented areas, ecologically significant areas, and wildlife corridors: the remaining large unfragmented areas, ecologically significant areas, and functional wildlife corridors should be protected immediately from further fragmentation. Critical areas should be identified where further fragmentation is an imminent threat and their rapid preservation is crucial before they would be lost to fragmentation by roads and railroads. This task is particularly urgent in regions with a rapid pace of development, such as large parts of the Eastern and Central European countries. Ongoing increase in landscape fragmentation will also increase the future costs for the restoration of wildlife corridors and habitats and

for the rescue of endangered wildlife populations. Policymaking on the European level has an important responsibility to link the provision of funds for transportation to the requirement of protecting unfragmented areas in these regions.

2. Monitoring of landscape fragmentation: landscape fragmentation is an essential indicator of threats to biodiversity, to the sustainability of human LU, and to landscape quality. It should be implemented in monitoring systems of biodiversity, sustainable development, and landscape quality, which is a precondition for being able to diagnose the rate of increase and changes in trends.

3. Application of fragmentation analysis as a tool in transportation planning and regional planning: the cumulative effects of new transportation infrastructure on the degree of landscape fragmentation should be analyzed quantitatively and in more detail in the planning process. The effective mesh density method should be included in the planning process as an instrument for this task, in combination with other relevant criteria (such as habitat amount and quality), for example, to compare alternative transportation corridors for new roads and railway lines. This should be a requirement for all transportation infrastructure to which the EU provides some financial support. The uncertain effects of landscape fragmentation need to be considered more seriously and studied, for example, through the use of the Before-After-Control-Impact (BACI) study design (Roedenbeck et al., 2007).

Large unfragmented areas are a limited and nonrenewable resource. This fact is particularly important to consider in Europe, where high human population densities compete for land with biodiversity. As a consequence, mankind's growing demands for renewable energy, food, and land cannot be circumvented by any form of adaptation. Haber (2007) has called these growing demands the three major "ecological traps" that threaten mankind probably more severely than any other environmental problem. If endeavors for promoting sustainable development disregard these three ecological traps, they will inevitably miss their goals. Therefore, much higher efforts are now required to conserve unfragmented landscapes.

Acknowledgments

This book chapter is based on the publication of the original study by EEA and FOEN (2011). We thank the Swiss Federal Office for the Environment (FOEN) and the European Environment Agency (EEA) for funding the original study. We thank all partners from FOEN and EEA for the enjoyable collaboration.

References

Antrop, M. 2004. Landscape change and urbanization process in Europe. *Landscape and Urban Planning* 67:9–26.

Arbeitskreis Strassen im VCD-Kreisverband Ludwigsburg (Ed.). 1996. *Positionspapier Oasen-Konzept: Vision 2020* (T. Wolf, author). Ludwigsburg, Germany: Verkehrsclub Deutschland (VCD).

Bayne, E. M., Hobson, K. A. 2002. Apparent survival of male ovenbirds in fragmented and forested boreal landscapes. *Ecology* 83:1307–1316.

Bertiller, R., Schwick, C., Jaeger, J. 2007. *Degree of landscape fragmentation in Switzerland: Quantitative analysis 1885–2002 and implications for traffic planning and regional planning.* Project report (in German). FEDRO Report No. 1175. Berne: Swiss Federal Roads Authority.

Boitani, L. 2000. Action plan for the conservation of the wolves (*Canis lupus*) in Europe. *Nature and Environment*, No. 113. Strasbourg: Council of Europe Publishing. 84 pp.

Böttcher, M., Reck, H., Hänel, K., Winter, A. 2005. Habitat corridors for humans and nature in Germany (in German, Lebensraumkorridore für Mensch und Natur in Deutschland). *GAIA* 14(2):163–166 + 1 map.

Bundesverfassung der Schweizerischen Eidgenossenschaft vom 18. April 1999, Art. 84 Alpenquerender Transitverkehr. Available at: http://www.basisinformationen.diagnose-funk.org/downloads/bundesverfassung.pdf.

CIA (Central Intelligence Agency). 2010. *The world factbook—Netherlands.* Available at: https://www.cia.gov/library/publications/the-world-factbook/geos/nl.html.

Council of Europe. 1996. Pan-European biological and landscape diversity strategy. *Nature and Environment* 74. Strasbourg: Council of Europe Press.

Council of Europe. 2009. The Emerald Network—Network of areas of special conservation interest. Strasbourg: Council of Europe. Available at: http://www.coe.int/t/dg4/cultureheritage/nature/econetworks/default_en.asp.

Council of the European Communities. 1992. Council Directive 92/43/EEC on the conservation of natural habitats of wild fauna and flora. Official Journal of the European Communities, L206/7.

DeVelice, R. L., Martin, J. R. 2001. Assessing the extent to which roadless areas complement the conservation of biological diversity. *Ecological Applications* 11(4):1008–1018.

EC (European Commission). 1998. European Community biodiversity strategy (COM(98) 42). Commission of the European Communities, Office for Official Publications of the European Communities, Luxembourg. Available at: http://biodiversity-chm.eea.eu.int/convention/cbd_ec.

EC (European Commission). 2007. *Growing regions, growing Europe.* Fourth Report on Economic and Social Cohesion. Luxembourg: Office for Official Publications of the European Communities.

EC (European Commission). 2010. *2009 environment policy review.* Luxembourg: Office for Official Publications of the European Communities.

EC (European Commission). 2013. *Building a green infrastructure for Europe.* Luxembourg: Publications Office of the European Union.

EEA (European Environment Agency). 2001. *Late lessons from early warnings: The precautionary principle 1896–2000.* Environmental Issue Report No. 22. Copenhagen: European Environment Agency.

EEA (European Environment Agency). 2006a. *Urban sprawl in Europe—The ignored challenge.* EEA Report No. 10/2006. Luxembourg: Office for Official Publications of the European Communities. Available at: http://www.eea.europa.eu/publications/eea_report_2006_10/eea_report_10_2006.pdf.

EEA (European Environment Agency). 2006b. *Land accounts for Europe 1990–2000: Towards integrated land and ecosystem accounting.* EEA Report No. 11/2006. Luxembourg: Office for Official Publications of the European Communities.

EEA (European Environment Agency). 2010a. *The European environment—State and outlook 2010: Land use.* Luxembourg: Publications Office of the European Union.

EEA (European Environment Agency). 2010b. SOER 2010—The European environment—State and outlook 2010—Country assessments—Land use (Germany). Available at: http://www.eea.europa.eu/soer/countries/de/soertopic_view?topic=land.

EEA (European Environment Agency). 2010c. *Towards a resource-efficient transport system—TERM 2009: Indicators tracking transport and environment in the European Union.* EEA Report No. 2/2010: Luxembourg: Office for Official Publications of the European Union.

EEA (European Environment Agency) and FOEN (Swiss Federal Office for the Environment). 2011. *Landscape fragmentation in Europe. Joint EEA-FOEN report.* EEA Report No. 2/2011 (J. A. G. Jaeger, T. Soukup, L. F. Madriñán, C. Schwick, and F. Kienast, authors). Luxembourg: Publications Office of the European Union.

Esswein, H., Schwarz-von Raumer, H.-G. 2008. *Landschaftszerschneidung in Baden-Württemberg: Neuberechnung des Landschaftszerschneidungsgrades 2008. Verbindungsräume geringer Zerschneidung.* Universität Stuttgart, im Auftrag der Landesanstalt für Umwelt, Messungen und Naturschutz Baden-Württemberg. Karlsruhe.

ETC (European Travel Commission). 2010. European tourism in 2010: Trends and prospects (Q3/2010). Brussels: European Travel Commission.

EuroNatur. 2010. *Trans-European Wildlife Networks project: TEWN, TEWN manual: Recommendations for the reduction of habitat fragmentation caused by transport infrastructure development.* Radolfzell, Germany: EuroNatur Foundation.

Ewald, K. C., Klaus G. 2009. *Die ausgewechselte Landschaft: vom Umgang der Schweiz mit ihrer wichtigsten Ressource.* Bern: Haupt Verlag.

Fahrig, L. 2001. How much habitat is enough? *Biological Conservation* 100:65–74.

Fahrig, L. 2002. Effect of habitat fragmentation on the extinction threshold: A synthesis. *Ecological Applications* 12:346–353.

Fahrig, L., Rytwinski, T. 2009. Effects of roads on animal abundance: An empirical review and synthesis. *Ecology and Society* 14(1):21. Available at: http://www.ecologyandsociety.org/vol14/iss1/art21/.

Federal Ministry for the Environment, Nature Conservation and Nuclear Safety Bundesministerium für Umwelt, Naturschutz und Reaktorsicherheit, (BMU). 2007. *National strategy on biological diversity.* Berlin and Bonn, Germany: BMU.

Findlay, C. S., Bourdages, J. 2000. Response time of wetland biodiversity to road construction on adjacent lands. *Conservation Biology* 14:86–94.

Forman, R. T. T. 1995. *Land mosaics—The ecology of landscapes and regions.* Cambridge, UK/New York, NY, USA: Cambridge University Press.

Forman, R. T. T., Alexander, L. E. 1998. Roads and their major ecological effects. *Annual Review of Ecology and Systematics* 29:207–231.

Forman, R. T. T, Sperling, D., Bissonette, J. A., Clevenger, A. P., Cutshall, C. D., Dale, V. H., Fahrig, L., France, R., Goldman, C. R., Heanue, K., Jones, J. A. et al. 2003. *Road ecology—Science and solutions.* Washington, DC: Island Press.

Fraser, L. H., Henry, H. A., Carlyle, C. N., White, S. R., Beierkuhnlein, C., Cahill, J. F., Jr., Casper, B. B., Cleland, E., S. L. Collins, J. S. Dukes, A. K. Knapp et al. 2013. Coordinated distributed experiments: An emerging tool for testing global hypotheses in ecology and environmental science. *Frontiers in Ecology and the Environment* 11:147–155.

Froment, A., Wildmann, B. 1987. Landscape ecology and rural restructuring in Belgium. *Landscape and Urban Planning* 14:415–426.

Galliker, R. 2009. Stand und Ausbau des schweizerischen Nationalstrassennetzes. *Die Volkswirtschaft* 5-2009:1215.

Geist, H. J., Lambin, E. F. 2001. *What drives tropical deforestation?—A meta-analysis of proximate and underlying causes of deforestation based on subnational case study evidence.* International Human Dimensions Programme on Global Environmental Change (IHDP), International Geosphere-Biosphere Programme (IGBP). LUCC Report Series No. 4, Louvain-la-Neuve, Belgium LUCC International Project Office, University of Louvain, Dpt. of Geography.

Girvetz, E. H., Jaeger, J. A. G., Thorne, J. H. 2007. Comment on "Roadless space of the conterminous United States." *Science* 318(5854):1240b.

Green Infrastructure. 2007. *Towards a green infrastructure for Europe—Developing new concepts for integration of Natura 2000 network into a broader countryside.* EC study ENV.B.2/SER/2007/0076. Available at: http://ec.europa.eu/environment/nature/ecosystems/docs/green_infrastructure_integration.pdf, see also www.green-infrastructure-europe.org.

Gustafson, E. J. 1998. Quantifying landscape spatial pattern: What is the state of the art? *Ecosystems* 1:143–156.

Haber, W. 2007. Energy, food and land—The ecological traps of humankind. *Environmental Science and Pollution Research* 14(6):359–365.

Hanski, I. 1999. *Metapopulation ecology.* Oxford: Oxford University Press.

Holzgang, O., Pfister H. P., Heynen, D., Blant, M., Righetti, A., Berthoud, G., Marchesi, P., Maddalena, T., Müri, H., Wendelspiess, M., Dändliker, G. et al. 2001. *Korridore für Wildtiere in der Schweiz— Grundlagen zur überregionalen Vernetzung von Lebensräumen.* Bundesamt für Umwelt, Wald und Landschaft (BUWAL), Schweizerische Gesellschaft für Wildlterbiologie (SGW) and Vogelwarte Sempach, Schriftenreihe Umwelt Nr. 326. Bern, Switzerland.

IUCN (International Union for Conservation of Nature and Natural Resources). 2001. *IUCN red list.* Gland, Switzerland: IUCN.

Iuell, B., Bekker, G. J., Cuperus, R., Dufek, J., Fry, G., Hicks, C., Hlavac, V., Keller, V., Rosell, C., Sangwine, T., Torslov, N., Wandall, B. M. (Eds.). 2003. *COST 341: Habitat fragmentation due to transportation infrastructure—Wildlife and traffic—A European handbook for identifying conflicts and designing solutions.* Zeist, the Netherlands: KNNV Publishers.

Jaeger, J. A. G. 2000. Landscape division, splitting index, and effective mesh size: New measures of landscape fragmentation. *Landscape Ecology* 15(2):115–130.

Jaeger, J. A. G. 2002. *Landscape fragmentation—A transdisciplinary study according to the concept of environmental threat* (in German: *Landschaftszerschneidung—Eine transdisziplinäre Studie gemäß dem Konzept der Umweltgefährdung*). Stuttgart: Verlag Eugen Ulmer.

Jaeger, J. 2003. II-5.3 Landschaftszerschneidung. In W. Konold, R. Böcker, and U. Hampicke (Eds.), *Handbuch Naturschutz und Landschaftspflege* (1999 ff.), 11. Erg.-Lieferung. Landsberg: Ecomed-Verlag.

Jaeger, J. A. G. 2007. Effects of the configuration of road networks on landscape connectivity. In C. Irwin, D. Nelson, and K. P. McDermott (Eds.), *Proceedings of the 2007 International Conference on Ecology and Transportation (ICOET)* (pp. 267–280). Center for Transportation and the Environment, North Carolina State University, Raleigh.

Jaeger, J., Holderegger, R. 2005. Thresholds of landscape fragmentation (in German: Schwellenwerte der Landschaftszerschneidung). *GAIA* 14(2):113–118.

Jaeger, J. A. G., Bowman, J., Brennan, J., Fahrig, L., Bert, D., Bouchard, J., Charbonneau, N., Frank, K., Gruber, B., Tluk von Toschanowitz, K. 2005. Predicting when animal populations are at risk from roads: An interactive model of road avoidance behavior. *Ecological Modelling* 185: 329–348.

Jaeger, J. A. G., Fahrig, L., Ewald, K. 2006. Does the configuration of road networks influence the degree to which roads affect wildlife populations? In C. L. Irwin, P. Garrett, and K. P. McDermott (Eds.), *Proceedings of the 2005 International Conference on Ecology and Transportation (ICOET)* (pp. 151–163).Center for Transportation and the Environment, North Carolina State University, Raleigh.

Jaeger, J., Bertiller, R., Schwick, C. 2007. *Degree of landscape fragmentation in Switzerland—Quantitative analysis 1885–2002 and implications for traffic planning and regional planning—Condensed version.* Order No. 868-0200, Bundesamt für Statistik, Neuchâtel. Also available in French (Order No. 867-0200) and German (Order No. 866-0200). Available at: http://www.bfs.admin.ch/bfs/por tal/en/index/themen/02/22/publ.html?publicationID=2992.

Jaeger, J. A. G., Bertiller, R., Schwick, C., Müller, K., Steinmeier, C., Ewald, K. C., Ghazoul, J. 2008. Implementing landscape fragmentation as an indicator in the Swiss Monitoring System of Sustainable Development (MONET). *Journal of Environmental Management* 88(4):737–751.

Jongman, R. H. G., Bouwma, I. M., Griffioen, A., Jones-Walters, L., Van Doorn, A. M.. 2011. The Pan European Ecological Network: PEEN. *Landscape Ecology* 26:311–326.

Kienast, F., Frick, J., van Strien, M. J., Hunziker, M. 2015. The Swiss Landscape Monitoring Program—A comprehensive indicator set to measure landscape change. *Ecological Modelling* 295:136–150.

Kriebel, D., Tickner, J., Epstein, P., Lemon, J., Levins, R., Loechler, E. L., Quinn, M., Rudel, R., Schettler, T., Stoto, M. 2001. The precautionary principle in environmental science. *Environmental Health Perspectives* 109(9):871–876.

Laurance, W. F. 1999. Reflections on the tropical deforestation crisis. *Biological Conservation* 91:109–117.

Leitão, A. B., Miller, J., Ahern, J., McGarigal, K. 2006. *Measuring landscapes—A planner's handbook.* Washington, DC: Island Press.

Li, H., Wu, J. 2004. Use and misuse of landscape indices. *Landscape Ecology* 19:389–399.

Moser, B., Jaeger, J. A. G., Tappeiner, U., Tasser, E., Eiselt, B. 2007. Modification of the effective mesh size for measuring landscape fragmentation to solve the boundary problem. *Landscape Ecology* 22(3):447–459.

Mysterud, A., Bartoń, K. A. Jędrzejewska, B., Krasiński, Z. A., Niedziałkowska, M., Kamler, J. F., Yoccoz, N. G., Stenseth, N. C. 2007. Population ecology and conservation of endangered megafauna: The case of European bison in Białowieża Primeval Forest, Poland. *Animal Conservation* 10:77–87.

Nijkamp, P., Goede, E. 2002. Urban development in the Netherlands: New perspectives. In H. S. Geyer (Ed.), *International handbook of urban systems: Studies of urbanization and migration in advanced and developing countries* (pp. 185–213). Cheltenham: Edward Elgar.

Penn-Bressel, G. 2005. Begrenzung der Landschaftszerschneidung bei der Planung von Verkehrswegen. *GAIA* 14(2):130–134.

Racicot, A., Babin-Roussel, V., Dauphinais, J.-F., Joly, J.-S., Noël, P., Lavoie, C. 2014. A framework to predict the impacts of shale gas infrastructures on the forest fragmentation of an agroforest region. *Environmental Management* 53:1023–1033.

Reck, H., Hänel, K., Jeßberger, J., Lorenzen, D. 2008. *UZVR, UFR + Biologische Vielfalt*. Naturschutz und Biologische Vielfalt 62, German Bundesamt für Naturschutz, Münster, Landwirtschaftsverlag.

Roedenbeck, I. A., Voser, P. 2008. Effects of roads on spatial distribution, abundance and mortality of brown hare (*Lepus europaeus*) in Switzerland. *European Journal of Wildlife Research* 54:425–437.

Roedenbeck, I. A., Fahrig, L., Findlay, C. S., Houlahan, J., Jaeger, J. A. G., Klar, N., Kramer-Schadt, S., van der Grift, E. A. 2007. The Rauischholzhausen-Agenda for road ecology. *Ecology and Society* 12(1):11. Available at: http://www.ecologyandsociety.org/vol12/iss1/art11/.

Romanowicz, A., Daffner, F., Uhel, R., Weber, J.-L. 2007. European Environment Agency developments of land and ecosystem accounts: General overview. In *Conference Information: 11th World Multi-Conference on Systemics, Cybernetics and Informatics (WMSCI)/13th International Conference on Information Systems Analysis and Synthesis*, July 8–11, 2007 Orlando, FL, Vol. IV, Proceedings (pp. 327–332).

Roth, U., Schwick, C., Spichtig, F. 2010. *L'état du paysage en Suisse—Rapport intermédiaire du programme Observation du paysage suisse (OPS)*. *État de l'environnement 2010*. Berne: Office Fédéral de l'Environnement.

Saunders D. A., Hobbs, R. J., Margules, C. R. 1991. Biological consequences of ecosystem fragmentation: A review. *Conservation Biology* 5(1):18–32.

Scheringer, M. 2002. *Persistence and spatial range of environmental chemicals*. Weinheim: Wiley-VCH.

Schupp, D. 2005. Umweltindikator Landschaftszerschneidung—Ein zentrales Element zur Verknüpfung von Wissenschaft und Politik. *GAIA* 14(2):101–106.

Schwick, C., Jaeger, J. A. G., Bertiller, R., Kienast, F. 2012. *Urban sprawl in Switzerland—Unstoppable? Quantitative analysis 1935–2002 and implications for regional planning*. Bristol Series 30, Bristol-Stiftung. Berne/Stuttgart/Vienna: Haupt-Verlag (in English and French).

Selva, N., Kreft, S., Kati, V., Schluck, M., Jonsson, B.-G., Mihok, B., Okarma, H., Ibisch, P. L. 2011. Roadless and low-traffic areas as conservation targets in Europe. *Environmental Management* 48:865–877.

State Institute for Environment, Measurements and Nature Conservation Baden-Württemberg. 2006. *Umweltdaten 2006 Baden-Württemberg*. Landesanstalt für Umwelt, Messungen und Naturschutz Baden-Württemberg (LUBW). Karlsruhe, Germany: JVA Mannheim.

Streffer, C., Bücker, J., Cansier, A., Cansier, D., Gethmann, C. F., Guderian, R., Hanekamp, G., Henschler, D., Pöch, G., Rehbinder, E., Renn, O. et al. 2003. Environmental Standards: Combined Exposures and Their Effects on Human Beings and Their Environment. Berlin/Heidelberg, Germany: Springer-Verlag.

Tasser, E., Sternbach, E., Tappeiner, U. 2008. Biodiversity indicators for sustainability monitoring at municipality level: An example of implementation in an alpine region. *Ecological Indicators* 8:204–223.

Taylor, P. D., Fahrig, L., Henein, K., Merriam, G. 1993. Connectivity is a vital element of landscape structure. *Oikos* 68(3):571–573.

Tillmann, J. E. 2005. Habitat fragmentation and ecological networks in Europe. *GAIA* 14(2):119–123.

Tilman, D., May, R. M., Lehman, C. L., Nowak, M. A. 1994. Habitat destruction and the extinction debt. *Nature* 371:65–66.

Tischendorf, L., Fahrig, L. 2000. On the usage and measurement of landscape connectivity. *Oikos* 90:7–19.

Umweltbundesamt (UBA) (Ed.). 2003. *Reduzierung der Flächeninanspruchnahme durch Siedlung und Verkehr—Materialienband.* UBA-Texte 90/03, Berlin.

United Nations. 2010. UNdata: Composition of macro geographical (continental) regions, geographical sub-regions, and selected economic and other groupings, December 2010. Available at: http://unstats.un.org/unsd/methods/m49/m49regin.htm.

van der Grift, E. A. 2005. Defragmentation in the Netherlands: A success story? *GAIA* 14(2):144–147.

van der Ree, R., Jaeger, J. A. G., van der Grift, E. A., Clevenger, A. P. 2011. Guest editorial: Effects of roads and traffic on wildlife populations and landscape function: Road ecology is moving towards larger scales. *Ecology and Society* 16(1):48. Available at: http://www.ecologyandsociety.org/vol16/iss1/art48/.

van der Ree, R., Jaeger, J. A. G., Rytwinski, T., van der Grift, E. A. 2015. Good science and experimentation are needed in road ecology. In R. van der Ree, D. J. Smith, and C. Grilo (Eds.), *Handbook of road ecology* (pp. 71–81). Hoboken, NJ: Wiley-Blackwell.

Vogt, J. V., Soille, P., de Jaeger, A., Rimaviciute, E., Mehl, W., Foisneau, S., Bodis, K., Dusart, J., Paracchini, M. L., Haastrup, P., Bamps, C. 2007. *A pan-European river and catchment database.* Report EUR 22920 EN. Luxembourg: European Commission–Joint Research Centre.

Waterstraat, A., Baier, H., Holz, R., Spieß, H. J., Ulbricht, J. 1996. *Unzerschnittene, störungsarme Landschaftsräume—Versuch der Beschreibung eines Schutzgutes. Die Bedeutung unzerschnittener, störungsarmer Landschaftsräume für Wirbeltierarten mit großen Raumansprüchen—Ein Forschungsprojekt.* Schriftenreihe des Landesamtes für Umwelt und Natur Mecklenburg-Vorpommern, Heft 1: 5–24.

Weber, J.-L. 2007. Implementation of land and ecosystem accounts at the European Environment Agency. *Ecological Economics* 61(4):695–707.

21

Ecosystem Mapping and Assessment

Markus Erhard, Branislav Olah, Gebhard Banko,
Stefan Kleeschulte, and Dania Abdul-Malak

CONTENTS

21.1 Policy Background

In May 2011 the European Commission and Council adopted the "Communication for the Implementation of a Biodiversity Strategy to 2020" (European Commission [EC], 2011), in line with the time lines to meet the Aichi targets of the Convention of Biodiversity (CBD, 2010). The Headline target for 2020 states "Halting the loss of biodiversity and the degradation of ecosystem services in the European Union (EU) by 2020, and restoring them in so far as feasible, while stepping up the EU contribution to averting global biodiversity loss." The Strategy translates this central objective into six specific targets, "Conserving and restoring nature," "Maintaining and restoring ecosystems and their services," "Sustainable agriculture and forestry," "Sustainable fishery," "Combatting invasive alien species," and "Addressing the global biodiversity crisis" and 20 concrete actions to achieve them.

The Strategy implies two timelines for its targets. For the medium term it states: "By 2020, ecosystems and their services are maintained and enhanced by establishing green infrastructure and restoring at least 15% of degraded ecosystems." As a long-term target, it declares "By 2050, EU biodiversity and the ecosystem services it provides—its natural capital—are protected, valued and appropriately restored for biodiversity's intrinsic value and for their essential contribution to human wellbeing and economic prosperity, so that catastrophic changes caused by the loss of biodiversity are avoided."

Action 5 of Target 2 "Mapping and Assessment of Ecosystems and their Services" (MAES) foresees that "Member States (MS), with the assistance of the Commission, will map and assess the state of ecosystems and their services in their national territory by 2014, assess the economic value of such services, and promote the integration of these values into accounting and reporting systems at EU and national level by 2020." It is currently implemented by the Working Group MAES (a joint body of Commission Services and MS with support of the European Environment Agency [EEA]). Assessments and valuation of Action 5 are closely linked to the other Actions of Target 2, mainly Actions 6a: Restoration Prioritization Framework; 6b: Green Infrastructure Strategy; 7a: Biodiversity

Proofing Methodology; and 7b: No Net Loss Initiative. There is also need to establish links to the other five Targets of the Communication. This is important for an integrative view in decision-making processes and key to address synergies and trade-offs of policy impacts on ecosystems and their services. EEA and Joint Research Centre (JRC) committed to provide an EU-wide assessment complementary to the MS activities. EEA is focusing on mapping and assessing ecosystems, their condition, the pressures they are subjected to, and the impacts on their biodiversity, while JRC is mapping ecosystem services. Both parts will then be merged to analyze how ecosystem condition affects biodiversity and ecosystem services. The activities are based on existing data and information making use of the reporting obligations of the European environmental legislation, namely the Nature and Habitat Directives, the Water Framework Directive (WFD), the Marine Strategy Framework Directive (MSFD), and other relevant data flows. The approach is documented in the MAES analytical framework (Maes et al., 2013) and data availability is evaluated in a second report (Maes et al., 2014).

21.2 Land Cover and Ecosystems

An ecosystem in the definition of the CBD is "a dynamic complex of plant, animal and micro-organism communities and their non-living environment interacting as a functional unit" (CBD, 1992). This definition applies to all hierarchical levels, from a single water drop and its microorganisms to Earth's major vegetation zones (biomes). For the practical purposes of policy-relevant mapping and assessment at the European level and in view of the available information, ecosystems are considered here at the scale of land cover (LC) related units such as urban, cropland, grassland, forests, rivers, and lakes. These units also represent the key elements for human management, for example, by nature protection, territorial planning, agriculture, forestry, fisheries, and water management to make the best use of their services and the respective European Environmental Directives, mainly the Habitat (HD), Bird (BD), WFD, and MSFD.

Ecosystems are multifunctional. Each system provides a series of services for human well-being, either directly such as food and fiber or more indirectly by, for example, providing clean air and water. At this scale, ecosystems also provide many habitats for major groups of species or taxa such as plants, butterflies, amphibians, reptiles, birds, or mammals in different stages of their life cycles. The functioning of an ecosystem, its capacity to provide habitats and services, is triggered by its condition. Its condition is steered by the exposure to direct pressures such as land take, fragmentation, use, and management including application of fertilizers and pesticides, and indirect pressures, mainly air pollution, climate change, invasive alien species, and so forth. The assessment of the condition requires spatially explicit mapping to capture the different gradients and variations of all relevant components in space and time affecting ecosystem condition and subsequent functions (Maes et al., 2014).

The MAES Working Group agreed on a list of European wide ecosystem types to be feasible for aggregation of national and local data and disaggregation of European data (Table 21.1) and to reflect main policy lines and environmental reporting. The typology should also allow for zonal classifications such as "mountainous areas" or "coastal zones" which can be generated using the proposed ecosystem types in combination with reference data such as elevation or coast lines. A detailed description is available in Maes et al. (2013).

TABLE 21.1

Typology of Ecosystems as Used for the EU-Wide Mapping and Assessment Based on CLC and Bathymetry Data

Major Ecosystem Category (Level 1)	Ecosystem Type for Mapping and Assessment (Level 2)	Description
Terrestrial	*Urban*[a]	Urban ecosystems are areas where most of the human population lives: This class includes urban, industrial, commercial, and transport areas, urban green areas, mines, dumping and construction sites.
	Cropland	Cropland is the main food production area including both intensively managed ecosystems and multifunctional areas supporting many semi- and natural species along with food production (lower intensity management). It includes regularly or recently cultivated agricultural, horticultural and domestic habitats and agro-ecosystems with significant coverage of natural vegetation (agricultural mosaics).
	Grassland	Grasslands are areas covered by a mix of annual and perennial grass and herbaceous non-woody species (including tall forbs, mosses, and lichens) with little or no tree cover. The two main types are managed pastures and (semi-)natural (extensively managed) grasslands.
	Forest and Woodlands	Woodland and forest are areas dominated by woody vegetation of various age or they have succession climax vegetation types on most of the area supporting many ecosystem services. Information on ecosystem structure (age class, species diversity, etc.) is especially important for this ecosystem type.
	Heathland and Shrub	Heathland and shrub are areas with vegetation dominated by shrubs or dwarf shrubs. They are mostly secondary ecosystems with unfavourable natural conditions. They include moors, heathland, and sclerophyllous vegetation.
	Sparsely Vegetated Land[a]	Sparsely vegetated lands often have extreme natural conditions that might support particular species. They include bare rocks, glaciers and dunes, beaches, and sand plains.
	Wetlands	Inland wetlands are predominantly water-logged specific plant and animal communities supporting water regulation and peat-related processes. This class includes natural or modified mires, bogs and fens, as well as peat extraction sites.
Freshwater	*Rivers and lakes*	Rivers and lakes which are the permanent freshwater inland surface waters. This class includes water courses and water bodies.
Marine[b]	*Marine inlets and transitional waters*	Marine inlets and transitional waters are ecosystems on the land-water interface under the influence of tides and with salinity higher than 0.5%. They include coastal wetlands, lagoons, estuaries and other transitional waters, fjords and sea lochs, as well as embayments.
	Coastal	The coastal ecosystems include coastal, shallow, marine systems that experience significant land-based influences. These systems undergo diurnal fluctuations in temperature, salinity, and turbidity, and are subject to wave disturbance. Depth is between 50 and 70 m.

(Continued)

TABLE 21.1 (CONTINUED)

Typology of Ecosystems as Used for the EU-Wide Mapping and Assessment Based on CLC and Bathymetry Data

Major Ecosystem Category (Level 1)	Ecosystem Type for Mapping and Assessment (Level 2)	Description
	Shelf	The shelf refers to marine systems away from coastal influence, down to the shelf break. They experience more stable temperature and salinity regimes than coastal systems, and their seabed is below wave disturbance. They are usually up to 200 m deep.
	Open ocean	The open ocean refers to marine systems beyond the shelf break with very stable temperature and salinity regimes, in particular in the deep seabed. Depth is beyond 200 m.

[a] Currently not implemented in ecosystem/ecosystem service assessment.

[b] The current zonal classification refers to the reporting units of the MSFD and might be replaced by a more ecosystem-based approach in the future.

21.3 Producing a European Ecosystem Map

According to its intended application in various EU environmental policies (as described in the EU Biodiversity Strategy), the production of a European ecosystem (ES) map must follow several necessary conditions. The mapping scale must allow for capturing ecosystems at the landscape level and for covering pan-European territory (preferably EU28+). The ES map must be spatially and thematically consistent allowing for aggregation (bottom-up) or disaggregation (top-down) across geographical scales (nested scale approach). Further, the ecosystem classification must reflect Europe's nature specifics and allow for interoperability with existing policies and relevant European (such as habitats in HD) or national ecosystem classifications. The underpinning data for ecosystem mapping and assessment must be updated regularly to allow for change detection and assessment of meeting targets set by the environmental policies.

Because no spatially referenced European ES map existed so far, a pilot version was produced using CLC 2006 for ecosystem delineation and European Nature Information System (EUNIS; http://eunis.eea.europa.eu/) for ecosystem classification. EUNIS is a pan-European hierarchical classification system that provides non-georeferenced information about habitats and their species. Both datasets fulfill the conditions described in the preceding text. CoORdination of Information on the Environment (CORINE) Land Cover (CLC) has proven itself as a solid and operational pan-European dataset with a secured update cycle (Copernicus) and fully developed applications for change assessment such as land accounting (EEA, 2006). The EUNIS classification was developed to categorize hierarchically all types of ecosystems that occur in the European territory. In addition, the EUNIS classification system allows for translation of its classes into other ecosystem classification systems such as habitat types in the HD or syntaxonomic units of the European Vegetation Survey (EVS; http://euroveg.org/) as well as to CLC classes. So-called cross-walk tables provide the thematic links between the different classification systems for attribution.

Despite the overall suitability of the CLC dataset, its application in producing a European ES map was not straightforward and it required several spatial and thematic refinements.

Spatial refinements had to provide more detailed information that is not delineated in the CLC map due to the minimum mapping units (MMUs) of CLC (omission of areas <25 ha or <100 m wide) and the structural heterogeneity of the so-called mixed classes (i.e., agricultural mosaics of croplands, grasslands and permanent crops, or mixtures of agricultural land and natural vegetation). Using more detailed datasets such as High-Resolution Layers (HRLs) supported the refinement process. Thematic refinements were based on using ancillary reference data to translate CLC polygons into EUNIS habitats (ecosystem) types.

The production implied four major steps (European Topic Centre on Spatial Information and Analysis [ETC/SIA], 2014a):

1. Reclassification of CLC2006 and geometrical refinement
2. Development and application of the CLC–EUNIS–MAES ecosystem typology cross-walk table
3. Thematic refinement delineating EUNIS habitats using CLC and reference data
4. Assessment of ecosystem condition (experimental)

Additional refinement will be available as soon as the HRLs are fully developed as enhancements of major CLC classes, which include the layers on artificial surfaces, urban, forest, agriculture, wetland, and water bodies (http://land.copernicus.eu/pan-european). Precursors such as the soil sealing degree or the JRC-forest layer (2006) already provide enhancements of the current CLC geometric resolution. Beside the HRLs, other thematic data also provide useful estimators for an enriched and integrated LC map with a resolution of 100×100 m (e.g., open street map roads and open street map land use [LU], Riparian Areas, or Small Linear Features). Those data that do not provide explicit spatial delineations (e.g., linear vectors of water courses) will not be taken into consideration for the refined LC maps. To be integrated in a spatial explicit value, the minimum size of a specific LC type should cover at least 50% of a 100×100 m grid cell. This approach was widely discussed within the Harmonized European Land Monitoring (HELM) Project (http://www.helm-project.eu/) as the "grid approach." Using the grid as a basic spatial data element not only defines ES map resolution and fixes its exact location but also allows for a precise change assessment using ecosystem accounting methodology, based on the change of the LU type as proposed in EC et al. (2013) and applied in Remme et al. (2014).

In a second step, a MAES ecosystem typology–EUNIS cross-walk table was developed to attribute habitat information to the MAES ecosystems (Table 21.2). The MAES typology is based on CLC nomenclature (mainly Level 2) and its approach links MAES Level 2 with EUNIS Levels 1 and 2 for the ecosystem mapping approach. Whenever a better regional differentiation was possible, EUNIS Level 3 was used.

As these two nomenclatures are not mutually exclusive, the relation between them is modeled as a multiple-to-multiple relation. This means that one MAES/CLC class can contain multiple EUNIS classes and vice versa. To resolve these multiple relationships additional data have to be integrated for thematic refinement. This third step is based on an approach used for selected habitats of the EU Habitat Directive (HD) as described in Mücher et al. (2004).

For the implementation of this method, important and available environmental datasets with the highest possible accuracy for Europe were identified based on a list of ancillary data (Figure 21.1). Also a set of knowledge rules was identified for each habitat. In contrast to Mücher et al. (2004), where three likelihood classes were modeled for each grid cell, only one habitat class was attributed in this study based on a Geographical Information System (GIS) based decision tree approach. Each attributed habitat class was labeled with

TABLE 21.2

Typology of Ecosystems Based on the Refinement with EUNIS Habitat Classes, Area of Coverage, and Percentage of Area per EUNIS Level 1 Class

Ecosystem Type	EUNIS Level 1	EUNIS Level 2	Total Ecosystem Coverage	
			Area (km²)	% Area EUNIS Level 2 per Level 1
Urban	J Constructed, industrial and other artificial habitats	J1 Buildings of cities, towns and villages	102,151	46.08
		J2 Low-density buildings	94,150	42.47
		J3 Extractive industrial sites	6453	2.91
		J4 Transport networks and other constructed hard-surface areas	16,100	7.26
		J5 Highly artificial man-made waters and associated structures	1828	0.82
		J6 Waste deposits	998	0.45
Cropland	I Regularly or recently cultivated agricultural, horticultural, and domestic habitats	I1 Arable land and market gardens	1,243,168	99.18
		I2 Cultivated areas of gardens and parks	10,292	0.82
Grassland	E Grasslands and land dominated by forbs, mosses, or lichens	E1 Dry grasslands	9330	1.35
		E2 Mesic grasslands	571,931	82.48
		E3 Seasonally wet and wet grasslands	55,771	8.04
		E4 Alpine and subalpine grasslands	21,128	3.05
		E5 Woodland fringes, clearings, and tall forbs stands	0	0.00
		E6 Inland salt steppes	3043	0.44
		E7 Sparsely wooded grasslands	32,195	4.64
Woodland and Forest	G Woodland, forest, and other wooded land	G1 Broad-leaved deciduous woodland	487,970	28.29
		G2 Broad-leaved evergreen woodland	49,248	2.86
		G3 Coniferous woodland	695,907	40.35
		G4 Mixed woodland	291,687	16.91
		G5 Lines of trees, small woodlands, recently felled woodlands, early stage woodland, coppice	199,784	11.58
Heathland and Shrub	F Heathland, scrub, and tundra	F1 Tundra	0	0.00
		F2 Arctic, alpine, and subalpine scrub	34,524	14.88
		F3 Temperate and mediterraneo-montane scrub	52,824	22.76
		F4 Temperate shrub heathland	691	0.30
		F5 Maquis, arborescent matorral and thermo-Mediterranean brushes	50,162	21.61

(Continued)

TABLE 21.2 (CONTINUED)

Typology of Ecosystems Based on the Refinement with EUNIS Habitat Classes, Area of Coverage, and Percentage of Area per EUNIS Level 1 Class

Ecosystem Type	EUNIS Level 1	EUNIS Level 2	Total Ecosystem Coverage	
			Area (km²)	% Area EUNIS Level 2 per Level 1
		F6 Garrigue	10,135	4.37
		F7 Spiny Mediterranean heaths	19,485	8.40
		F8 Thermo-Atlantic xerophytic scrub	3233	1.39
		F9 Riverine and fen shrubs	140	0.06
		FA Hedgerows	0	0.00
		FB Shrub plantations	60,890	26.24
Attributed to Sparsely Vegetated Land	B Coastal habitats	B1 Coastal dunes and sandy shores	n/a	n/a
		B2 Coastal shingle	n/a	n/a
		B3 Rock, cliffs, ledges and shores, including supralittoral	n/a	n/a
Wetlands	D Mires, bogs and fens	D1 Raised and blanked bogs	28,246	35.27
		D2 Valley mires, poor fens, and transition mires	1856	2.32
		D3 Aapa, palsa, and polygon mires	44,981	56.16
		D4 Base-rich fens and calcareolus spring mires	590	0.74
		D5 Sedge and reedbeds, normally without free-standing water	4373	5.46
		D6 Inland saline and brackish marshes and reedbeds	42	0.05
Rivers and Lakes	C Inland surface waters	C1 Surface standing waters	92,690	87.66
		C2 Surface running waters	9694	9.17
		C3 Littoral zone of inland surface water bodies	3356	3.17
Marine Inlets and Transitional Waters, Coastal, Shelf, Open Ocean	A Marine habitats	A1 Littoral rock and other hard substrate	n/a	n/a
		A2 Littoral sediment	n/a	n/a
		A3 Infralittoral rock and other hard substrate	n/a	n/a
		A4 Circalittoral rock and other hard substrate	n/a	n/a
		A5 Sub-littoral sediment	n/a	n/a
		A6 Deep sea-bed	n/a	n/a
		A7 Pelagic water column	n/a	n/a
		A8 Ice-associated marine habitats	n/a	n/a

a reliability for its geometric and thematic accuracy. Using the ancillary datasets, an expert rule system is developed to define differentiations within the multiple relations between MAES/CLC classes and EUNIS classes.

Either using geographic delineations (occurrence of special EUNIS classes in special geographic areas) or using attributive environmental specifications (occurrence of special

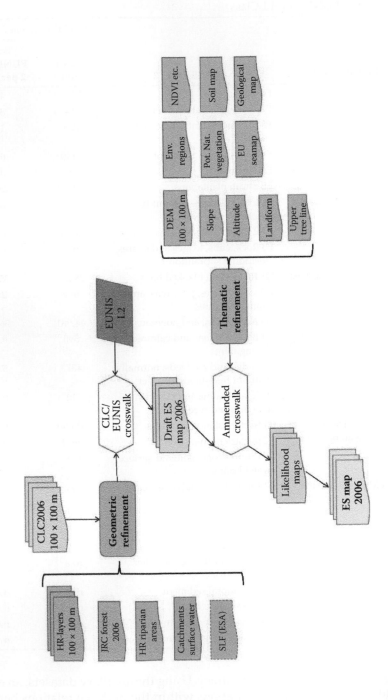

FIGURE 21.1
Work flow: Ecosystem (ES) map development. (From ETC/SIA [European Topic Centre on Spatial Information and Analysis]. 2014a. Developing conceptual framework for ecosystem mapping. ETC/SIA draft internal report. Available at: http://projects.eionet.europa.eu/eea-ecosystem-assessments/library/draft -ecosystem-map-europe.)

EUNIS classes only under special environmental conditions) made it possible to make these differentiations.

The application of these rules by performing a series of overlays and selection rules using the ancillary data provided a first version of the ES map for Europe (see Figure 21.2).*

21.4 Application and Outlook

The map provides a first overview about the potential spatial distribution of major habitats across Europe. As such, it can build the reference for numerous applications for assessing ecosystem condition that triggers their habitat quality, structure, extension, biodiversity, and capacity to provide services.

All CLC-based assessments such as land take and LC change, as well as local, regional, and national assessments, can be linked to the map via the cross-walks and can be related to habitat quality and underneath biodiversity. This offers a wide variety of information sources for assessments. The exploration of its potential just started. A first very simple application is the provision of the level of habitat protection, namely NATURA 2000 sites (http://ec.europa.eu/environment/nature/natura2000/index_en.htm) under the HD and BD in Europe. Figure 21.3 shows the distribution of main ecosystem types over Europe and the outcome of a simple overlay of NATURA 2000 and the ES map. Habitats that are relatively rare such as freshwater bodies, heathlands, and bogs have a relatively high level of protection (expressed as % heathlands % of bogs, etc.). Refinement of the analysis on MS or regional level using EUNIS Level 2 and if available Level 3 data will provide very valuable information about the spatial distribution of the protected habitats across Europe and in the MS and potential gaps that will support decision processes including meeting the targets set in the EU Biodiversity Strategy to 2020.

Another area for applications is the description of ecosystem conditions. As illustrated in Figure 21.4, the ES map can be populated with habitat quality and species assessments using information from European Environmental Directives to characterize the current ecosystem conditions.

To assess the condition of biodiversity in Europe, datasets from the HD are key. Namely, Article 17 data of the HD focuses on protecting species and habitats considered to be most at risk across the EU. The data of the first assessment of conservation status reported by member states and reported on the EU level in 2009 (EEA, 2009) provided information on aggregated national level not feasible to be used for detailed analyses at ecosystem level across Europe. The potential of this approach will be evaluated as soon as data for the current reporting cycle become available.

Another method implies the combination of the ES map with maps of drivers and pressures that indicate how much ecosystems and their habitats and biodiversity are affected by human activities. Among other applications, nitrogen has a high impact on the biodiversity in agro-ecosystems (Kleijn et al., 2009) and is commonly used in LU intensity indicators (Herzog et al., 2006). Figure 21.5 illustrates the level of pressure from nitrogen fertilizer application and its variation across Europe, where the highest pressures are expected to have the most severe impacts on biodiversity and ecosystem services such as water purification.

---

* The map and related information is available on http://biodiversity.europa.eu/maes and can be downloaded from http://www.eea.europa.eu/data-and-maps/data/ecosystem-types-of-europe.

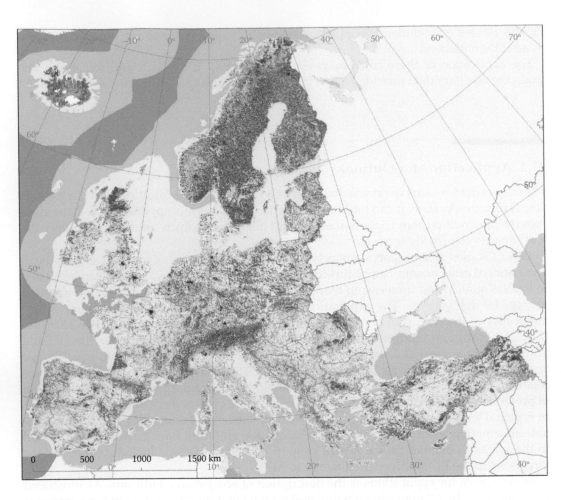

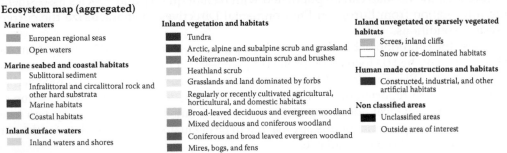

Ecosystem map (aggregated)

Marine waters
- European regional seas
- Open waters

Marine seabed and coastal habitats
- Sublittoral sediment
- Infralittoral and circalittoral rock and other hard substrata
- Marine habitats
- Coastal habitats

Inland surface waters
- Inland waters and shores

Inland vegetation and habitats
- Tundra
- Arctic, alpine and subalpine scrub and grassland
- Mediterranean-mountain scrub and brushes
- Heathland scrub
- Grasslands and land dominated by forbs
- Regularly or recently cultivated agricultural, horticultural, and domestic habitats
- Broad-leaved deciduous and evergreen woodland
- Mixed deciduous and coniferous woodland
- Coniferous and broad leaved evergreen woodland
- Mires, bogs, and fens

Inland unvegetated or sparsely vegetated habitats
- Screes, inland cliffs
- Snow or ice-dominated habitats

Human made constructions and habitats
- Constructed, industrial, and other artificial habitats

Non classified areas
- Unclassified areas
- Outside area of interest

FIGURE 21.2

ES map version 2.1. Note: The classes have been aggregated for better presentation. (From ETC/SIA [European Topic Centre on Spatial Information and Analysis]. 2014a. Developing conceptual framework for ecosystem mapping. ETC/SIA draft internal report. Available at: http://projects.eionet.europa.eu/eea-ecosystem-assessments/library/draft-ecosystem-map-europe.)

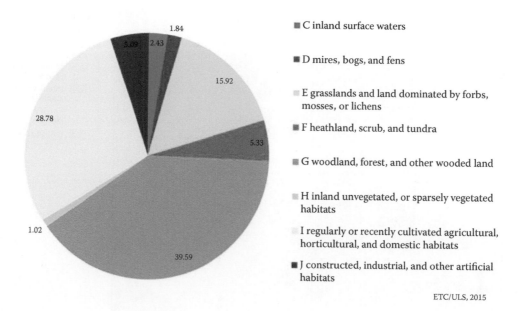

C inland surface waters

D mires, bogs, and fens

E grasslands and land dominated by forbs, mosses, or lichens

F heathland, scrub, and tundra

G woodland, forest, and other wooded land

H inland unvegetated, or sparsely vegetated habitats

I regularly or recently cultivated agricultural, horticultural, and domestic habitats

J constructed, industrial, and other artificial habitats

ETC/ULS, 2015

FIGURE 21.3
Percentage of EUNIS habitat types Level 1 in EU28 and their protection status by Natura 2000 (%) for terrestrial and freshwater ecosystems.

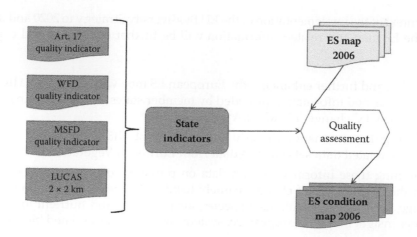

FIGURE 21.4
Work flow: Mapping ecosystem condition using the ES map. (From ETC/SIA [European Topic Centre on Spatial Information and Analysis]. 2014a. Developing conceptual framework for ecosystem mapping. ETC/SIA draft internal report. Available at: http://projects.eionet.europa.eu/eea-ecosystem-assessments/library/draft -ecosystem-map-europe.)

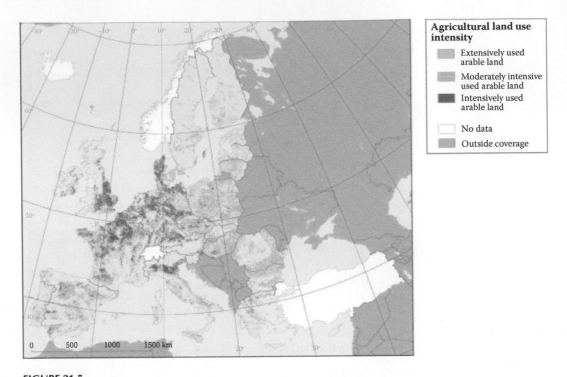

FIGURE 21.5
Agricultural LU intensity as function of nitrogen fertilizer application in Europe. (From ETC/SIA [European Topic Centre on Spatial Information and Analysis]. 2014b. Indicators to assess condition and major pressures on agro-ecosystems in Europe. ETC/SIA draft internal report. Available at: http://projects.eionet.europa.eu/eea -ecosystem-assessments/library/working-documents-and-maps-ecosystem-pressures.)

In the context of the implementation of the EU Biodiversity Strategy to 2020 and its related Actions, the ES map and related information will be further elaborated and explored. It includes

- Updating and further enhancing the European ES map with CLC 2012, HRLs, reference data, and information provided by member states including change detection using CLC change (LEAC, EEA 2006)
- Exploring the information directly linked to the EUNIS habitats and other biodiversity related data sets such as data from European Vegetation Survey (EVS)
- Combining these information with data on pressures on ecosystems generated from the main driver of change, namely habitat change, fragmentation, climate change, land management, alien species, air pollution, and nutrient loads to get further insight into the European ecosystem conditions and related biodiversity

CLC data provide the spatially referenced baseline to map and assess ecosystems and their condition across Europe. As an operational data flow, it provides very valuable information about changes in LC and as such indicates subsequent gain and loss of habitats over time. The MAES process will provide the baseline for a number of European environmental policies related to ecosystems and their services with CLC as one of the key input datasets for mapping and assessment.

References

CBD (Convention on Biological Diversity). 1992. Article 2. Use of terms: Ecosystem. Available at: http://www.cbd.int/ecosystem/.

CBD (Convention on Biological Diversity). 2010. Aichi biodiversity targets. Available at: http://www.cbd.int/sp/targets/.

EC (European Commission). 2011. Our life insurance, our natural capital: An EU biodiversity strategy to 2020. COM (2011) 244. Brussels. Available at: http://ec.europa.eu/environment/nature/bio diversity/comm2006/pdf/2020/1_EN_ACT_part1_v7[1].pdf.

EC (European Commission), OECD (Organisation for Economic Co-operation and Development), UN (United Nations), WB (World Bank). 2013. System of environmental-economic accounting 2012, Experimental ecosystem accounting. Available at: http://unstats.un.org/unsd/envac counting/eea_white_cover.pdf.

EEA (European Environment Agency). 2006. Land accounts for Europe 1990–2000. Towards integrated land and ecosystem accounting. EEA Report 11/2006. Copenhagen. Available at: http://www.eea.europa.eu/publications/eea_report_2006_11.

EEA (European Environment Agency). 2009. Progress towards the European 2010 biodiversity target. EEA Report 4/2009. Copenhagen. Available at: http://www.eea.europa.eu/publications/prog ress-towards-the-european-2010-biodiversity-target/.

ETC/SIA (European Topic Centre on Spatial Information and Analysis). 2014a. Developing conceptual framework for ecosystem mapping. ETC/SIA draft internal report. Available at: http://projects.eionet.europa.eu/eea-ecosystem-assessments/library/draft-ecosystem-map-europe.

ETC/SIA (European Topic Centre on Spatial Information and Analysis). 2014b. Indicators to assess condition and major pressures on agro-ecosystems in Europe. ETC/SIA draft internal report. Available at: http://projects.eionet.europa.eu/eea-ecosystem-assessments/library/working -documents-and-maps-ecosystem-pressures.

Herzog, F., Steiner, B., Bailey, D., Baudry, J., Billeter, R., Bukáček, R., De Blust, G., De Cock, R., Dirksen, J., Dormann, C. F., De Filippi, R. et al. 2006. Assessing the intensity of temperate European agriculture at the landscape scale. *European Journal of Agronomy* 24:165–181.

Kleijn, D., Kohler, F., Báldi, A., Batáry, P., Concepción, E. D., Clough, Y., Díaz, M., Gabriel, D., Holzschuh, A., Knop, E., Kovács, A. et al. 2009. On the relationship between farmland biodiversity and land-use intensity in Europe. *Proceedings of the Royal Society of London B: Biological Sciences* 276:903–909.

Maes, J., Teller, A., Erhard, M., Liquete, C., Braat, L., Berry, P., Egoh, B., Puydarrieux, P., Fiorina, C., Santos, F., Paracchini, M. L. et al. 2013. Mapping and Assessment of Ecosystems and their Services. An analytical framework for ecosystem assessments under action 5 of the EU biodiversity strategy to 2020. Luxembourg: Publications Office of the European Union. Available at: http://ec.europa.eu/environment/nature/knowledge/ecosystem_assessment/pdf/MAES WorkingPaper2013.pdf.

Maes, J., Teller, A., Erhard, M., Murphy, P., Paracchini, M. L., Barredo, J. I., Grizzetti, B., Cardoso, A., Somma, F., Petersen, J. E., Meiner, A. et al. 2014. Mapping and Assessment of Ecosystems and their Services Indicators for ecosystem assessments under Action 5 of the EU Biodiversity Strategy to 2020. 2nd Report–Final, Technical Report–080. Available at: http://ec.europa.eu/envi ronment/nature/knowledge/ecosystem_assessment/pdf/2ndMAESWorkingPaper.pdf.

Mücher, C. A., Hennekens, S. M., Bunce, R. G. H., Schaminée, J. H. J. 2004. Mapping European Habitat to support the design and implementation of a Pan-European Network. The PEENHAB project. Alterra Report 952. Wageningen: Alterra.

Remme, R. P., Schröter, M., Hein, L. 2014. Developing spatial biophysical accounting for multiple ecosystem services. *Ecosystem Services* 10:6–18.

22

High Nature Value Farmland and the Common Agricultural Policy

Ivone Pereira Martins, Katarzyna Biala, and Ana Maria Ribeiro de Sousa

CONTENTS

22.1 High Nature Value: Introducing the Concept

High Nature Value (HNV) is a concept used to describe some of the oldest and most biodiversity rich farming and forestry systems in Europe, many of which are now under serious threat. HNV farming and farmland areas contribute to biodiversity of European agricultural landscapes. On this HNV land, semi-natural habitats and wild species have been coexisting with low-intensity management by local rural communities for hundreds or even thousands of years. The high nature value proposed in 1993 in an influential report from the Institute for European Environmental Policy (Baldock et al., 1993) and since then has played an important role in research and policy analysis on farming practices and resulting land uses (LUs).

Across Europe, certain types of farming (e.g., subsistence, extensive, low input) are especially valuable for the environment. These HNV farming systems vary from country to country, but all provide important environmental and ecosystems services (European Environment Agency [EEA], 2015a). Not only do they support the conservation of some of Europe's most threatened habitats and species, but they also contribute to other ecosystem services such as soil carbon storage, the protection of water resources, and fire prevention. HNV farming systems also contribute to other more intangible services, as they are part of a rich cultural heritage (Baldock and Beaufoy, 1994). They are as well supporting an active land managing rural population, often sustaining a unique way of life in some of the most economically fragile areas in Europe, prone to land abandonment and depopulation. HNV farming systems, and their benefits, not only support the Common Agricultural Policy (CAP), but also play an important role on the Regional and Cohesion policies and the EU2020 Biodiversity Strategy.

HNV farmland areas and respective features have been widely recognized as a valuable element and asset of European agricultural and broader landscapes. They provide and support highly varied ecological conditions sustaining important habitats of European relevance for nature conservation and for a wide range of species and thereby contribute to maintaining and even halting biodiversity loss, the objectives of the Habitats and Birds Directives and the targets of the EU2020 Biodiversity Strategy and the Strategic Plan for Biodiversity 2011–2020 at global level.

HNV farmland results from a combination of specific LU and farming systems. The dominant feature of HNV farming is nonetheless low-intensity management, with a significant presence of semi-natural vegetation, in particular extensive grasslands. Diversity of land cover (LC), and the presence of green infrastructure features such as ponds, hedges, and woodland, is also a characteristic. Typical HNV farmland areas—which occur throughout Europe, but more common on the peripheries where less intensive agriculture has been practiced until very recently—are extensively grazed uplands, alpine meadows and pasture, steppic areas in Eastern and southern Europe, and dehesas and montados in Spain and Portugal, respectively. More intensively farmed areas in Western Europe lowlands or in the Iberian Peninsula can also host features enabling concentrations of species of particular conservation interest, such as migratory waterfowl or steppary birds such as the great bustard.

22.2 HNV: Applying the Concept

The concept of HNV farmland emerged as a policy consideration at the European Union (EU) level some years ago. From its first inclusion in the original set of agri-environmental indicators developed by the European Commission (EC) following the June 1998 Cardiff European Council it has remained part of the Agro-Environment Indicator (AEI) set coordinated by Eurostat. For the 2007–2013 programming period, the Community strategic guidelines for rural development highlighted the preservation and development of HNV farmland systems as a priority (Council Decision 2006/144/EC). This priority was reinforced through the introduction of biodiversity concerns as one of the new challenges for the CAP within its "Health Check" in 2009 (Council Regulation EC No. 73/2009). The rural development legal proposal for 2014–2020 includes restoring and preserving biodiversity in areas of high nature value farmland as one of the six Union priorities for rural development.

"HNV farming and farmland" was proposed as one of the impact indicators to be included in the CAP monitoring and evaluation framework for 2014–2020. The evaluation process is undertaken under a common framework, that is, Common Monitoring and Evaluation Framework [CMEF]) (DG AGRI, 2006). As such it will fall under the provisions of Article 110 of the proposed CAP Horizontal Regulation (Council Regulation [EC] No. 1259/1999), associated implementing rules, and the legislative framework for rural development. EU member states will therefore be required to supply values for this indicator (a baseline situation, plus updates at specific points during the period) in the context of the CAP monitoring and evaluation framework. In particular it will be needed for the baseline description of each Rural Development Programme (RDP) territory, and the subsequent evaluation of RDPs.

This concept of HNV developed in the early 1990s got growing recognition for conservation of biodiversity in Europe based on low-intensity farming systems across large areas of countryside (Baldock and Beaufoy, 1994; Bignal and McCracken, 1996a,b, 2000).

The EEA has, since 2004 (EEA, 2004), conducted work around this concept together with its European Topic Centre for Spatial Information and Analysis (ETC/SIA, which ended its activities in 2014) and the Joint Research Centre (JRC).

All of this work conducted for more than 20 years had the objective to influence the distribution of CAP funds through a shift from more intensive farming to a more sustainable LU.

To assess the extent of HNV farming and farmland in Europe a wide variety of approaches and combinations of methods are currently being used. Although good progress has been made in assessing the extent of HNV farming, the assessment of its condition or quality still presents a considerable challenge (EEA, 2009, 2010).

With increasing pressure on resources, environmental objectives become increasingly important for the CAP. Streamlining, complementarity and convergence between the AEIs and the monitoring and evaluation framework being developed for the CAP (CMEF) on one hand and the common implementation framework under the 2020 Biodiversity Strategy on the other are being sought to the extent possible. Two main options have been pursued that require explanation and most of the time are misunderstood:

1. HNV farming and farmland data produced at national level are based on the best national information available both in situ or spatial. These national HNV data are used under the CMEF, which assumes and supports the principle that CAP support should be weighted in favor of low-intensity farming systems throughout the European Union (EU) territory, and that this would in turn provide a robust basis for biodiversity conservation in Europe. Elegibility for support payments would depend on simple criteria applied at the level of the farm holding (e.g., such as livestock density). It was not intended that support payments for low-intensity farming should be provided only in delineated "HNV areas." An indicator definition based on the previous version of agri-environmental indicators (AEI 23; http://ec.europa.eu/eurostat/statistics-explained/index.php/Agri -environmental_indicator_-_High_Nature_Value_farmland) published by the statistical office of the EU (Eurostat) together with the experience gained through using HNV farmland as an indicator within the CMEF for rural development for 2007–2013, has been used within the CAP framework and based on country-submitted information.

2. HNV farmland produced as a European map and indicator, based on LC data, namely the CoORdination of Information on the Environment (CORINE) Land Cover (CLC) datasets, conducted at the EEA level. The data are geographically comparable and used for state of the environment assessment.

Owing to the variation in data availability across the EU member states and the range of physical and ecological situations (territory size, farm structure and systems, predominant land and habitat types), the implementation of one common methodology for the assessment of HNV farming within the CMEF framework has not been considered suitable (Cooper et al., 2007). As some EU member states have a rich array of data sources and have developed more refined methods for using them, the use of one single method would restrict the analysis of available data and would impose on those member states a reduction in the quality and accuracy of the assessments.

At European level, distribution patterns of HNV farmland are based on LC and biodiversity data approach, developed by EEA and JRC (Paracchini et al., 2006, 2008). The HNV farmland map produced by EEA provides comparable estimates across Europe, shows an overview of the likely spatial distribution of HNV farmland across European countries, and gives a rough indication of the share of HNV farmland in the agricultural land across Europe. On its last version available (2012) based on CLC2006 it includes information from 39 countries, excluding Greece.

22.3 The 2012 HNV Farmland Map

HNV farmland and farming refers to the causality link established between certain types of farming activities and corresponding positive environmental outcomes (now commonly designated as ecosystems services), including high levels of biodiversity and the presence of diverse and valuable habitats and species. HNV farmland and farming are therefore key indicators for the assessment of the impact of policy interventions with respect to the preservation and enhancement of biodiversity, habitats and ecosystems dependent on agriculture, and the preservation of traditional rural landscapes for their intrinsic and functional values and roles.

Though the HNV farming and farmland concepts had been established in the early 1990s, there was no map on the distribution of HNV farmland. In 2003 the first HNV map of Europe was presented by Andersen (2003) in a report for the EEA that shows the approximate distribution of HNV farmland in Europe. Two approaches for identifying HNV farmland were proposed: based on (1) farm system data (derived from the Farm Accountancy Data Network [FADN]; this approach is not developed further in the context of this chapter) and (2) LC data (represented by the CLC data) were considered to provide an indication of the broad distribution pattern of HNV farmland, whereas farm system data can give information about the types and characteristics of the farms concerned and their estimated share in the farm population.

The objective of EEA was to estimate the likelihood of presence of HNV farmland within Europe using existing European-wide datasets. The CLC European dataset was used as it was available consistently across the great majority of European countries. It was, though, noted that these data have several drawbacks related to geometric limitations (minimum mapping unit [MMU] of 25 ha), thematic resolution (particular class assessment) and updates dynamic (10 and 6 years), which had an impact in the quality/reliability of the results, which can be seen only as approximate.

EEA and the JRC continued to work to enhance and update the map and the results of a major update were presented in 2008 (Paracchini et al., 2008). The basis of the mapping exercise is provided by the CLC data for the year 2000 and the Environmental Stratification of Europe (Metzger et al., 2005). To enhance the map, additional data sources were used: NATURA 2000 network, Important Bird Areas (IBAs), Prime Butterfly Areas (PBAs), and national biodiversity datasets.

In 2009 the EEA started a new update of the HNV farmland map based on the 2006 update of CLC, new data on NATURA 2000 network, IBA data, PBA data, and including missing countries such as Switzerland, Iceland, Norway, Turkey, and the West Balkan states. This update had in view to gain a better insight into the current distribution and extent of farmland that is of inherent biodiversity value, and to develop a more effective tool for testing out further analyses on spatial and time trends and review the suitability

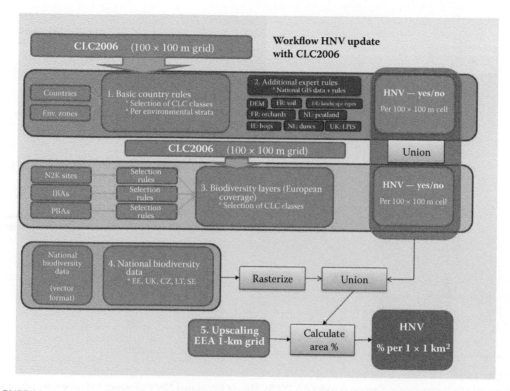

FIGURE 22.1
Workflow of the HNV farmland update procedure. (From ETC/SIA [European Topic Centre for Spatial Information and Analysis] 2012. Updated High Nature Value Farmland in Europe, An estimate of the distribution patterns on the basis of CLC2006 and biodiversity data [unpublished report].)

of policy measures (targeting) for supporting HNV farmland and farming systems. Figure 22.1 represents the work flow supporting such exercise.

According to Paracchini et al. (2008), the basic mapping steps (summarized in Figure 22.1) are

1. Selection of relevant CLC classes in the different environmental zones in Europe per country
2. Refinement of the draft LC map on the basis of additional expert rules (e.g., relating to altitude, soil quality) and country-specific information
3. Addition of the biodiversity data layers with European coverage
4. Addition of national biodiversity datasets
5. Upscaling to 1×1 km^2 INSPIRE grid

Figure 22.2, produced according to the aforementioned steps, shows the estimated HNV farmland presence in Europe for the reference year 2012 (using CLC2006).

Figure 22.3 shows the estimated share of HNV farmland at level 2 of the European Nomenclature of Territorial Units for Statistics (NUTS 2) level, based on a consistent methodology for all countries. To compare data holding the same characteristics, the estimated

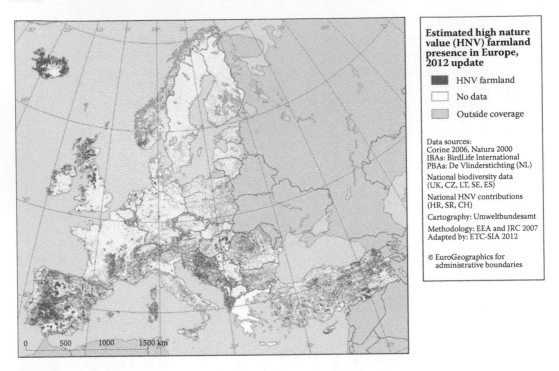

FIGURE 22.2
Estimated HNV presence in Europe. (From EEA, 2015b. Map of Estimated HNV presence in Europe. Available at: http://www.eea.europa.eu/data-and-maps/figures/estimated-high-nature-hnv-presence [published February 17, 2015].)

share of HNV farmland is calculated on the basis of total agricultural area as derived from CLC2006 agricultural classes plus identified HNV areas outside these classes.

Green areas show likelihood to contain primarily HNV land, on the basis of a stratified selection of CLC classes per country and environmental zone, and national biodiversity data when available. The values in the map are a proxy for the proportion of HNV in each 1-km^2 cell (reference year 2012, based on CLC2006).

The estimated share of HNV for each NUTS 2 area in the EU-27 was calculated according to the methodology described in ETC/SIA (2012). In this study data on the estimated HNV farmland are also available for the following (at the time) non EU-27 countries: Albania, Bosnia and Herzegovina, Croatia, Iceland, Kosovo, Liechtenstein, Macedonia FYR, Montenegro, Norway, Serbia, Switzerland, and Turkey.

The use of CLC data leads to certain data artefacts in some countries or regions, in spite of refined selection criteria and the inclusion of additional biodiversity data sets. Further refinements on the basis of national datasets would be advantageous in several regions. In general, this approach faces two crucial constraints as also indicated in Paracchini et al. (2008). The one is the uncertainty in the data on the distribution and extent of HNV farmland in different countries and the other issue is to find comparable data for agricultural land.

In the context of the monitoring and evaluation framework of Rural Development 2007–2013 Programmes, DG AGRI has issued guidelines for reporting on HNV farmland and forestry indicators (http://ec.europa.eu/agriculture/rurdev/eval/hnv/guidance_en.pdf), to support member states wishing to make use of a national definition for this indicator, and to develop the indicator further to include aspects of the HNV concept not covered so far.

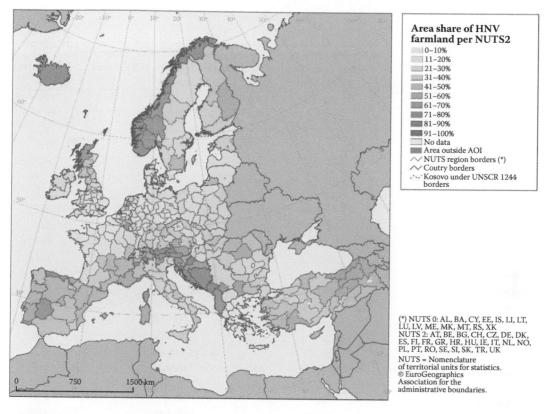

FIGURE 22.3
Share of HNV farmland per utilized agricultural area (UAA) per NUTS 2 area (in 10% gradation). (From ETC/ SIA, 2012. Updated High Nature Value Farmland in Europe, An estimate of the distribution patterns on the basis of CLC2006 and biodiversity data [unpublished report].)

22.4 HNV as a SEBI Indicator

The Streamlining European Biodiversity Indicators (SEBI) process was set up in response to a request from the EU Environment Council. Its aim was to streamline national, regional, and global indicators and, crucially, to develop a simple and workable set of indicators to measure progress and help reach the 2010 target. More information about the SEBI process and its role in monitoring progress toward the new 2020 biodiversity targets can be found in EEA (2012).

One of the indicators in the SEBI set is SEBI 020: Agriculture: area under management practices potentially supporting biodiversity. A key component of this indicator is information on the estimated distribution of HNV farmland in Europe (http://www.eea .europa.eu/data-and-maps/indicators/agriculture-area-under-management-practices /agriculture-area-under-management-practices-2).

Of Europe's farmland area, an estimated 37.7% share is HNV farmland (see Table 22.1). The highest estimated share of HNV farmland in the agricultural area (more than 60%) is observed in Austria, Norway and Slovenia as well as Albania, Bosnia and Herzegovina, Croatia, and Montenegro. In Cyprus, Finland, Portugal, Spain, Switzerland, and Turkey HNV farming systems represent between 41% and 60%. On the other hand, the lowest share

TABLE 22.1

HNV Farmland Share per Country Assessed in 2012

Country	HNV 2006 Farmland Area (million ha)	Agricultural Land (CLC 2006 Agricultural Classes) + HNV Areas (million ha)	Area Share of HNV (%)
Albania	1,231,102	1,531,030	80.4
Austria	2,140,879	3,340,014	64.1
Belgium	435,153	1,780,001	24.4
Bosnia and Herzegovina	2,056,975	2,209,639	93.1
Bulgaria	2,578,005	6,745,380	38.2
Croatia	2,955,012	3,285,969	89.9
Cyprus	343,209	629,220	54.5
Czech Republic	1,190,319	4,625,715	25.7
Denmark	191,262	3,433,650	5.6
Estonia	531,554	1,605,575	33.1
Finland	1,268,980	2,992,428	42.4
France	8,023,118	35,204,602	22.8
Germany	3,248,177	21,508,158	15.1
Hungary	1,935,454	6,768,833	28.6
Iceland	6,155,976	6,156,311	100
Ireland	1,154,495	5,729,074	20.2
Italy	6,196,451	18,393,993	33.7
Kosovo (under UNSCR1244/99)	497,705	610,960	81.5
Latvia	569,534	2,848,613	20
Liechtenstein	81	5706	1.4
Lithuania	640,277	4,011,830	16
Luxembourg	13,637	140,317	9.7
Macedonia, FYR	196,688	1,158,067	17
Malta	1034	15,666	6.6
Montenegro	465,414	469,682	99.1
Netherlands	390,551	2,570,614	15.2
Norway	5,966,735	6,601,527	90.4
Poland	4,488,811	19,750,026	22.7
Portugal	2,854,853	4,881,341	58.5
Romania	5,221,251	14,401,453	36.3
Serbia	1,003,818	4,867,569	20.6
Slovakia	479,205	2,413,272	19.9
Slovenia	570,551	754,220	75.6
Spain	18,820,501	33,698,696	55.8
Sweden	1,166,103	4,311,707	27
Switzerland	884,778	1,924,777	46
Turkey	19,810,869	42,973,114	46.1
United Kingdom	5,376,637	19,281,655	27.9
Total			
EEA-38	110,557,449	293,630,404	37.7
EU-27	69,830,001	221,836,053	31.5
EU-15	51,280,797	157,266,250	32.6
EU-N12	18,549,204	64,569,803	28.7

of HNV areas is estimated to be in Denmark, Germany, Latvia, Liechtenstein, Lithuania, Luxembourg, Macedonia FYR, Malta, the Netherlands, and Slovakia, where it is between 0 and 20% (see Table 22.1).

22.5 Steps Forward

Steps forward are to be considered at very divergent scales: on a very technical level, through the development of the knowledge base both at European and country/local level; and at an advocacy and practice level.

On the first, around the knowledge base it is relevant to improve the current HNV map with the new European datasets developed under the Copernicus program (see Chapter 1). As more and more data are being made available through this program, the spatial and thematic detail of the HNV farmland map could be enriched, wherever possible, using those harmonized and standardized European products.

In 2013, a study requested by EEA (ETC SIA, 2013) investigated whether the existing HNV farmland layer could be improved using High-Resolution Layers (HRLs) (see Chapter 9) and Landscape Linear Features (LLFs). The assessment showed that HNV farmland maps can be improved using higher geometrical details compared to the CLC mapping considering the limitation of the MMU of 25 ha (the HRLs are derived at 20 m pixel and then aggregated to 100 m).

Unfortunately, the grasslands layer does not distinguish between management intensity and covers intensively used grassland as well, not supporting the HNV farmland approach, which underlines the importance of including intensity measures for the classification of HNV farmlands. Exclusion of more intensively used grassland areas from the existing HNV farmland map would produce a better map, a topic still to be developed under the umbrella of the Copernicus program.

In the study referenced in the preceding text, the following conceptual ideas were applied:

Exclusion of areas from current HNV mask that are very unlikely to represent agriculture at all:

- HRL imperviousness
  - Areas >30% sealing
- HRL forest
  - Outside of CLC class 31x (forest), areas >75% forest
- HRL water
  - Outside the CLC class 5xx (water); areas >75% water

Inclusion of additional areas that are very likely to contain high nature value farmland:

- LLFs
  - High percentage of edges
  - Include within CLC class 211 those areas that are characterized by a very high landscape diversity
- HRL grassland
  - Include within CLC class 211 those areas that represent grassland

Further improvements are expected from the data layers being developed under the local component of Copernicus: Urban Atlas and Riparian Zones mapping represent an improvement geometrically and thematically speaking. Though, the usefulness of the thematic details of Urban Atlas for HNV mapping is quite limited, Riparian Zones is expected to provide valuable input for the future enhancement of HNV farmland.

Tracking changes overtime is another issue of key relevance for policy evaluation. It has been recognized that owing to the variation in data availability across the member states and regions of the EU and the range of physical situations (territory size, farm structure and systems, predominant land and habitat types), it is not appropriate to impose a common methodology for the assessment of HNV farming as a CMEF indicator. Use of one single method would restrict the analysis to data available throughout the EU, which would exclude the richest and most relevant data sources, and preclude those member states that have developed more refined methods from using them, with a consequent reduction in the quality and accuracy of the assessment. Therefore, a unique definition embracing all types of HNV farming areas across Europe is not possible, given the variation in HNV farming in member states and regions. Nor it is possible to derive an aggregate value for the EU of the extent in hectares of the HNV area.

On the contrary, agreement on the common parameters being measured, transparency and acceptance of the various methodologies, does allow for comparability and aggregation. This is possible provided that appropriate methodology to identify land fulfilling the criteria for one of the three HNV types (Hazeu et al., 2014) in that biophysical situation is used., that is type 1—Farmland with a high proportion of semi-natural vegetation; type 2—Farmland with a mosaic of low intensity agriculture and natural and structural elements, such as field margins, hedgerows, stone walls, patches of woodland or scrub, and so forth; and type 3—Farmland supporting rare species or a high proportion of European or world wildlife populations. It is, however, requested that in each territory the same methodology is used for successive assessments, so that trends can be correctly estimated.

22.6 General Considerations

From a practical perspective it is clear that member states and the EU should continue to recognize their HNV farming systems, analyze the socioeconomic issues they face, and identify the specific practices that provide the highest environmental and cohesion benefits. More targeted agri-environment schemes are needed to support HNV farming systems, as these will continue to play a vital role in ensuring that all farming systems can deliver environmental benefits.

Although the importance of HNV farming systems has been officially recognized in Europe since the 1990s, there is still registered a poor understanding and low awareness on the relevance of using this concept.

Difficult climatic conditions, poor land quality, and distance from markets make many of these areas difficult to farm, yet HNV farmers/crofters play a significant role in the process of producing high quality food alongside securing a range of public goods—storing vast amounts of carbon, supporting clean water supplies, holding water to prevent flooding downstream, providing areas for recreation, protecting cultural and historic heritage, and supporting rural communities and economies. Despite all this, many receive inadequate recognition and often go unrewarded at market leaving many farmers struggling to survive.

Although large amounts of public money are spent on the CAP, most of this is not targeted to the farms that deliver the most for society. Existing support mechanisms such as agri-environment schemes have helped to slow the loss of these systems, but alone are insufficient to make HNV farms commercially viable.

Market forces and social pressures are leaving farmers with a stark choice between intensifying, abandoning parts of their farms and in some cases ceasing to farm altogether, all of which will have disastrous consequences for wildlife dependent on these systems and the survival of these rural communities. In situ and proactive working with farming communities and projects to advice at the local level is needed to improve economic sustainability, while maintaining their environmental and social benefits and cohesion as the socioeconomic causes and environmental effects of abandonment should be researched and assessed as part of the evaluation of both the 2020 EU Biodiversity Strategy and the 7th Environmental Action Programme (EAP).

Coordinating the voice of HNV farming and demonstrating the relevance of its results to support these vulnerable farming systems to ensure farming is a viable economic option for future generations, keeping people on the land to maintain a vibrant rural community and protecting our special wildlife and landscapes, is an important element in the future of Europe in the context of global food security, tackling aspects of soil performance, nutrient recycling, and biodiversity. In turn it will reinforce the role of nature in the transition to a Green Economy in the context of sustainable development and poverty eradication. As human and societal well-being depends on nature and as all sectors of the economy benefit directly or indirectly from nature, making sure there is a clear understanding of its value as natural capital is of key importance. HNV as understood in Europe is a case study and a practical example of how nature and low-intensity farming are key for a greener, safer, and more resilient future.

References

Andersen, E. (Ed.). 2003. Developing a high nature value farming area indicator. Internal report. Copenhagen: European Environment Agency.

Baldock, D., Beaufoy, G. 1994. The nature of farming, low intensity farming systems in nine European countries. London: Institute for European Environmental Policy.

Baldock, D., Beaufoy, G., Benne, G., Clark, J. 1993. Nature conservation and new directions in the common agricultural policy. London: Institute for European Environmental Policy.

Bignal, E. M., McCracken, D. I. 1996a. The ecological resources of European farmland. In M. Whitby (Ed.), *The European environment and CAP reform: Policies and proposals for conservation* (pp. 26–42). Wallingford, UK: Centre for Agriculture and Biosciences International.

Bignal, E. M., McCracken, D. I. 1996b. Low-intensity farming systems in the conservation of the countryside. *Journal of Applied Ecology* 33:413–424.

Bignal, E. M., McCracken, D. I. 2000. The nature conservation value of European traditional farming systems. *Environmental Reviews* 8:149–171.

Cooper, T., Arblaster, K., Baldock, D., Farmer, M., Beaufoy, G., Jones, G., Poux, X., McCracken, D., Bignal, E., Elbersen, B., Wascher, D. et al. 2007. Final report for the study on HNV indicators for evaluation. London: Institute for European Environmental Policy.

Council Regulation (EC) No. 1259/1999. 1999. Official Journal of the European Communities, L 160.

Council Regulation (EC) No. 73/2009 establishing common rules for direct support schemes for farmers amending Regulations (EC) No. 1290/2005, (EC) No. 247/2006, (EC) No. 378/2007 and repealing Regulation (EC) No. 1782/2003, OJ L 30/16,31.01.2009. Available at: http://eur-lex .europa.eu/legal-content/EN/ALL/?uri=CELEX:32009R0073.

Directorate-General of Agriculture and Rural Development. 2006, *Handbook on Common Monitoring and Evaluation Framework* (guidance document). Available at: http://ec.europa.eu/agriculture /rurdev/eval/guidance/document_en.pdf.

EEA (European Environment Agency). 2004. High nature value farmland—Characteristics, trends and policy challenges. Copenhagen: European Environment Agency. Available at: http:// www.eea.europa.eu/publications/report_2004_1.

EEA (European Environment Agency). 2012. Streamlining European biodiversity indicators 2020: Building a future on lessons learnt from the SEBI 2010 process. Available at: http://www.eea .europa.eu/publications/streamlining-european-biodiversity-indicators-2020.

EEA (European Environment Agency). 2015a. State and outlook 2015.European briefing on agriculture: Available at: http://www.eea.europa.eu/soer-2015/europe/agriculture.

EEA (European Environment Agency). 2015b. Map of Estimated High Nature (HNV) presence in Europe. Available at: http://www.eea.europa.eu/data-and-maps/figures/estimated-high-nature -hnv-presence (published February 17, 2015).

ETC/SIA (European Topic Centre for Spatial Information and Analysis). 2012. Updated High Nature Value Farmland in Europe: An estimate of the distribution patterns on the basis of CORINE Land Cover 2006 and biodiversity data (unpublished report).

ETC/SIA (European Topic Centre for Spatial Information and Analysis). 2013. Agri-environmental indicator and assessment: Improvement of the HNV farmland indicator based on GIO HRL data and synergies with SLF products (unpublished report).

Evaluation Expert Network. 2009. The application of the High Nature Value impact indicator. Brussels: European Commission.

Evaluation Expert Network. 2010. Approaches for assessing the impacts of the Rural Development Programmes in the context of multiple intervening factors (working paper). Available at: http:// enrd.ec.europa.eu/enrd-static/fms/pdf/EB43A527-C292-F36C-FC51-9EA5B47CEDAE.pdf.

Hazeu, G., Milenov, P., Pedroli, B., Samoungi, V., Van Eupen, M., Vassilev, V. 2014. High Nature Value farmland identification from satellite imagery: A comparison of two methodological approaches. *International Journal of Applied Earth Observation and Geoinformation* 30:98–112.

Metzger, M. J., Bunce, R. G. H., Jongman, R. H. G., Mucher, C. A., Watkins, J. W. 2005. A climatic stratification of the environment of Europe. *Global Ecological. Biogeography* 14:549–563.

Parachini, M. L., Terres, J. M., Petersen, J. E., Hoogeveen, Y. 2006. Background document on the Methodology for Mapping High Nature Value Farmland in EU27. Internal Report EEA/JRC. Copenhagen: European Environment Agency.

Paracchini, M. L., Peterson, J.-E., Hoogeveen, Y., Bamps, C., Burfield, I., van Swaay, C. 2008. High Nature Value Farmland in Europe. EEA and JRC: Available at: http://agrienv.jrc.it/publications /pdfs/HNV_Final_Report.pdf.

23

CLC for National Accounting:
Land and Ecosystem Natural Capital Accounts

Jean-Louis Weber and Gabriel Jaffrain

CONTENTS

23.1 Introduction

CoORdination of Information on the Environment (CORINE) Land Cover (CLC) data can be used in many domains, principally for assessing the environment and its relation to human activities. The development of CLC has indeed made possible the development of ecosystem accounts. The first update of CLC has been followed by the publication by the European Environment Agency (EEA) of land accounts for Europe 1990–2000 (Haines-Young, 2006) and the launch of a project of simplified ecosystem capital accounts in 2010. In 2013, the United Nations (jointly with the European Commission, the Organisation for Economic Co-operation and Development, and the World Bank) edited a manual called *System of Environmental-Economic Accounting 2012, Experimental Ecosystem Accounting* that states that "For ecosystem accounting purposes, the economic territory is disaggregated into spatial units [...]. These spatial units form a focus for ecosystem accounting performing a similar role to economic units (such as enterprises, households and governments) in national accounting" (SEEA-EEA, 2013, paragraph 1.27). The technical guidelines published in 2014 by the Secretariat of the Convention on Biological Diversity put the SEEA-EEA broad recommendations to work and achieve their target on "Integration of Biodiversity Values in National Accounting Systems." The guidelines provide detailed references to the use of land cover (LC) data and contain a full chapter on LC accounting (Weber, 2014a).

Associating CLC and the National Accounts may appear very novel and even a strange idea as long as the two realms seem so different in terms of representations, methods, and purpose: maps and geographical objects on the one hand and statistical units defined from legal properties and economic functions on the other hand; monitoring, image processing, and cartography versus collection of statistics via surveys and administrative data; and analysis of locations and spatial patterns for land-based policies versus aggregation of economic flows for macroeconomic policies.

However, from an historical perspective, one of the origins of CLC is the attempt to counterbalance the domination of national accounts based on financial transactions with accounts of the natural resource measured in physical units.

23.2 LC Mapping and the Development of Ecosystem Accounting: A Historical Reminder

The first framework of environmental accounts based on spatial entities was presented in the French Natural Patrimony Accounts (CICPN, 1986). This methodology stated that the material balances of assets used by the economy such as oil, coal, ores, timber, or water should be supplemented for the living resource by accounts of the ecosystems that reproduce them. At this stage, the objective was to sketch a statistical and accounting framework for the ecosystems. Because of their extreme diversity and complexity, starting from their component and/or functions and services was not considered as a realistic option. Monetary valuation, often proposed as a way of aggregating heterogeneous things, was not considered as relevant here as long as the purpose was an independent measurement of the impacts of market-driven activities on natural systems. To define statistical units for ecosystem accounting, it was decided to use the LC maps that started to be produced in the late 1970s.

In the early analyses of the newly sensed satellite images, two pioneer experts working for the French Ministry of Environment, Michel Lenco and Yves Heymann, came to the conclusion that the most robust and reliable methodology to classify satellite images (in early years Landsat MSS) was visual photointerpretation (Lenco, 1979). The outcome of the process was composed of mapping units called at that time "ecozones" to which were attributed characteristics related to biophysical LC. This LC image, which at the same time reflected aspects of ecosystems and land use (LU), appeared finally as an acceptable representation of land ecosystems, the best available at that time—and still the best today. It has inspired the framework of ecozones (the name given to the ecosystem accounting units) accounts in the Natural Patrimony Accounts (Figure 23.1).

During the same period, the European Commission's DG XI (later renamed Directorate General of Environment [DG ENV]) was implementing the first European geographical information system (CORINE*). It was decided that the visual photointerpretation methodology developed and standardized for the Natural Patrimony

---

* COoRdination de l'INformation sur l'Environnement; the officers in charge of CORINE at DG XI were Günther Schneider and Michel Cornaert.

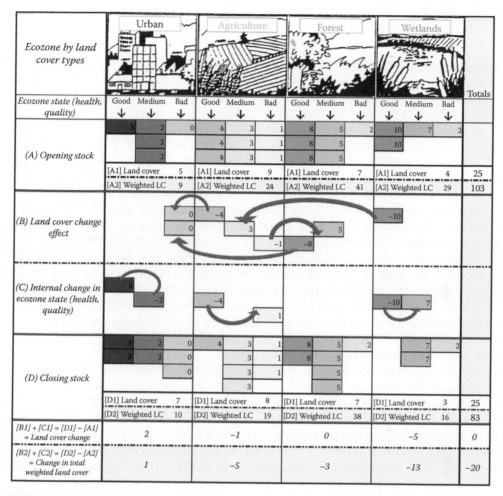

FIGURE 23.1
Illustration of ecozones accounts based on LC. (From Weber, adapted from CICPN [Commission Interministérielle des Comptes du Patrimoine Naturel]. 1986. *Les Comptes du Patrimoine Naturel*. Collections de l'INSEE, Série C 137–138. Paris: Institut National de la Statistique et des Etudes Economiques.)

Accounts in France would be tested in 10 European countries. It was done in 1985 under the leadership of Yves Heymann. As tests results proved to be positive, CLC was launched.

Land and ecosystem accounting was developed further during the period between 1986 and the early 2000s. This development took place in the context of a project of the UN Economic Commission for Europe (UNECE) that continued later on with the support of Eurostat (Parker, 1996). It is not a surprise then that ecosystem accounts have been developed on the basis of CLC (Europe) or equivalent LC data. One consequence of this dependence of ecosystem accounting in Europe on CLC is that the methodology could not be implemented before CLC's regular production and update. The first applications of Land and Ecosystem Accounts (LEAC) carried out in 2003 by the EEA with the support of Eurostat were made possible by production of LC 1975 and 1990 data carried out in four Central and Eastern European countries (Soukup, 2003) on the one hand and for a coastal strip of 10 km on the other hand (Weber, 2003).

23.3 Land Ecosystem Accounting: A Follow-Up in Space and Time

To a large extent, the subsequent development of ecosystem accounts was made possible when CLC1990 was updated to 2000. The outcome was the publication of the EEA report on *"Land accounts for Europe 1990–2000: Towards integrated land and ecosystem accounting"* (Haines-Young, 2006).

Following this first step, the EEA started with the process of producing simplified Ecosystem Capital Accounts (ECA) for the EU member states. The policy context at that time was that of the Millennium Ecosystem Assessment of 2005, the TEEB (The Economics of Ecosystems and Biodiversity) study,[*] and the "Beyond Gross Domestic Products (GDP)" process launched by the European Commission end of 2007.[†] TEEB was initiated by Germany and the European Commission in response to a proposal by the G8+5 Environment Ministers in Potsdam, 2007, to develop a global study on the economics of biodiversity loss.

On the statistical side, the United Nations Committee of experts on Economic-Environmental Accounting (UNCEEA), which steered from 2006 the revision of the SEEA 2003, accepted the idea of a special volume on ecosystem accounting (SEEA-EEA, 2013). This was approved by the UN Statistical Commission on an "experimental" basis. Taking stock of the experience gained in Europe as well as through its active participation into TEEB and the SEEA revision, the EEA published in 2011 "An experimental framework for ecosystem capital accounting in Europe" (Weber, 2011) for the requirements of its own activity as well as an input to the SEEA revision.

In 2013, the Secretariat of the Convention on Biological Diversity (CBD) commissioned the writing of the practical guidelines for "Ecosystem Natural Capital Accounting: A Quick Start Package" (ENCA-QSP) (Weber, 2014a) in view of stimulating their implementation considering the CBD target of integrating by 2020 ecosystem and biodiversity values into the mainstream of decision making and policies, including into national accounts. The novel aspect of these developments is that spatial information in general and LC data for terrestrial ecosystems have been recognized as a fundamental component of ecosystem assessment and accounting.

At present, ECA (or SECA) are the acronyms commonly used for the project of (Simplified) Ecosystem Capital Accounts implemented at the EEA. ENCA (for Ecosystem Natural Capital Accounts) is the acronym used for the 2013 Mauritius case study as well as for the CBD report of 2014 (Weber, 2014b). There are no major conceptual differences between ECA, SECA, and ENCA. The differences are in the terms of focus. ECA (SECA) is a EU-wide program making use of the data available in Europe at the EEA, Eurostat, and the Joint Research Centre (JRC) of the European Commission (EC). ENCA has a global focus and needs being adapted to various available data sources. LEAC, the EEA land and ecosystem accounting framework of 2006, is the base for ECA (SECA) implementation. Although the acronym is not used, the LEAC methodology transposed to the international context is at the core of ENCA as well. In the CBD report, the adjunction of QSP (Quick Start Package) to ENCA indicates that the aim is to contribute to a quick implementation (by 2020), which means that priorities have been set: first core physical accounts and

---

[*] TEEB is now hosted by the United Nations Environment Programme (UNEP).

[†] The Beyond GDP initiative of the European Commission aimed at supplementing the broadly used macroeconomic aggregate with other measurements in order to correct well known biases. BGDP was launched at the end of 2007 in a high-level conference convened by the European Commission, European Parliament, Club of Rome, OECD, and WWF.

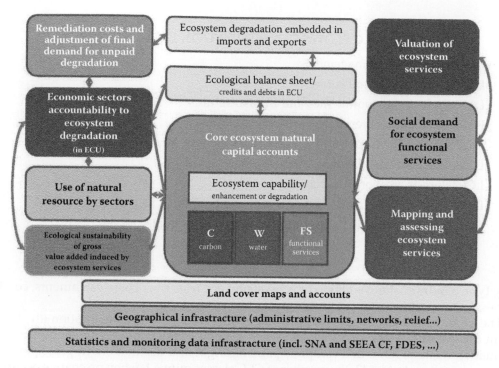

FIGURE 23.2

Structure of ecosystem natural capital accounts. (From Weber, J.-L. 2014a. *Ecosystem Natural Capital Accounts: A Quick Start Package*. Technical Series 77. Montreal: Secretariat of the Convention on Biological Diversity, Montreal. Available at: http://www.cbd.int/doc/publications/cbd-ts-77-en.pdf.)

second, once the base is in place, more specific ecosystem services assessments and monetary valuation (Figure 23.2).

23.4 LEAC: The European Framework

In early 2000s the EEA started the implementation of a LEAC program. The purpose is to integrate information across the various ecosystem components and to support further assessments and modeling of these components and their interactions with economic and social developments. This program reflects the increasing demand for environmental policy integration in Europe, both vertically through thematic policies as well as horizontally across policies in those sectors that contribute most to environmental impacts.

The construction of land and ecosystem accounts has been made feasible as a result of continuous improvements in monitoring, collecting, and processing data. Progress with the development of statistical methods that facilitate data assimilation and integration is also part of this process. In Europe, CLC is at the core of the program. The accounts produced in Burkina Faso in 2006 rely also on a LC database, that is, the so-called BDOT (Base de Données d'Occupation des Terres du Burkina Faso; see Chapter 25; Jaffrain et al., 2007), produced and organized in a way similar to CLC. Comprehensive LC accounts are based on explicit spatial patterns and can be scaled up and down using a grid 1 × 1 km to

BOX 23.1   LC FLOWS CLASSIFICATION USED FOR LEAC

LCF1 Urban land management

LCF2 Urban residential sprawl

LCF3 Sprawl of economic sites and infrastructures

LCF4 Agriculture internal conversions

LCF5 Conversion from forested and natural land to agriculture

LCF6 Withdrawal of farming

LCF7 Forests creation and management

LCF8 Water bodies creation and management

LCF9 Changes of Land Cover due to natural and multiple causes

any type of administrative region or ecosystem zone (e.g., river basin catchments, coastal zones, or biogeographic areas).

The LEAC methodology structures the accounts of change into consumption (the losses from the initial LC) and formation (the gains leading to the final LC). Consumption and formation relate to flows that comprise pairs of origin and destination. In principle, CLC changes count up to 44 × 43 (the number of CLC classes minus 1, when no change occurs) = 1982. If change is computed from the aggregated level, of course the number of possibilities shrinks sharply but the consequence is that many of the actual changes drop into the "no change" category because only changes between aggregated classes are recorded then. The LEAC flows methodology allows summarizing the total observed changes without eliminating any one of them. It means that flows will reflect changes that are internal to broad categories as well as changes between them. A clear example of flows aggregated classification is *LCF4 Agriculture internal conversions*. There are many others of these internal flows that are listed in the detailed LCF Levels 2 and 3 used in the European LEAC (Haines-Young, 2006) (Box 23.1).

The LEAC accounts for Europe are accessible on the EEA website using the *Land accounts data viewer 1990, 2000, 2006** (Figure 23.3). The viewer will be updated for year 2012 with the CLC update, which was finalized in 2015. This viewer allows visualization and download of LC statistics derived from land accounts applied methodology.

On the internal scene, the CLC and the LEAC methodology have been an important input to the development of ecosystem accounting. The discussions that took place in the course of the SEEA revision resulted in the better understanding of the relations between CLC and the Food and Agriculture Organization of the United Nations (FAO) Land Cover Classification System. For some, a basic issue seemed to relate to the idea of "pure" LC classes as opposed to the impure or "mixed" classes of CLC and its LU character. The clarification carried out by FAO and EEA highlighted the confusion between elementary or "pure" physical objects that are at the base of a LC classification (e.g., grass, shrubs, trees, mineral objects, ice, water) and the LC classes themselves, which depict the way these objects are distributed over space. Establishing an LC classification is defining rules

---

* EEA's Land accounts data viewer 1990, 2000, 2006. http://www.eea.europa.eu/data-and-maps/data/data-viewers /land-accounts.

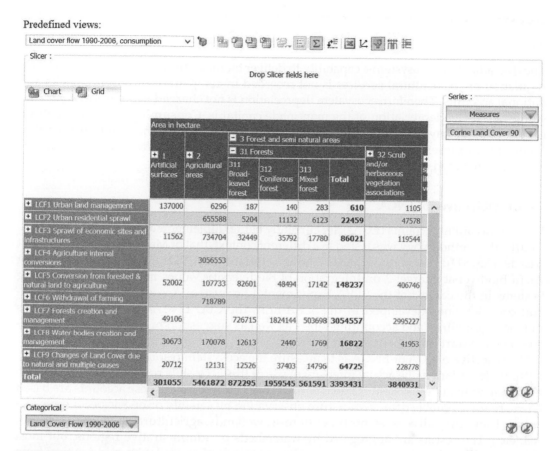

FIGURE 23.3
The LEAC viewer. (From EEA, 2014. http://www.eea.europa.eu/data-and-maps/data/data-viewers/land-accounts.)

to combine these pure objects to obtain a correct characterization of the way they appear in the real world, with their spatial combinations and patterns. This may include characteristics influenced by LU, such as in the case of the distinction between managed pastures and natural grassland, which is relevant in many regions. It was therefore possible to define LC ecosystem units (LCEUs) as one of the essential categories for accounting and to characterize them in terms of the ISO Land Cover Meta Language (LCML) promoted by FAO. The LCEU classification is much aggregated, made of 15 classes only. More detailed LC classification is left to the countries, considering their specific landscapes and priorities. For this purpose they can use the LCCS3 software package developed by FAO to define detailed classes in correct LCML terminology. For ecosystem natural capital accounting, however, it should be noted that the boundaries of land are broader than those commonly addressed by FAO LCCS and EEA CLC, which limit LC mapping to the shoreline. Instead, the LCEU classification includes land below the sea coastal water as long as its cover can be mapped in biophysical terms: algae and sea grass beds and coral reefs.

Based on the LCEU classification and the LEAC approach, an aggregated classification of LC flows is proposed in the CBD *Ecosystem Natural Capital Accounts—A Quick Start Package* manual (Weber, 2014). Here again, the idea is to start with a very limited number of items and to develop additional levels according to local conditions.

23.5 LC in Ecosystem Natural Capital Accounting

The degradation of ecosystems' capability to deliver biomass, freshwater, and natural cycle regulation or sociocultural services is not recorded in companies' accounting books and national accounts. Therefore ecosystem depreciation is not charged in the price of our consumption, while depreciation of the man-made capital is. Consuming ecosystem natural capital without paying is equivalent to creating ecological debts that are transmitted to others, to our present and future generations, or to those countries from which we import products produced under unsustainable conditions.

23.5.1 Objectives of ENCA

Ecosystem capital accounts (ECAs) are currently being implemented in Europe by the EEA. In 2013, the methodology was tested in a different context, that of Mauritius (Weber, 2014). It was developed further at the request of the Secretariat of the CBD in view of "Implementing Aichi Biodiversity Target 2 on Integration of Biodiversity Values in National Accounting Systems in the context of the SEEA Experimental Ecosystem Accounts." Beyond the general considerations developed in the SEEA-EEA, the CBD ENCA-QSP manual presents practical guidelines and accounting tables as well as suggestion of data sources for countries keen to start implementation now.

The objective of ECA or ENCA is to measure the ecosystem resources that are accessible without degradation, the actual intensity of use of this accessible resource, and the change in the capability of ecosystems to deliver their services over time. These accounts are based on currently available data from nature observation by satellite or on socioeconomic statistics. They cover all ecosystems types (forests, wetlands, agricultural and urban systems, sea, etc.). The results are aggregated by watersheds or administrative regions, but most data are collected or disaggregated according to standard assimilation grid ranging from 100×100 m to 1×1 km according to the dimension of the country or the study area or the appropriate scale for ecosystem assessment. This geographic data management is required to analyze short-term degradation of different ecosystems and to articulate programs of national, regional, or local initiative with broader assessments such as those made at the European level.

ENCA are established as two blocks of accounts: integrated core accounts of the basic resources of all ecosystems (ecosystem carbon, ecosystem water, and ecosystem infrastructure functional services) on the one hand and on the other hand functional analysis of restoration costs and of key ecosystem services, addressed one by one in physical terms and valued in money (Figure 23.2).

23.5.2 From Mapping LC to Accounting for Ecosystem Degradation and Ecological Debts

In physical ecosystem natural capital accounts, in a first step measurements are made in basic units (tons, joules, m³, or ha) and then converted to a special composite currency named Ecosystem Capability Unit (ECU) used to quantify ecosystem degradation or enhancement. The price of one physical unit (e.g., 1 ton of biomass) in ECU expresses at the same time the intensity of use of the resource in terms of maximum sustainable yield and the direct and indirect impacts on ecosystem condition (e.g., contamination or biodiversity loss). While money measures the economic value of ecosystem resources, ECU measures their ecological value. ECU plays for ecosystems and biodiversity the same role as the CO_2-equivalent used to measure the contribution of economic sectors to global warming (Figure 23.4).

Because the costs of using nature have not been paid, loss of ecosystem capability in ECU (the measurement of ecosystem capital degradation) is a measurement of an ecological debt. It is a debt to present and future generations as they lose (or will lose) benefits connected with the ecosystem that are considered a public good. In addition to negative impacts on the national territory, the economy can induce ecosystem degradation in the rest of the world either by exporting pollution or by importing commodities produced under unsustainable conditions. In the latter case, ecosystem degradation is embodied within international trade. A country's ecological balance sheet would therefore record both internal and external debts. Ecological debt in ECU (and symmetrically credits when improvements are verified) could be incorporated into portfolios of financial instruments. Physical degradation or ecological debts can be converted in a second step into money on the basis of the costs necessary to restore ecosystem capability. Various mechanisms could be envisaged to recover ecological debts: ecological taxes, insurance systems against ecological risks, offset credit–debit systems (again, as for the Kyoto Protocol) or broader compensation mechanisms based on ECU.

23.5.3 Specific Role of LC Data in Ecosystem Natural Capital Accounting

LC is an essential input to ecosystem accounting. In the CBD ENCA-QSP manual, LC appears in mostly all chapters under two aspects: as a way to characterize the units for which accounts are produced and frequently as an input for calculating or downscaling various variables. LC is one of the elements used for defining the ecosystem accounting units. A first category is constituted of the usual LC mapping units, the so-called LC ecosystem units. LCEUs are closely correlated to the provision of ecosystem-based goods such as food, timber, or other fiber products. Other ecosystem units reflect the complexity of interactions. They are generally called socio-ecosystems and defined primarily according to their functions. For accounting theoretical socio-ecosystems must be defined as entities to which variables can be attributed; they need to have clear borders similarly to the institutional units used for national accounting (see Section 23.1). These units, called socio-ecological landscape units (SELUs), are produced by integrating the image of biophysical cover and LU with physical entities such as river basins or classes of altitude. CLC is particularly fit for supporting this construction, and the analytical methodologies presented in the EEA LEAC report of 2006 (smoothing of LC classes and creation of dominant LC types) can be used for this modeling.

LC is also frequently used in ENCA to calibrate or downscale datasets to feed accounts by SELUs (e.g., in view of establishing a spatially explicit biocarbon balance). Other typical uses are calculating biomass productivity or evapotranspiration; downscaling agriculture statistics of crops; or producing indicators of landscape greenness, integrity, or potential.

These major contribution of CLC to ecosystem natural capital accounting is due to its capacity at describing changes for large areas (more than 30 European countries) for a long period of time (from 1990 to 2012, and in some cases, even from 1975). This is a unique and invaluable property of CLC that makes it possible to develop a long-term vision (required for ecosystem assessments) in a world dominated by the short term. Based on this CLC experience of monitoring change, the CBD ENCA-QSP report devotes a full chapter to LC accounts where LC classification methodologies are assessed

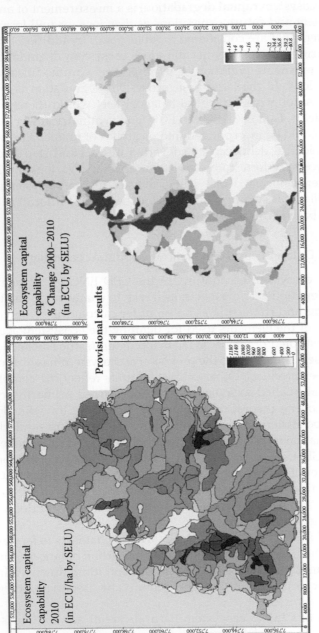

FIGURE 23.4

Illustration of ecosystem accounts in ECU—Experimental results from the Mauritius case study. (From Weber, J.-L. 2014b. *Experimental Ecosystems Natural Capital Accounts Mauritius Case Study, Methodology and preliminary results 2000–2010.* Indian Ocean Commission. Available at: http://commissionoceanindien.org/fileadmin /resources/ISLANDSpdf/Experimental_Ecosystems_Natural_Capital_Accounts_Mauritius.pdf.)

regarding their appropriateness to be used for accounting, starting with the basic LC consumption and formation.

ENCA-QSP is not the only project in which CLC (or similar LC data) plays a crucial role. A few years ago, the JRC of the EC undertook a project of "Mapping and Assessment of Ecosystem Services" (Maes, 2014) in response to EC Directorate General of Environment demand for support in the implementation of the EU Biodiversity Strategy to 2020. The principle of MAES is to assess services one by one and to map and quantify them in reference to the CLC classes likely to deliver them. In a few cases, higher resolution data are used when available at the JRC but in most cases, CLC is the starting point. As long as MAES (developed by the JRC) and ECA (developed by the EEA) have this common CLC reference a synergy will take place in which assessment of some specific services will be used as input in the ECA diagnosis while the latter will provide useful information on the state of the ecosystems delivering a certain service, that is, on their sustainability.

23.6 Conclusion

The goal of ecosystem natural capital accounting is to quantify the responsibility toward the goods state of the ecosystem (assessed in quantity and quality) of market economies and their actors and sectors that are targeted at the maximization of private profits. Ecosystems are at the same time resources privately exploitable for profit (economic assets) and a public good from which nobody, present and future generations altogether, should be excluded. Their maintenance requires that economic actors as well as governments and citizens are given the tools for assessing the trade-off between the benefits provided by the economy and all the other values that we are enjoying from the market system.

Ecosystem natural capital accounting architecture has been made possible by the fast development of monitoring systems, in situ and spatial, and the capacity of data management systems, in particular Geographical Information System (GIS). This development has been essential for the scientific communities involved, as much as for people who are now able to access in one click or touch of their smartphone to information and its localization on a map or a satellite image. In this context the role of land and ecosystem accounting is not to mimic what already exists but to summarize huge amounts of micro data into a small number of reliable and verifiable aggregated indicators likely to be used in decision making on par with well-established economic variables. In the economic realm, market transactions agglomerate individual preferences, and finally, statisticians just need to add and subtract to produce national accounts. In the ecological realm, this agglomeration has to be done carefully, respecting space and time scales, the complexity of the things, and the variability of conditions and issues. It could not be possible without referents, of which LC, the "skin of ecosystems,"* is a key one. The implementation of CLC has been a decisive milestone in the assessment of the liability of the economy to the natural systems that not only provide useful services but also make life possible.

---

* A nice expression by Pierre Calame in his *Essai sur l'Œconomie*, Editions Charles-Leopold Mayer, Paris 2009.

References

CICPN (Commission Interministérielle des Comptes du Patrimoine Naturel). 1986. *Les Comptes du Patrimoine Naturel*. Collections de l'INSEE, Série C 137–138. Paris: Institut National de la Statistique et des Etudes Economiques. Available at: http://projects.eionet.europa.eu/leac/library/background-papers-and-publications/comptes_patrimoine_naturel_insee_1986.

Haines-Young, R., Weber, J.-L. 2006. *Land accounts for Europe 1990–2000. Towards integrated land and ecosystem accounting*. EEA Report 11. Available at: http://www.eea.europa.eu/publications/eea_report_2006_11.

Jaffrain, G., Diallo, A., Rolland, N. 2007. *Deuxième Programme National de Gestion des Terroirs- Burkina Faso,/Base de données d'occupation des terres (BDOT): Evolution de l'occupation des terres entre 1992 et 2002 au Burkina Faso*. Available at: http://www.fidafrique.net/IMG/pdf/BDOT_Analyse _Comptes_Langage_accessible__Janvier_2007_-2.pdf.

Lenco, M. 1979. La télédétection: Une nouvelle source d'information pour l'environnement. *Journal de la Société de Statistique de Paris* 120(1):32–50. Available at: http://www.numdam.org/item?id =JSFS_1979__120_1_32_0.

Maes, J., Teller, A., Erhard, M., Murphy, P., Paracchini, M. L., Barredo, J. I., Grizzetti, B., Cardoso, A., Somma, F., Petersen, J.-E., Meiner, A. et al. 2014. *Mapping and Assessment of Ecosystems and their Services (MAES), Indicators for ecosystem assessments under Action 5 of the EU Biodiversity Strategy to 2020*. Final Report. Available at: http://ec.europa.eu/environment/nature/knowledge/eco system_assessment/pdf/2ndMAESWorkingPaper.pdf.

Parker, J., Steurer, A., Uhel, R., Weber J.-L. 1996. A general model for land cover and land use accounting. Invited paper drafted from the report of the UN-ECE Task Force on Physical Environmental Accounting—Special Conference on "Environmental Accounting in Theory and Practice" in Tokyo. Available at: http://www.ecosystemaccounting.net.

Soukup, T., Kupkova, L., Weber, J.-L., Paramo, F. 2003. Integration of geographical and statistical data in the environmental accounting framework; methodological development based on two case studies: Action 1: Accounts of the impacts on Forest and Biodiversity of Land Cover/Land Use changes; case from the land cover changes 1975–90 in the 4 Central and Eastern European countries. Prague: Report of the European Topic Centre on Terrestrial Environment for Eurostat and the EEA. Available at: http://projects.eionet.europa.eu/leac/library/reportsposters /reports_notes/leacforestfinalreportjun.

System of Environmental-Economic Accounting. 2013. *Experimental Ecosystem Accounting* (SEEA-EEA). White cover publication, preedited text subject to official editing. New York: European Commission, Organisation for Economic Co-operation and Development, United Nations, World Bank. Available at: http://unstats.un.org/unsd/envaccounting/eea_white_cover.pdf.

Weber, J.-L., Paramo, F., Breton, F., Haines-Young, R. 2003. Integration of geographical and statistical data in the environmental accounting framework; methodological development based on two case studies: Action 2: Integration of environmental accounts in coastal zones; case study of tourism. Barcelona Bellaterra: Report of the European Topic Centre on Terrestrial Environment for Eurostat and the EEA. Available at: http://projects.eionet.europa.eu/leac/library /reportsposters/reports_notes/leacfinalreport.

Weber, J.-L. 2011. *An experimental framework for ecosystem capital accounting in Europe*. EEA Technical Report 13. Available at: http://www.eea.europa.eu/publications/an-experimental-framework -for-ecosystem.

Weber, J.-L. 2014a. *Ecosystem Natural Capital Accounts: A Quick Start Package*. Technical Series 77. Montreal: Secretariat of the Convention on Biological Diversity, Montreal. Available at: http:// www.cbd.int/doc/publications/cbd-ts-77-en.pdf.

Weber, J.-L. 2014b. *Experimental Ecosystems Natural Capital Accounts Mauritius Case Study, Methodology and preliminary results 2000–2010*. Indian Ocean Commission. Available at: http://commission oceanindien.org/fileadmin/resources/ISLANDSpdf/Experimental_Ecosystems_Natural _Capital_Accounts_Mauritius.pdf.

24

Land Use and Scenario Modeling for Integrated Sustainability Assessment

Carlo Lavalle, Filipe Batista e Silva, Claudia Baranzelli, Chris Jacobs-Crisioni,
Ine Vandecasteele, Ana Luisa Barbosa, Joachim Maes, Grazia Zulian,
Carolina Perpiña Castillo, Ricardo Barranco, and Sara Vallecillo

CONTENTS

24.1 Introduction: Integrated Impact Assessment and the LUISA Concept

Land change models are a key means for understanding how humans are reshaping the Earth's surface in the past and present, for forecasting future landscape conditions, and for developing policies to manage our use of resources and the environment at scales ranging from an individual parcel of land in a city to vast expanses of forests around the world.

National Research Council (2013)

Beyond a direct impact on land use (LU) patterns, European Commission (EC) policies often have substantial impacts on various domains. To uphold and improve the quality of legislation, a thorough ex ante impact assessment of EC policies should therefore evaluate policy proposals on all three domains of sustainability: economy, environment, and society (Organisation for Economic Co-operation and Development [OECD], 2010). However, straightforward impact assessment of EC policies is often not possible because various policy domains are inherently intertwined. For example, a trade policy can have a direct impact on the agricultural sector, and new transport infrastructures can influence economic growth, while both policies, together and individually, may affect LU. To make the impact assessment challenge even greater, various policies may either propagate or compensate impacts, thus causing nonlinear interactions. Such complex dynamics can be grasped only by an elaborate impact assessment effort. In 2002 the EC introduced an Impact Assessment (IA) procedure to provide "evidence for political decision-makers on the advantages and disadvantages of possible policy options by assessing their potential impacts" (EC, 2002a). This procedure has to be applied to all Commission initiatives that aim to quantify economic, social, and environmental impact. More recently, the EC has also published guidelines for Territorial Impact Assessment, which should be carried out when policies target specific territories, or when policies may produce uneven impacts throughout the territory (EC, 2013a).

The EC's Joint Research Centre (JRC) employs LU modeling to execute territorial impact assessments following EC guidelines. Early LU modeling activities at the JRC focused mainly on case studies at the regional and urban scale, using a cellular automata model and detailed LU data produced in the context of the MOnitoring LANd use/cover Dynamics (MOLAND) project (Lavalle et al., 2002; Barredo et al., 2003; Engelen et al., 2007). One of the aims of the MOLAND project was to "provide a spatial planning tool for monitoring, modeling and assessing the development of urban and regional systems" (Engelen et al., 2007). The supporting LU data of the MOLAND project were characterized by remarkably high thematic, spatial, and temporal resolutions, but covering only a very limited sample of urban areas across Europe. With increasing demand for LU-based, ex ante impact assessments for the EC, the JRC, with the support of the Directorate General for Environment, started developing an improved LU modeling framework, the Land Use Modeling Platform (LUMP). This platform is capable of performing more integrated assessments to fulfill the emerging policy needs of different services of the EC; for full details we refer readers to Lavalle et al. (2011a). In short, that new modeling framework was designed to simulate LU changes for the entire EU extent. An early version of LUMP operated at a spatial resolution of 1 km, just before the 100-m version became available. LUMP relied on CoORdination of Information on the Environment (CORINE) Land Cover (CLC) datasets for complete and consistent information on LU across Europe. At the core of LUMP was a statistically calibrated LU change model, EU-ClueScanner100, that integrated top-down and bottom-up drivers of LU change, while taking into account policies with territorial impact at both the macro and the micro level. Since its inception, LUMP has been changed substantially to meet ever more demanding policy requirements. To emphasize these changes the platform has been renamed Land Use–based Integrated Sustainability Assessment (LUISA).

Essentially, LUISA still entails the EU-ClueScanner100 LU model and is still based on CLC data, but has been redeveloped to incorporate additional information on "Land Functions." Those Land Functions are a new concept for cross-sector integration and for the representation of complex system dynamics. They are instrumental to better understand LU change processes and to better inform on the impacts of policy options. LUISA simulates future LU changes, and land functions related to the resulting LU patterns are then inferred and described by means of spatially explicit indicators. A land function can, for example, be physical (e.g., related to hydrology or topography), ecological (e.g., related to landscape or phenology), social (e.g., related to housing or recreation), economic (e.g., related to employment or production or to an infrastructural asset), or political (e.g., consequence of policy decisions). Commonly, one portion of land is perceived to exercise many functions. Land functions are temporally dynamic, depend on the characteristics of land parcels, and are constrained and driven by natural, socioeconomic, and technological processes. As it is centered on this novel concept, LUISA is far beyond a single, stand-alone model. It can be best described as a platform with an LU model at its core, linked to other upstream and downstream models. LUISA was designed to yield, ultimately, a comprehensive, consistent, and harmonized analysis of the impacts of environmental, socioeconomic, and policy changes in Europe.

LUISA has already been used to provide contributions to impact assessments (either in a formal or informal framework) related to the Integrated Coastal Zone Management (Lavalle et al., 2011c), Common Agricultural Policy (Lavalle et al., 2011b), Energy (shale gas and energy package) (Lavalle et al., 2013c), EU Water Blueprint (Burek et al., 2012; De Roo et al., 2012), Regional Policy (Batista e Silva et al., 2013a), and the Resource Efficiency Roadmap (Lavalle et al., 2013b). In this chapter we attempt to describe the state of the art of the LUISA model, with emphasis on the policy evaluations in which LUISA has been used, how LUISA employs CLC maps, and LUISA's future development. In Section 24.2 the main technical and structural characteristics of LUISA are summarized. The role of the CLC map in the processing chain is further explained there as well. Examples of applications in European policy domains are illustrated, emphasizing the integrative aspect of the adopted methodology (Section 24.3). A review of the opportunities and limitations of working with CLC data and an overview of future applications and future improvements conclude the chapter in Section 24.4.

24.2 LUISA in a Nutshell

24.2.1 Overview

The LUISA platform has been specifically designed to assess LU impacts of European policies by providing a vision of possible futures and quantitative comparisons between policy options. The platform accommodates multipolicy scenarios, so that several interacting and complementary dimensions of the EU are represented. At the core of LUISA is a computationally dynamic spatial model that simulates discrete LU changes based on biophysical and socioeconomic drivers. This LU model receives direct input from several external models covering demography, economy, agriculture, forestry, and hydrology, which define the main macro assumptions that drive the model. LUISA is also compliant with given energy and climate scenarios, which are modeled further upstream and link directly to economy, forestry, or hydrology models. The model was initially based on other LU models, namely the Land Use Scanner and Conversion of Land Use and its Effects (CLUE) models (Hilferink and Rietveld, 1999; Dekkers and Koomen, 2007; Verburg and Overmars, 2009), but in its current form LUISA is the result of a continuous development effort by the JRC (Lavalle et al., 2011a). The model is written in GeoDMS, an open source, high-level programming language. The model projects future LU changes at the relatively fine spatial resolution of 1 ha (100 × 100 m), with the most relevant groups of LU types being represented (see Section 24.2.4). LUISA is usually run for all EU countries, but can be used for more detailed case studies or, on the contrary, be expanded to cover pan-European territory. Table 24.1 provides an overview of LUISA's keynote characteristics.

As with many modeling tools, LUISA is not a forecasting model. The most meaningful and useful way to use it is by simulating two or more comparable scenarios. Typically, a "baseline" scenario captures the policies already in place, assuming the most likely socioeconomic trends and "business-as-usual" dynamics (i.e., as observed in the recent past). Such a baseline serves as a benchmark to compare other scenarios in which future conditions or policies are assumed to change. This approach to impact assessment provides relevant elements to structure discussion and debate in a decision-making process. Two elements are crucial when performing an assessment with the LUISA integrated modeling framework: (1) the definition of a coherent multisector baseline scenario to be used as the benchmark for the evaluation of alternative options and (2) a consistent and comprehensive database covering socioeconomic, environmental, and infrastructural themes.

The baseline scenario provides the basis for comparing policy options and should ideally include the full scope of relevant policies at European level. A comprehensive baseline integrated in a modeling platform such as LUISA serves to capture the aggregated impact of the drivers and policies that it covers. Sensitivity analysis can be helpful to identify linkages, feedbacks, mutual benefits, and trade-offs between policies. The definition of the baseline should be the result of agreements between the main stakeholders and experts involved. Ideally, the baseline's assumptions should be shared and used by different

TABLE 24.1

Main Model Characteristics

LU Model Characteristics	In LUISA
Spatial extent	All EU countries
Spatial resolution	100 m
Thematic resolution	Eight main LU classes (+ agricultural breakdown + "abandoned" LUs)
Temporal resolution	Yearly
Time span	2006–2050
Primary outputs	LU maps, LU changes, potential accessibility, population distribution map
Secondary outputs	Spatially explicit thematic indicators

models in integrated impact assessment. Since 2013, LUISA has been configured and updated to be in line with the EC's "Reference scenario" (Lavalle et al., 2013c), which has been used as a baseline in subsequent impact assessments. Various aspects of the model, such as sector forecasts and land suitability definitions, are updated whenever pertinent.

The second element refers to the wealth of data that are needed to cope with the European-wide coverage and multithematic nature of a territorial impact assessment. The principal input datasets required by LUISA must comply with the following set of characteristics:

- EU-wide (ideally pan-European) coverage
- Geographically referenced to bring information together and infer relationships from diverse sources
- Consistency of data nomenclature, quality, and resolution to allow cross-country/region comparison
- Adjustable spatial and thematic resolutions to resolve local features and provide continental patterns

CLC data are the backbone of the LUISA platform. First and foremost, they describe the reference situation, based on which all model results are computed. Besides providing the base map for the LU projections, the CLC data are used for calibration and validation. Here, knowledge of observed LU patterns is essential to establish statistical relationships between the occurrence of particular LU types observed in CLC2006 and a set of bio-physical, socioeconomic and neighborhood factors. In addition, some other input maps of LUISA, such as the population distribution maps (Batista e Silva et al., 2013b), could not be constructed without CLC data. Next to CLC a number of other data sources are used. LUISA relies as much as possible on widely accepted, institutional sources for its input datasets, namely the European Statistical Office (Eurostat), the European Environment Agency (EEA), and, as already mentioned, exogenous models.

LUISA is structured into three main modules: a "demand module," an "LU allocation module," and an "indicator module." The main, final output of the allocation module is an LU map. Potential accessibility and population distribution maps are also endogenously computed by the model as a result of the simulation, and are themselves important factors for the final projected LU map. From these outputs, and in conjunction with other modeling tools that have been coupled with LUISA (e.g., for a concrete example, see Section 24.3.7), a number of relevant indicators can be computed in the indicator module. The indicators capture policy-relevant information from the model's outputs for specific LU functions, such as water retention or accessibility. When computed for various scenarios, differences in the indicators can be geographically identified, sensitive regions can be pin-pointed, and impacts can be related to certain driving factors assumed in the definition of the scenarios. In Sections 24.2.2 through 24.2.4 LUISA's, demand module, land allocation module and LU classification are discussed respectively. The indicators available in the LUISA platform are not reviewed systematically in this chapter, but some are elaborated on in the project descriptions in Section 24.3.

24.2.2 The Demand Module

The demand module captures top-down or macro drivers of LU change that limit the regional quantities of the modeled LU types. The demands for different LU categories are modeled by specialized upstream models. For example, regional land demands for agricultural commodities are taken from the Common Agricultural Policy Regionalised Impact (CAPRI) model (Britz and Witzke, 2008), which simulates the consequences of the Common Agricultural Policy; demographic projections from Eurostat are used to derive future demands for additional residential areas in each region; and land demands for industrial and commercial areas are driven primarily by the growth of different economic sectors. It is clear that LUISA is linked to several thematic models, and thus it also inherits the scenario configurations and assumptions of those models. Special care is therefore taken when integrating the input data from multiple source models to ensure that inputs are mutually consistent in terms of scenario assumptions.

In the case of urban, industrial, and commercial areas the link between macro driving forces and land demands are modeled within LUISA's demand module. Urban LU demands are obtained from combining demand for residences and tourist accommodations. The demand for residential urban areas is a function of the number of households and an LU intensity parameter that indicates the number of households per hectare of residential urban land. The number of households is a function of the regional population and of an average household size that is assumed to converge across European regions. The LU intensity parameter can either be extrapolated from observed past trends in a business-as-usual approach, or can be modified to depend on specific urban policies. The demand for touristic LU is a function of the number of beds in a region and another LU intensity parameter that indicates the number of beds in tourist accommodations per hectare of touristic urban land. The number of beds is a function of the projected number of tourist arrivals, which are in turn obtained from the United Nations World Tourism Organization. Finally, demand for industrial, commercial, and services (ICS) LU is a function of economic growth in those three sectors of activity, and, again, an LU intensity parameter that in this case indicates gross value added per hectare of ICS land (Batista e Silva et al., 2014). Here the LU intensity parameter responds to Gross Domestic Product (GDP) per capita because it has been found that economic LU intensity depends foremost on that factor.

24.2.3 The Land Allocation Module

The LU allocation module is based on the principle that competing LU classes vie for most suitable locations, given available land and the demand for various LU classes. Given that assumption, the actual allocation of LUs to space is governed by an LU optimization approach, in which discrete LU transitions per grid cell occur in each discrete time step. The suitability of locations for various LU types is based on both rules and statistically inferred transition probabilities that are derived from the following factors: terrain factors such as slope, orientation, and elevation; socioeconomic factors such as potential accessibility, accessibility to towns, and distance to roads; and neighborhood interactions between LU. The association between these factors and each LU type is obtained from past LU observations by means of statistical regressions. In addition to exogenous suitability factors, spatial planning, regulatory constraints (e.g., protected areas), and exogenous incentives influencing specific LU conversions can also be taken into account in the model. Furthermore, two matrices govern the occurrence of LU transitions. A "transition cost matrix" informs the model on the likelihood of pairwise transitions. This transition cost matrix is obtained from observed LU transitions recorded in the CLC time series (1990–2006); for example, indicating that in general an LU transition from agriculture to urban is more likely than from forest to urban. An "allow matrix" informs the model on which transitions are permitted, and can also be specified to define the number of years required for a transition to take place. Both matrices can be used as either calibration or scenario parameters, and contribute, in addition to the aforementioned factors, to the overall suitability of grid cells for each LU type.

The LU allocation is done independently for each Nomenclature of Statistical Territorial Units 2 (NUTS2) region. Spillover effects between regions are not yet dealt with in an integrated manner, which is still a limitation of the modeling approach. Although more elegant and integrated ways of dealing with cross-border effects are not developed, problems arising from fixed border systems have been minimized after a revision of the regional system of LUISA, whereby small NUTS2 regions comprising a large conurbation but little hinterland were merged with the adjacent NUTS2. This has been done for Berlin, Prague, Brussels, and Vienna, to cite just a few.

Recent developments are shifting LUISA from traditional, land cover (LC)–based modeling approaches to activity-based modeling. The foremost developments entail the endogenous computation of accessibility levels and population distributions for each grid cell as part of the LU modeling exercise; this is explained exhaustively in Batista e Silva et al. (2013a). Essentially these developments add that for each year, potential accessibility levels are computed given a road network and population distribution (Jacobs-Crisioni et al., in press), while the population allocation module in the model allocates people (newcomers and internal migrants) across each region based on a range of factors. With regard to population distributions the model assumes that people are, among others, driven by high accessibility, vicinity to other people (a proxy for economies of scale), and preferences to build housing on certain LUs. A proxy for housing supply at each location limits the number of people who can be accommodated without further development. Finally, whether one grid cell will be urban no longer depends on the discrete LU allocation process, but instead is obtained from population distributions given straightforward threshold rules.

24.2.4 LU Classification

In the LUISA framework, each LU class belongs to one of the following types: active, passive, or fixed. Active classes are those that are fully simulated by the LU model, and for which land demands are determined. Passive classes are LU classes that are allocated by the model, but for which demands are not predefined. These classes expand or shrink depending on pressures exerted by the active classes. The spatial arrangement of both active and passive classes is governed by the previously discussed land allocation mechanism. Lastly, fixed classes are those for which processes, demands, and suitability criteria are not known or cannot be modeled with current knowledge and/or data, and thus remain unchanged in terms of both quantity and location. This categorization of LU classes is similar to that adopted in other modeling frameworks (Barredo et al., 2003; Verburg and Overmars, 2009).

CLC LU categories have been grouped in a smaller number of relevant classes for use in LUISA, while other classes have been added to meet specific project requirements. For example, a set of "abandoned" classes have been introduced to better capture land abandonment processes (a relevant issue for environmental impact assessment). Arable land has also been broken down into more specific crop classes to better assess the impacts of the Common Agricultural Policy (CAP) reforms (see Section 24.3.2). Table 24.2 shows, for each of LUISA's LU classes, its correspondence with the CLC nomenclature, how it is modeled in LUISA, and its overall share in LUISA's base map. In 2006, the active classes represent the vast majority of the EU's territory (73.8%), while the passive classes represent 14.5%. Fixed LU classes represent only 11.7%.

TABLE 24.2

LU Classification in LUISA

Label in LUISA	Corresponding Class(es) in CLC	Type of Simulation	Share in Base Map (%)
Urban	Urban fabric (11X), built-up component of the sport and leisure facilities (142)	Active	2.8
Industry, commerce, services	Industrial or commercial units (121)	Active	0.6
Infrastructure	Road and rail networks (122), ports (123), airports (124), mine, dump, and construction sites (13X)	Fixed	1.0
Green urban areas	Green urban areas (141), green component of the sport and leisure facilities (142)	Fixed	0.2
Arable land	Arable land (21X), heterogeneous agricultural areas (24X, except agroforestry areas)	Active	32.8
Cereals	LUISA specific class	Active	–
Maize	LUISA specific class	Active	–
Root crops	LUISA specific class	Active	–
New energy crops	LUISA specific class	Active	–
Other arable	LUISA specific class	Active	–
Permanent crops	Permanent crops (22X), agroforestry areas (244)	Active	2.5
Pasture	Pastures (231)	Active	6.7
Forest	Forest (31X)	Active	28.4
Transitional woodland shrub	Transitional woodland shrub (324)	Passive	6.1
Scrub and/or herbaceous vegetation associations	Natural grasslands (321), moors and heathland (322), sclerophyllous vegetation (323)	Passive	8.4
Other nature	Open spaces with little or no vegetation (33X)	Fixed	6.1
Wetlands	Wetlands (4XX)	Fixed	2.1
Water bodies	Water bodies (5XX)	Fixed	2.3
Abandoned urban	LUISA specific class	Passive	–
Abandoned industry, commerce, services	LUISA specific class	Passive	–
Abandoned arable	LUISA specific class	Passive	–

24.3 Recent Applications of LUISA

LUISA has been employed in a number of different policy assessments projects. As can be seen in Sections 24.3.1 to 24.3.7 the scope and goals of those projects varies substantially, thus emphasizing the breadth of topics where LU impacts are relevant as well as the flexibility of the LUISA platform. In all project descriptions we start with laying out the policy context, scope, and objectives. Subsequently some methodological issues are highlighted. Finally, the main conclusions of the projects are given.

24.3.1 Impacts of Policy Alternatives for Coastal Zone Management

24.3.1.1 *Policy Context, Scope, and Objectives*

Since the early 1970s, the EC has been consistently active in promoting the creation and implementation of an integrated approach to manage coastal zones. Recent initiatives include the European Demonstration Programme on Integrated Coastal Zone Management (EC, 1999a) and the publication of the Communication on "Integrated Coastal Zone Management: A Strategy for Europe" in 2000 (EC, 2000). The proposed Directive on maritime spatial planning and integrated coastal management, drafted in 2013 (EC, 2013b), is the latest in a long history of actions taken.

In line with the cited initiatives and other international coastal conventions, the JRC was commissioned to evaluate the impacts of two policy alternatives specially developed for European coastal zones, one of which was directly targeted to implementing recommendations related to Integrated Coastal Zone Management (ICZM). The two policy scenarios were assessed through a range of indicators to give support in the definition of future strategies that were currently evaluated in the framework of the follow-up proposal of the EU ICZM Recommendation (EC, 2002b).

24.3.1.2 Methodological Issues

The study was based on the use of an early version of the LUISA model at 1 km resolution. Considering the European scope of the study and the fact that the modeling framework chosen only focused on LU, it was necessary to define a European coastal zone that could not only embrace terrain heterogeneity, but also be wide enough to encompass the LU dynamics of these areas. In light of this and in compliance with the coastal zone definition as given in the EC Demonstration Programme on ICZM, the following criteria were adopted to define the geographical delimitation of coastal zones:

- A 10-km buffer from the coastline (derived from the Geographical Information System of the COmmission [GISCO] administrative boundaries)
- A 2-km buffer from the aggregation of five CLC classes
  - Coastal wetlands (salt marshes, salines, and intertidal flats)
  - Marine waters (coastal lagoons and estuaries)

Even though the coastal zones are the focus of these policies, the LU simulations were carried out for the entire territory of all EU countries.

The simulated policy alternatives represent two theoretical and rather extreme policy alternatives: they both refer to the simulation period of 2000–2050 and represent alternatives to the global B1 scenario (IPCC, 2000), with which they share the main socioeconomic assumptions. The two policy alternatives are characterized as follows:

- Uncontrolled development: Urban growth will continue, and is restricted only by the preexistent framework of environmental protection (e.g., NATURA 2000). The scenario reflects a lack of restrictions in urban planning and an increase in the potential of urban sprawl.
- Sustainable Environmental Friendly Planning: A more balanced development is promoted, that is, urban growth is constrained by increasing measures to protect vulnerable areas (e.g., erosion/flood sensitive areas) and the presence of natural areas is fostered (forest or semi-natural). Specific LU conversions are not allowed (i.e., change of semi-natural and forest into built-up areas). Regarding built-up targeted policies, built-up areas are constrained within the immediate shoreline to control risks and provide access to coast, as well as to provide landscape protection in a broader band of 3 km from the shoreline. In addition, urban sprawl is

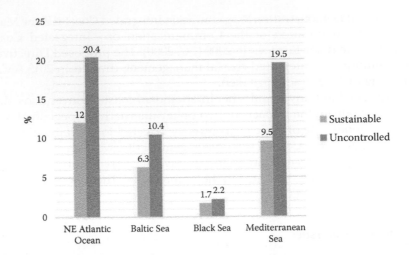

FIGURE 24.1
Change in built-up areas in coastal zones per marine region, according to the uncontrolled and sustainable policy alternatives, 2000–2050 (%).

discouraged. A policy targeted at clustering natural LU types toward large robust natural areas, and against landscape fragmentation (semi-natural and forest), is implemented.

24.3.1.3 Main Results and Conclusions

In order to evaluate the pressure on coastal zones, the policy alternatives were assessed by computing LU change indicators and a selection of thematic indicators: built-up pressure, soil sealing, vulnerability to coastal erosion (EC, 2004), vulnerability to coastal floods (Barredo et al., 2009), mean species abundance index (Alkemade et al., 2009), and green infrastructures connectivity (Saura and Rubio, 2010).

The two policy alternatives show greatly contrasting environmental impacts. The results show that the built-up pressure in coastal zones, if not managed with particular attention, could increase significantly (see Figure 24.1). According to the simulations, expansion of built-up areas in the "uncontrolled" scenario between 2000 and 2050 is ca. 8% more than under the "sustainable" policy alternative. The simulation results showed, in addition, that urbanization patterns within coastal areas are significantly more sprawled under the "uncontrolled" scenario, leading to additional landscape fragmentation, increased habitat loss, reduced biodiversity, and further imbalances of the water cycle. The analysis indicates as well that Europe has much to benefit from further protection measures of coastal zones. For example, lower urbanization pressure in coastal areas will limit the exposure of people and economic assets to coastal erosion and coastal flooding.

24.3.2 Impact of the Green Measures of CAP

24.3.2.1 Policy Context, Scope, and Objectives

In November 2010, the European Commission launched the revision of the CAP with the Communication "The CAP towards 2020" (EC, 2010), based on the outcome of a wide public debate that was initiated in April 2010. This document identified the challenges that should have been addressed in the forthcoming years, and in line with the "Europe 2020 Strategy" defines as main objectives of the reform (1) viable food production, (2) sustainable management of natural resources and climate action, and (3) balanced territorial development. To accomplish these targets, three policy options were outlined: "adjustment," "integration," and "re-focus." These options differ mainly in the weight that is given to a specific objective and present diverse ways to achieve these objectives.

In this context, and in the framework of the impact assessment procedure, the JRC was requested by Directorate General of Environment (DG ENV) to assess a range of environmental impacts that may be caused by the implementation of different policy settings foreseen under the CAP reform; and focus on the greening component of direct payments, as defined in the integration policy option.

24.3.2.2 Methodological Issues

A baseline scenario and a policy alternative were defined and implemented in LUISA, resulting in two different simulated LU maps for the year 2020:

- "Status quo" scenario: represents the current socioeconomic and environmental trends with existing policy provision maintained (business-as-usual)
- "Integration" policy option: builds on the present policy provisions but encompasses a specific set of greening measures

The "status quo" was considered to be the benchmark scenario, with which the impacts of the integration policy option were then compared. For the "integration" policy option, the following specific greening measures were considered:

- Ecological focus area
- Maintenance of permanent pastures
- Separate payment for NATURA 2000 areas

The implementation of the "integration" policy option settings in the LU model required both the introduction of new measures and the strengthening of measures already existent in the "status quo." For example, the green measure "ecological focus area" was implemented by encouraging the occurrence of natural vegetation in a 50-m strip width along water courses within currently mapped Nitrate Vulnerable Zones (CEC, 1991). In addition, in both scenarios, LU change to arable land and permanent crops was encouraged in Less Favoured Areas (see Articles 18 and 20 of EC, 1999b) and discouraged in environmentally sensitive areas. In this particular context, sensitive areas included the following components: (1) 50-m-wide strips along water courses in currently designated Nitrate Vulnerable Zones and (2) areas sensitive to erosion processes (>20 ton/ha/year). Under the "integration" policy alternative, however, the latter measures were more strongly enforced.

24.3.2.3 Main Results and Conclusions

To assess the impact of the projected changes in LU by 2020, a set of indicators was computed: LU shares and conversions, conservation of natural areas, conservation and connectivity of green infrastructure, homogeneity of agricultural areas (Estreguil and Mouton, 2009; Riitters et al., 2009; Estreguil et al., 2012) and soil organic carbon stocks (Intergovernmental

Panel on Climate Change [IPCC], 2003, 2006). Overall, the modeled greening options were found to reduce the pressure on naturally vegetated areas and on environmentally sensitive sites. Summarily for all EU countries, the "integration" policy option resulted in a lower level of environmental impact as compared to the "status quo" scenario. However, several indicators also showed pronounced regional differences in impact.

24.3.3 Water Blueprint

24.3.3.1 Policy Context, Scope, and Objectives

The Blueprint to Safeguard Europe's Waters (EC, 2012) was a commission communication looking into the current state of Europe's water management, and into options to improve it with emphasis on quantity management and efficiency. Within this framework, a request was made to evaluate the effectiveness of Natural Water Retention Measures (NWRMs) in terms of reducing river discharges in flood-prone areas, and maintaining environmental flows in drought-prone areas. NWRMs are all landscape management methodologies which can be applied without artificial construction. The results of the analysis were synthesized in the report by (Burek et al., 2012).

24.3.3.2 Methodological Issues

The study required a strong linkage with the hydrological model LISFLOOD. The model is driven in part by LU-sensitive parameters derived from EU-ClueScanner100, including the Leaf Area Index, Rooting Depth, and Manning's coefficient. The fraction of land covered by forest, sealed area, water, or other LU is also a derived input—each of these LUs has their own adapted algorithm (e.g., for infiltration rates and surface runoff) within the hydrological model.

The NWRMs were individually translated into necessary model adaptations, either in the LU or hydrological modeling, or in both. Table 24.3 provides an overview of the model adaptations per scenario considered. Most measures were modeled by altering the settings of EU-ClueScanner100 to allow or encourage certain LUs to be allocated over others, usually following a Geographical Information System (GIS) procedure to outline the area in which the conversion was encouraged. Some additional scenarios (e.g., re-meandering of rivers, polders, buffer ponds) were assessed using only the LISFLOOD model and are not further described here.

Each NWRM was modeled as a separate LU scenario, which was run from the base year 2006 up to 2030. For each resulting LU map, the necessary input parameters were derived, and LISFLOOD was run to simulate the expected river discharge. The scenario results were then compared to the discharge results obtained from the baseline 2030 run (no NWRMs taken into account) for the whole of Europe, divided into 21 hydrological regions.

24.3.3.3 Main Results and Conclusions

The scenarios were compared in terms of their effectiveness in

- Reducing flood peaks: Re-meandering was the most effective measure in reducing flood peaks in Scandinavia, green urban in the United Kingdom, polders in central Europe, afforestation in Germany, and crop practices for the rest of the continent.

TABLE 24.3

Model Adjustments for Scenarios Ran for the Water Blueprint Project

Scenario	Description	Model Adjustments
Afforestation		
CAP areas	Afforestation of areas from LUISA-CAP scenarios	Conversion to forest encouraged in CAP areas
Riparian areas	Afforestation of areas alongside rivers 100 m, 200 m	Conversion to forest encouraged within a 100 m and 200 m buffer along major rivers
Green Urban Areas		
25%, 50% green	Green infrastructure, green roofs, rain gardens, park depressions	Decrease of sealed surface in all urban areas by 25% and 50%; in these "green" areas the direct runoff fraction ↓ 50% and evapotranspiration ↑ 50%
Agriculture		
Grassland	Conversion to grassland in CAP areas	Conversion to grassland encouraged in CAP areas
Buffer strips	5 m wide grass buffer strips within arable fields, on slopes <10%, every 200 m	2.5% of arable land converted to grassland, only on slopes <10%
Grassed waterways	10 m wide grass-covered areas in valley bottom	1% of arable land converted to grassland, in valley bottoms with slope >5%
Crop practices	Adapted crop practices to reverse organic matter decline and increase mulching	Increased infiltration, porosity, hydraulic parameters

- Increasing low flow: The most effective measures were grassland in Scandinavia and the United Kingdom, buffer ponds in southern Spain, green urban in Italy, and re-meandering for the rest of Europe.
- Increasing groundwater recharge: For most of Europe, the 50% urban green scenario was the most effective in improving recharge; in Scandinavia the grassland scenario was most effective.
- Reducing water stress: Where crop practices were applied all over Europe there was a significant reduction in the number of water stressed days per year.

LISFLOOD is indeed sensitive to changes in LU, and there was a wide variety of discharge response depending on the scenario evaluated. NWRMs can indeed contribute to increased low flows and reduced flood peaks, improve groundwater recharge, and decrease water stress. In each of the 21 macro-regions, however, the most effective set of measures will differ depending on the climate, flow regime, LU, and socioeconomic situation, and the application of measures should therefore be tailored to the specific site location. Using a combination of measures may also be beneficial.

24.3.4 Impacts of Shale Gas Development in Europe

24.3.4.1 Policy Context, Scope, and Objectives

The extraction of shale gas by hydraulic fracturing (fracking) in Europe remains controversial. There are several environmental concerns involved with this extraction procedure, including impacts on water and air quality, noise and visual pollution, potential impacts on biodiversity and nature conservation objectives, and even seismic triggering (Gény, 2010; Rutqvist et al., 2013). Competition for surface area with other LU is also of concern, as the actual take of land for shale gas exploitation may have serious and lasting impacts on the landscape, especially in densely populated areas (Wood et al., 2011). On request of the EC's DG ENV, a modeling exercise was carried out to assess the possible land and water use implications.

An extensive literature study was carried out (Kavalov and Pelletier, 2012), and Germany and Poland were selected as case sites because of the availability of necessary data and the fact that commercialization of shale gas extraction was proceeding in both countries.

The modeling exercise addressed the following questions:

- What are the discrete and potential aggregate LU requirements associated with individual shale gas sites, fields, and potential development scenarios?
- Are there potential LU conflicts associated with shale gas development in member states, including competition with alternative LUs?
- To what degree does shale gas extraction have to compete for water with other water-intensive sectors?

24.3.4.2 Methodological Issues

To answer these questions, variables were identified that may influence the land and water requirements associated with shale gas development (e.g., the range of actual land area and freshwater required, spacing of well pads, rate of development of the resource, recycling ratio of water, etc.). A range of representative values spanning worst- and best-case scenarios for each variable were derived, and thus two specific development scenarios were defined with maximal and minimal expected environmental impact. Those scenarios were then compared to the baseline land and water use scenarios that assume no shale gas extraction. The parameters used are summarized in Table 24.4.

To assess the related land and water use of well pads (drilling sites), they were modeled as a new LU class, based initially on the settings used for the allocation of industrial land. A Water Exploitation Index (WEI), indicating the ratio of water consumption to water availability per subcatchment area (EEA, 2010) has been added to the model. Furthermore, a new suitability layer was created specifically for the allocation of well pads. Areas with favorable conditions for well pads (having low WEI values or beneficial geological characteristics, resource availability, and connectivity) were given higher suitability scores. Areas where the allocation of exploitation sites was strictly forbidden were excluded (e.g., protected areas, floodplains, minimal distance from urban zones),

TABLE 24.4

Summary of Parameters Used to Model Shale Gas Development in Scenarios with Minimum and Maximum Expected Environmental Impacts

Scenario	Low	High
Construction land use per well pad (ha)	9.93	3.55
Operation land use per well pad (ha)	3.75	1.06
Lifespan per well pad (years)	10	10
Minimum distance between well pads (m)	3200	600
Number of fracks per 10 years	1	5
Recycling scenario (%)	70%	0%
Water consumption per well per frack (m³)	3000	45,000

and the allocation was assumed more likely to take place with increasing distance from sensitive areas.

The LU simulation was carried out for the time period 2013–2028, with well pads allocated every 5 years. For each allocation year, four main steps were carried out:

- LU change simulation (EU-ClueScanner100, taking into account locations of previously placed well pads)
- Update of the WEI and suitability layer (as the LU-based parameters are dynamic in time)
- Allocation of well pads based on the suitability score and a neighborhood effect (which takes into account the location advantage of placing extraction sites in areas where the necessary infrastructure has already been developed)

24.3.4.3 Main Results and Conclusions

The modeled high- and low-impact scenarios vary substantially in terms of both projected land and water consumption and allocation patterns of the well pads. Highly complex, multiuse landscapes imply the presence of numerous barriers to drilling activities. The land taken up for shale gas extraction as a percentage of the total land converted to industrial purposes within the shale play area considered in the period 2006–2028 ranges from 19.0% to 38.2% for the low- versus high-impact scenarios respectively. Water consumed by the extraction of shale gas as compared to the total consumption of water per shale play reaches a maximum of 1%. Although these values may not seem significant at the country level, they may have large impacts locally.

24.3.5 Resource Efficiency

24.3.5.1 Policy Context, Scope, and Objectives

The concept of "land take" refers to the expansion of built-up areas or, more precisely, the loss of agriculture, forest and other semi-natural land due to land development (EEA, 1997). The Communication from the European Commission on the Roadmap to Resource Efficient Europe (RERM) proposed a milestone objective for land to contain land take and prevent related environmental problems. Overall for Europe, the objective is to achieve no net land take by 2050 (EC, 2011a).

The LUISA platform was used to assess the spatial impacts that this land take milestone may have in all EU countries by 2020. Two land take scenarios were developed: a "Reference" scenario, where future land take is driven by demographic and economic trends as defined in the Energy Roadmap 2050 (EC, 2011b); and a "Target 0" scenario where the future land take is forced to a "no net land take" by 2050. The RERM milestone proposed for land was transformed into specific urban and industrial/commercial future land demands. The simulated scenarios were compared by means of an average annual land take indicator; see Lavalle et al. (2013b).

24.3.5.2 Methodological Issues

In the reference scenario, future LU changes follow a business-as-usual trend and consider all legislation and EU initiatives currently in place. The growth of urban areas is driven by population projections provided by Eurostat (2010). The future demand for industrial and commercial areas is driven by the growth of the sector Gross Value Added (GVA), derived from long term economic projections (Batista e Silva et al., 2013a).

The 2020 milestone proposed for land take was incorporated in the demand module through the urban and industrial/commercial land requirements. The configuration of the Target 0 scenario was done by adjusting the urban and industrial/commercial land requirements necessary to be on track to reach the 2020 land take milestone. Therefore, in the Target 0 scenario, the annual land take between 2006 and 2050 decreases gradually until reaching zero land take by 2050.

24.3.5.3 Results and Conclusions

The method presented here allows the translation of a descriptive milestone into a quantitative goal, both at European and at member state level. Between 2000 and 2006 the average annual land take in the whole EU was roughly 100,031 ha/year. To achieve the 2020 land take milestone proposed in the RERM, annual land take would need to be reduced to an average of 82,980 ha yearly in the period 2006–2020, which translates into a 17% overall reduction (Lavalle et al., 2013b).

According to the modeling exercise, the proposed objectives for land take reduction appear to be ambitious and would require considerable effort to be achieved in certain regions, considering the large "distance to target" (difference between land take in the Reference and Target 0 scenarios; red and orange hues in Figure 24.2). On the other hand, the simulation results suggested that, in some regions, the 2020 land take milestone values can be reached even without any further action, as the land take in the Reference scenario already complies with the objectives set by the milestone (blue and green hues in Figure 24.2). Such an assessment can only be made through a modeling approach, capable of simulating local implications of more generic spatial planning policies.

24.3.6 Assessing the Direct and Indirect LU Impacts of the Cohesion Policy

24.3.6.1 Policy Context, Scope, and Objectives

Social and economic disparities between European regions have long been a preoccupation of the European Commission. In the late 1980s the EC initiated a series of multiannual investment programmes—known as Regional or Cohesion policies—to promote overall growth and convergence of the less developed countries and regions in the Union. Currently the Cohesion policy is one of the most important policy instruments of the EU. The 2014–2020 program allocates ca. 322 billion EUR (2011 prices), roughly one-third of the total EU budget, to promote competitiveness, economic growth and job creation, while reducing economic, social, and territorial disparities between regions. The actual investment is broken down into the following broad categories: research and development (12%), aid to the private sector (12%), environment (17%), infrastructure (32%), human resources (22%), and technical assistance (5%).

In support of the EC's Directorate General for Regional and Urban Policy (DG REGIO), the JRC was asked to investigate the potential future territorial and environmental impacts of the Cohesion Policy 2014–2020. The main goal of the study was to identify regional trade-offs between investments and LU, and provide insights on how potentially detrimental LU impacts could be minimized. Such a large scale and wide-scope assessment of

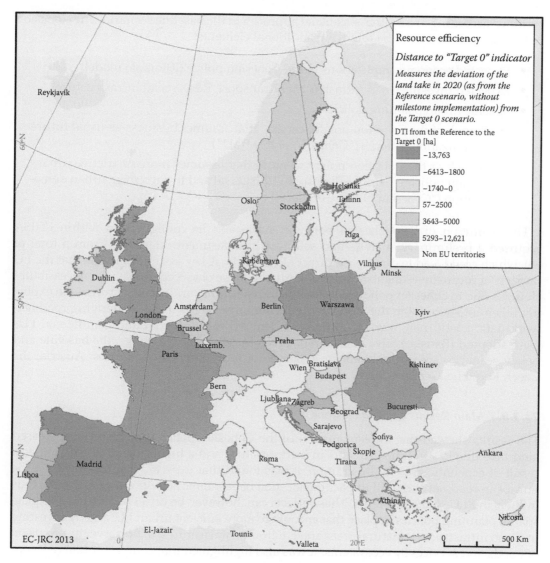

FIGURE 24.2
Distances of the annual land take from the Reference to the Target 0 Scenario at member state level (2006–2020). Red hues show a greater deviation from this target; blues and greens indicate that this target is on track.

the overall Cohesion policy has never been performed before. All details can be found in Batista e Silva et al. (2013a).

24.3.6.2 Methodological Issues

To determine the potential territorial impacts of the Cohesion policy, LUISA was configured to simulate three scenarios—one ignoring the Cohesion policy (baseline scenario), the others accounting for it (Cohesion policy scenarios). All scenarios were configured assuming the accomplishment of the "Europe 2020" targets regarding Energy and Climate, and compliance with major approved environmental legislation. Agriculture, forestry, and demographics projections were the same in both scenarios. Urbanization was assumed to continue according to a business-as-usual approach in the baseline scenario. The Cohesion policy scenario assumed the following additional elements:

- Expected economic implications of the Cohesion policy (Rhomolo model)
- Expected investments in infrastructure (transport, R&D, social infrastructure)
- Specific urban planning policies:
  - Branch of the Cohesion policy scenario that assumes business-as-usual future urbanization patterns ("Cohesion policy BAU")
  - Branch of the Cohesion policy scenario that assumes more restrictive urbanization policies to limit urban sprawl/dispersal and to encourage urban densification ("Cohesion policy Compact")

The definition of the Cohesion policy scenario and its configuration within LUISA required a preliminary analysis and selection of investment categories (from a total of 86), which could lead to direct or indirect LU impacts. A key assumption was that the LU impacts of future investments would be similar to those observed in the past. Accessibility changes due to cohesion policy investments in road transport infrastructure were explicitly taken into account in the model by reducing impedances in the underlying road network data. Lastly, new ways of estimating demand for industrial and commercial LUs were adopted (Batista e Silva et al., 2014). Given these scenario settings, the baseline and policy scenarios were computed for a sample of four European countries: Austria, the Czech Republic, Germany, and Poland.

24.3.6.3 Main Results and Conclusions

The results showed that implementation of the Cohesion policy may result in increased land take due to direct investments in infrastructure and a higher demand for land as a result of economic growth. In addition, it was found that expected improvements to the European road network will impact the location of economic activities and residents, thus influencing LU change patterns. These effects can, however, be offset by putting adequate spatial planning policies in place that encourage more efficient urban LU; and by investing in green infrastructure (natural areas and environmental features) to preserve the provision of goods and services by ecosystems. It was also highlighted that compact urban development and investment in urban green infrastructure have positive effects on the environment and air quality, notably regarding the estimated removal rate of nitrogen dioxide.

24.3.7 ESTIMAP: Linking Dynamic Land Use to the Assessment of Ecosystem Services

Ecosystems services are defined as the contribution that ecosystems make to human well-being, and include, among many others, food and water provision, climate regulation, pollination, and cultural and recreational enjoyment (Haines-Young and Potschin, 2012). Biodiversity is essential to maintain the basic ecosystem processes and support efficient ecosystem functions (Cardinale et al., 2011; Maes et al., 2014). The provision of ecosystem services is strongly supported by the EU policies; it is included in the EU Biodiversity strategy, which, under "Target 2," aims to maintain and enhance ecosystems and their services (EC, 2011a). Ecosystem services are also referred to in the Sixth Report on economic, social, and territorial cohesion as a means of tackling potentially changing conditions in future years (EC, 2014).

Mapping ecosystem services is a way to support decision making processes at different scales and policy levels (Maes et al., 2013). To be effective, the models and methods underlying the mapping process need to be robust, reliable, and comparable (Zulian et al., 2013). Moreover, spatially explicit models should be part of a complete assessment framework. LU changes are acknowledged to be one of the major drivers of biodiversity loss, causing pressure on habitats and species that consequently impacts on the capacity of ecosystems to provide goods and services. The assessment of ecosystem services should therefore be intrinsically linked to LU. ESTIMAP (Ecosystem Services Mapping tool) is a component of LUISA that represents a collection of spatially explicit models for the assessment of ecosystem services. It is framed on the cascade model from Haines-Young and Potschin (2010) and consistent with the Common International Classification of Ecosystem Services (CICES) classification (http://www.cices.eu; Haines-Young and Potschin, 2012), thereby providing an integrated evaluation of the capacity of ecosystems to deliver services using standardized output formats (Zulian et al., 2013).

Currently, ESTIMAP can model nine ecosystem services that are dynamically linked to the LU simulations. Table 24.5 lists the currently available models including units and output formats. The complete set of models can be aggregated into a single indicator ("TESI") by summing the normalized values per ecosystem service (Maes et al., 2012; Dick et al., 2014). An interesting recent application of ESTIMAP has been the analysis of the impact of LU change and the role of green infrastructure on the aggregated provision of ecosystem services (Maes et al., 2014). In this context, the models were used to estimate the required relative increment in Green Infrastructure to offset the expected future urban

TABLE 24.5

Currently Available Modules of ESTIMAP with Their Respective Indicators, Units, and Output Formats

Module	Indicators	Units
Pollination	Relative Pollinator Abundance (RPA)	Dimensionless indicator
Recreation	Recreation Potential (RP)	Dimensionless indicator
	Recreation Opportunity Spectrum (ROS)	Categories based on RP and proximity
	Potential trips	Share of the population that has access to ROS classes (%)
Air quality regulation[a]	Removal of pollutants by urban vegetation	Ton pollutant ha^{-1} $year^{-1}$ [b]
	Population exposure to threshold pollutant concentration	Share of the population exposed to different levels of pollutant concentrations (%)
Food production	Available agricultural land	Share of agricultural land (%)
Water retention	Relative capacity for retention of water in the landscape	Dimensionless indicator
Soil carbon stock	Soil organic carbon content	Ton/ha
Soil erosion control	Surface area of protective cover	%
Habitat quality	Habitat suitability map	Species richness
Coastal protection	Coastal protection capacity	Categorical indicator
	Coastal protection exposure	
	Coastal protection demand	

Note: Table updated from Zulian, G., Paracchini, M., Maes, J., Liquete, C. 2013. ESTIMAP: Ecosystem services mapping at European scale. EUR 26474 EN. Luxembourg: Publications Office of the European Union.

[a] The indicator is computed following Nowak, Crane, and Stevens (2006); concentration map is derived from a spatial regression model as presented by Beelen et al. (2009).

[b] The model considers NO_2 and PM_{10}.

expansion and maintain ecosystem service provision over time for the EU as a whole. The study estimated that for every percentage increase in built-up area, an increase of 2.2% in Green Infrastructure was required to maintain ecosystem service provision at the 2010 level. Efforts are currently being undertaken to further develop the ESTIMAP approach by improving the resolution of both inputs and outputs and by increasing the number of ecosystem services covered. The application of the ESTIMAP approach to assess ecosystem services specific to urban/metropolitan areas is also envisioned.

24.4 Lessons Learned from LU Modeling and the Way Forward

This chapter presented the LUISA framework, which in its current form projects discrete LU patterns based on CLC data, and ties to those LUs various land functions, which in turn often have implications that reach far beyond the grid cell. The projects presented in the preceding text have illustrated the usefulness of the LUISA concept in a wide range of policy impact assessment, and show the versatility and flexibility of its framework. Clearly, the inclusion of land functions such as water retention, ecosystem services, and residences in the modeling framework enable a more thorough exploration of the impacts that LU changing policies may bring forth.

The continuous development and employment of LUISA for impact assessments such as the preceding have given ample opportunity to reflect on the opportunities and limitations entailed with using CLC data. Such a reflection will be given in Section 24.4.1. We conclude this chapter by looking at the way LUISA will be further developed, keeping in mind both the ever changing requirements that policy impact assessments impose on Integrated Impact Assessment and the opportunities offered by newly available data sources.

24.4.1 The Use of CLC Data in LUISA: Opportunities and Limitations

CLC data have been used as the backbone and reference of JRC's LU modeling work since the inception of a full-fledged LU modeling framework at the JRC in 2011. The experiences with those data have shown that the CLC is a suitable data source for JRC's Integrated Impact Assessment. It has full EU coverage, with many non-EU countries also covered (e.g., EFTA—countries belonging to the European Free Trade Association Balkan countries and Turkey); has a high thematic detail with 44 different LU/categories; it records LU for a time series that includes the years 1990, 2000, 2006, and 2012 (under preparation); it is largely comparable across countries and dates due to common and stable mapping guidelines; and it is produced and validated by national experts. Throughout the many years of experience in using CLC data for LU modeling applications, we have come across a series of limitations (see Table 24.6). To overcome some of the problems arising from the spatial resolution of CLC, a modification of the 2006 version was produced by integrating data from more detailed thematic geo-sources, and enabling the reduction of the minimum mapping unit (MMU) to 1 ha for most of the artificial LU categories and water bodies. In addition, a more consistent classification of the urban areas into three comparable levels of density was achieved and the built-up component of the "leisure and sport facilities" was split from the green component. The modified version of the CLC, known as "CLC-refined," is described in a dedicated paper by Batista e Silva et al. (2013c). The JRC intends to undertake again a refinement procedure of the CLC as soon as the 2012 version

TABLE 24.6

Limitations of the CLC Encountered for LU Modeling Applications

Nature of Problem	Issue	Description	Perceived Severity
Temporal dimension	Broad time window for the CLC 1990	The CLC 1990 actually depict LU between 1985 and 1995, depending on the country. The data mismatches can, however, be taken into account in analysis.	Low
	Short time series	The CLC time series is still a relatively short one, encompassing ca. 16 years of LU change (or 22 years after the 2012 version is released).	Low
	Low frequency of updates	The frequency of the updates is still relatively low: 10 years between the first two versions (1990, 2000), and a more satisfactory 6 years between the subsequent versions.	Low
	Time lag	A large time lag still exists between the date of the data and the date of the release. An outdated base LU map strongly affects the quality of the baseline.	Medium
Spatial resolution	Large MMU	All LC features below the MMU are subject to a generalization procedure whereby the dominant surrounding LU class is kept. The large MMU chiefly affects LU types that occur in smaller, more dispersed and irregular patches, such as the urban areas, leading to underestimation of these LU types. This has substantial implications for both analysis and modeling tasks.	High
Thematic detail	Low thematic detail for certain artificial LU categories	It is challenging to model heterogeneous LU classes because of the different determinants involved. • The LU class "industrial or commercial units" lumps facilities as diverse as factories, health and education facilities, military fields, and large shopping malls. • The class "sport and leisure facilities" includes race tracks, amusement parks, and camping sites. • No distinction between monofunctional residential areas and urban areas with mixed LUs (services, business, residential).	High
	Inconsistency of urban definitions	Inconsistent definition and distinction between two different urban classes ("continuous" and "discontinuous urban fabric") across countries.	High
Geogr. coverage	Incomplete coverage in versions 1990 and 2006	The EU coverage is not complete for all the available time steps. For example, the Scandinavian countries and the United Kingdom are not covered in the 1990 version, and Greece is not yet covered in the 2006 version.	Medium

is released. Besides the already established improvements, a further breakdown of the "industrial and commercial sites" will be attempted by using nonconventional geo-data sources. This breakdown will allow a better modeling of the nonresidential urban areas, and will open the possibility for mapping activity distributions throughout the day.

24.4.2 The Way Forward

Besides the previously mentioned planned refinement of the CLC base data, the coming years will see much work to improve LUISA as a comprehensive tool for evaluating the effects of various policies on LU and associated indicators. The end goal of LUISA's development should be a modeling framework that closely approximates true economic land conversions, explicitly modeling all costs and benefits that are internalized in the LU change process, while broadly taking into account both the internal and external costs and benefits of LU changes when evaluating model results. This end goal includes a better grasp of the various economic activities that drive anthropogenic LU. Lastly, a number of efforts need to be undertaken to better underpin the validity of the model approach, variable selection, and model reliability. In the following sections we discern short-term plans, for which necessary data is available, and long-term plans, which will require data sources that are currently unavailable.

One of the most important planned improvements concerns integrating air quality indicators in the LUISA model. To do so, assumptions on activity levels have to be extended further from the population allocation model already in place. By integrating air quality levels in the model, the modeling platform gains a useful indicator necessary to understand the full range of external costs of LU change and also opens up possibilities to evaluate air quality improvement policies that aim at promoting behavioral changes and structural measures. Another important improvement involves redesigning the link between regional urban LU claims, the population allocation module and the discrete allocation method. This improvement aims to use population pressures as a driving force for grid-cell urbanization without allocating urban LUs parallel to EU-ClueScanner100 LU allocation method. Other works that will be undertaken on the short term aim to (1) underpin the conversion cost matrices currently used in the model with either empirically obtained probabilities or costs derived from an economic rationale, which serves to more closely link the model to real processes; and (2) include water scarcity levels as a suitability factor for particular LUs, to better assess direct and indirect effects of water policies.

One of the most substantial improvements planned for the long term is to fully integrate an economic rationale into the LU model—based on true utilities, true costs, and true willingness-to-pay data. This would better underpin the rationale of the model, and would allow inductive approaches in the model to evaluate the effect of policies on LU behavior (i.e., not starting from an assumed overall effect, but from a clearly defined added cost or financial incentive in the utilities of particular LU conversions). In this improvement, currently unavailable data on financial aspects of LU conversions are critical. Other planned improvements to the model are

- Feedback mechanisms with the macro agricultural and economic models, to have more accurate land demand forecasts for those sectors
- Modeling a wider range of urban activity levels to compute more accurate air quality levels, better assess land suitabilities for urbanization, and potentially to compute various social indicators that deal with urban activity levels and urban activity diversity as in Jacobs-Crisioni et al. (2014)
- Including public transport accessibility as an explanatory factor for urbanization, which is useful to more accurately model effects of EU-initiated public transport development
- Estimating transport use within the model to more accurately compute air quality levels and to assess transport consumption and potential congestion levels as a societally relevant effect of LU changes

Lastly, the frequent use of the LUISA framework in policy consultation presses the need to validate the model's output in terms of accuracy and reliability. In 2013 the JRC began a cross-validation exercise with Germany's Federal Institute for Research on Building, Urban Affairs and Spatial Development (BBSR) that also employs an LU model. It is expected that this validation exercise will yield useful insights into the importance of various model settings and factors that differ between the BBSR's and JRC's LU models. Furthermore, data to carry out an empirical validation of the model using historical trends are finally becoming available, in the form of a historical time series of municipal population counts and historical time series LU data (Barranco et al., 2014; EC, 2014). These historical data will be instrumental in empirical validation projects that are planned for the short to medium term.

References

Alkemade, R., Van Oorschot, M., Miles, L., Nellemann, C., Bakkenes, M., Ten Brink, B. 2009. GLOBIO3: A framework to investigate options for reducing global terrestrial biodiversity loss. *Ecosystems* 12:374–390.

Barranco, R., Batista e Silva, F., Marin Herrera, M., Lavalle, C. 2014. Integrating the MOLAND and the urban atlas geo-databases to analyze urban growth in European cities. *Journal of Map and Geography Libraries: Advances in Geospatial Information, Collections and Archives* 10(3):305–328.

Barredo, J. I., Kasanko, M., McCormick, N., Lavalle, C. 2003. Modelling dynamic spatial processes: Simulation of urban future scenarios through cellular automata. *Landscape and Urban Planning* 64(3):145–160.

Barredo, J. I., Salamon, P., Feyen, L., Nicholls, R. J. 2009. *Coastal and riverine flood damage potential in Europe.* Ispra, Italy: Joint Research Centre, Institute for Environment and Sustainability.

Batista e Silva, F., Lavalle, C., Jacobs-Crisioni, C., Barranco, R., Zulian, G., Maes, J., Baranzelli, C., Perpiña, C., Vandecasteele, I., Ustaoglu, E., Barbosa, A., Mubareka, S. 2013a. Direct and indirect land use impacts of the EU cohesion policy assessment with the Land Use Modelling Platform. Luxembourg: Publications Office of the European Union.

Batista e Silva, F., Gallego, J., Lavalle, C. 2013b. A high-resolution population grid map for Europe. *Journal of Maps* 9(1):16–28.

Batista e Silva, F., Lavalle, C., Koomen, E. 2013c. A procedure to obtain a refined European land use/cover map. *Journal of Land Use Science* 8(3):255–283.

Batista e Silva, F., Koomen, E., Diogo, V., Lavalle, C. 2014. Estimating demand for industrial and commercial land use given economic forecasts. *PLoS ONE* 9(3):e91991.

Britz, W., Witzke, H. P. 2008. Capri model documentation 2008: Version 2. Bonn: Institute for Food and Resource Economics, University of Bonn.

Burek, P. A., Mubareka, S., Rojas, R., De Roo, A., Bianchi, A., Baranzelli, C., Lavalle, C., Vandecasteele, I. 2012. Evaluation of the effectiveness of natural water retention measures—Support to the EU Blueprint to Safeguard Europe's Waters. EUR 25551 EN. Luxembourg: Publications Office of the European Union.

Cardinale, B. J., Matulich, K. L., Hooper, D. U., Byrnes, J. E., Duffy, E., Gamfeldt, L., Balvanera, P., O'Connor, M. I., Gonzalez, A. 2011. The functional role of producer diversity in ecosystems. *American Journal of Botany* 98:572–592.

CEC (Commission for Envionmental Cooperation). 1991. Council directive of 12 December 1991 concerning the protection of waters against pollution caused by nitrates from agricultural sources. 91/676/EEC.

Dekkers, J. E. C., Koomen, E. 2007. Land-use simulation for water management: Application of the Land Use Scanner model in two large-scale scenario-studies. In E. Koomen, J. Stillwell, A. Bakema, and H. J. Scholten (Eds.), *Modelling land-use change: Progress and applications* (pp. 355–373) Dordrecht: Springer.

De Roo, A., Burek, P. A., Gentile, A., Udias, A., Bouraoui, F., Aloe, A. Bianchi, A., La Notte, A., Kuik, O., Elorza Tenreiro, J., Vandecasteele, I. et al. 2012. A multi-criteria optimisation of scenarios for the protection of water resources in Europe: Support to the EU Blueprint to Safeguard Europe's Waters. EUR 25552 EN. Luxembourg: Publications Office of the European Union.

Dick, J., Maes, J., Smith, R. I., Paracchini, M., Zulian, G. 2014. Cross-scale analysis of ecosystem services identified and assessed at local and European level. *Ecological Indicators* 38:20–30.

EC (European Commission). 1999a. Towards a European integrated coastal zone management (ICZM) strategy: General principles and policy options—A reflection paper. Luxembourg: Publications Office of the European Union.

EC (European Commission). 1999b. On support for rural development from the European Agricultural Guidance and Guarantee Fund (EAGGF) and amending and repealing certain regulations. Council regulation (EC) 1257/1999.

EC (European Commission). 2000. Integrated coastal zone management: A strategy for Europe. COM(2000) 547 final.

EC (European Commission). 2002a. Communication from the Commission on Impact Assessment. In COM(2002) 276.

EC (European Commission). 2002b. Concerning the implementation of Integrated Coastal Zone Management in Europe (2002/413/EC). *Official Journal of the European Communities L* 148:24–27.

EC (European Commission). 2004. Living with coastal erosion in Europe: Sediment and space for sustainability. Report of EUROSION Project. Luxembourg: Publications Office of the European Union.

EC (European Commission). 2010. The CAP towards 2020: Meeting the food, natural resources and territorial challenges of the future. COM(2010) 672 final.

EC (European Commission). 2011a. Our life insurance, our natural capital: An EU biodiversity strategy to 2020. COM(2011) 0244 final.

EC (European Commission). 2011b. Roadmap to a resource efficient Europe. COM(2011) 571 final.

EC (European Commission). 2012. A blueprint to safeguard Europe's water resources. COM(2012) 673 final.

EC (European Commission). 2013a. Assessing territorial impacts: Operational guidance on how to assess regional and local impacts within the Commission Impact Assessment System.

EC (European Commission). 2013b. Proposal for a directive of the European parliament and of the council establishing a framework for maritime spatial planning and integrated coastal management. COM(2013) 133 final.

EC (European Commission). 2014. Investment for jobs and growth: Promoting development and good governance in EU regions and cities. Sixth Report on Economic, Social and Territorial Cohesion. Luxembourg: Publications Office of the European Union.

EEA (European Environment Agency). The concept of land consumption 1997 [cited 23/09/2014]. Available at: http://glossary.eea.europa.eu/terminology/concept_html?term=land%20consumption.

EEA (European Environment Agency). 2010. Use of freshwater resources. Copenhagen: European Environment Agency.

Engelen, G., Lavalle, C., Barredo, J. I., Meulen, M., White, R. 2007. The MOLAND modelling framework for urban and regional land-use dynamics. In E. Koomen, J. Stillwell, H. J. Scholten, and A. Bakema (Eds.), *Modelling land-use change: Progress and applications* (pp. 297–320) Dordrecht: Springer.

Estreguil, C., Mouton, C. 2009. Measuring and reporting on forest landscape pattern, fragmentation and connectivity in Europe: Methods and indicators. EUR 23841 EN. Luxembourg: Publications Office of the European Union.

Estreguil, C., Caudullo, G., De Rigo, D., San Miguel, J. 2012. Forest Landscape in Europe: Pattern, Fragmentation and Connectivity. EUR 28717 EN. Luxembourg: Publication Office of the European Union.

Eurostat. *EUROPOP2010—Convergence scenario, national level* 2010 [cited 30/07/2014]. Available at: http://epp.eurostat.ec.europa.eu/cache/ITY_SDDS/EN/proj_10c_esms.htm.

Gény, F. 2010. Can unconventional gas be a game changer in European gas markets? Oxford: Oxford Institute for Energy Studies.

Haines-Young, R., Potschin, M. 2010. The links between biodiversity, ecosystem services and human well-being. In D. G. Raffaelli and C. L. J. Frid (Eds.), *Ecosystem ecology: A new synthesis* (pp. 110–139) Cambridge, UK: Cambridge University Press.

Haines-Young, R., Potschin, M. 2012. Consultation on CICES version 4. Nottingham: University of Nottingham.

Hilferink, M., Rietveld, P. 1999. Land Use Scanner: An integrated GIS based model for long term projections of land use in urban and rural areas. *Journal of Geographical Systems* 1(2):155–177.

IPCC (Intergovernmental Panel on Climate Change). 2000. Emissions scenarios. Summary for policymakers. Geneva: Intergovernmental Panel on Climate Change.

IPCC (Intergovernmental Panel on Climate Change). 2003. *Good practice guidance for land use, land-use change and forestry*. Hayama: IPCC/OECD/IEA/IGES.

IPCC (Intergovernmental Panel on Climate Change). 2006. *Guidelines for national greenhouse gas inventories, Volume 4*. Hayama: IPCC/OECD/IEA/IGES.

Jacobs-Crisioni, C., Batista e Silva, F., Lavalle, C., Baranzelli, C., Barbosa, A., Perpiña Castillo, C. In press. Accessibility and territorial *cohesion* in a case of transport infrastructure improvements with endogenous population distributions. European Transport Research Review.

Jacobs-Crisioni, C., Rietveld, P., Koomen, E. 2014. Evaluating the impact of land-use density and mix on spatiotemporal urban activity patterns: An exploratory study using mobile phone data. *Environment and Planning A* 46(11):2769–2785.

Kavalov, B., Pelletier, N. 2012. Shale gas for Europe: Main environmental and social considerations (a literature review). EUR 25498 EN. Luxembourg: Publications Office of the European Union.

Lavalle, C., Demicelli, L., Kasanko, M., McCormick, N., Barredo, J. I., Turchini, M., da Graça Saraiva, M., Nunes da Silva, F., Loupa Ramos, I., Pinto Moreiro, F. 2002. Towards an urban atlas. Environmental Issue Report No. 30. Copenhagen, Denmark: European Environmental Agency.

Lavalle, C., Baranzelli, C., Batista e Silva, F., Mubareka, S., Rocha Gomes, C., Koomen, E., Hilferink, M. 2011a. A High Resolution Land use/cover Modelling Framework for Europe: Introducing the EU-ClueScanner100 model. In B. Murgante, O. Gervasi, A. Iglesias, D. Taniar, and B. O. Apduhan (Eds.), *Computational science and its applications ICCSA 2011*, Part I. Lecture Notes in Computer Science 6782. Berlin: Springer-Verlag.

Lavalle, C., Baranzelli, C., Mubareka, S, Rocha Gomes, C., Hiederer, R., Batista e Silva, F., Estreguil, C. 2011b. Implementation of the CAP policy options with the Land Use Modelling Platform—A first indicator-based analysis. EUR 24909 EN. Luxembourg: Publications Office of the European Union.

Lavalle, C., Rocha Gomes, C., Baranzelli, C., Batista e Silva, F. 2011c. Coastal zones—Policy alternatives impacts on European coastal zones 2000–2050. EUR 24792 EN. Luxembourg: Publications Office of the European Union.

Lavalle, C., Baranzelli, C., Vandecasteele, I., Ribeiro Barranco, R., Sala, S., Pelletier, N. 2013a. Spatially-resolved assessment of land and water use scenarios for shale gas development: Poland and Germany. EUR 26085 EN. Luxembourg: Publications Office of the European Union.

Lavalle, C., Barbosa, A., Mubareka, S., Jacobs-Crisioni, C., Baranzelli, C., Perpiña Castillo, C. 2013b. Land use related indicators for resource efficiency—Part I. Land take assessment. Luxembourg: Publications Office of the European Union.

Lavalle, C., Mubareka, S., Perpiña, C., Jacobs-Crisioni, C., Baranzelli, C., Batista e Silva, F., Vandecasteele, I. 2013c. Configuration of a reference scenario for the land use modelling platform. Luxembourg: Publications office of the European Union.

Maes, J., Paracchini, M., Zulian, G., Dunbar, M. B., Alkemade, R. 2012. Synergies and trade-offs between ecosystem service supply, biodiversity, and habitat conservation status in Europe. *Biological Conservation* 155:1–12.

Maes, J., Teller, A., Erhard, M., Liquete, C., Braat, L., Berry, P., Egoh, B., Puydarrieux, P., Fiorina, C., Santos, F., Paracchini, M. et al. 2013. Mapping and assessment of ecosystems and their services. An analytical framework for ecosystem assessments under Action 5 of the EU biodiversity strategy to 2020. Discussion paper. Luxembourg: Publications Office of the European Union.

Maes, J., Barbosa, A., Baranzelli, C., Zulian, G., Batista e Silva, F., Vandecasteele, I., Hiederer, R., Liquete, C., Paracchini, M., Mubareka, S., Jacobs-Crisioni, C. et al. 2014. More green infrastructure is required to maintain ecosystem services under current trends in land-use change in Europe. *Landscape Ecology* 30:517–534.

OECD (Organisation for Economic Co-operation and Development). 2010. *Guidance on sustainability impact assessment*. OECD Publishing.

Riitters, K. H., Wickham, J. D., Wade, T. G. 2009. An indicator of forest dynamics using a shifting landscape mosaic. *Ecological Indicators* 9:107–117.

Rutqvist, J., Rinaldi, A. P., Cappa, F., Moridis, G. J. 2013. Modeling of fault reactivation and induced seismicity during hydraulic fracturing of shale-gas reservoirs. *Journal of Petroleum Science and Engineering* 107:31–44.

Saura, S., Rubio, L. 2010. A common currency for the different ways in which patches and links can contribute to habitat availability and connectivity in the landscape. *Ecography* 33(3):523–537.

Verburg, P. H., Overmars, K. 2009. Combining top-down and bottom-up dynamics in land use modeling: Exploring the future of abandoned farmlands in Europe with the Dyna-CLUE model. *Landscape Ecology* 24(1167):1181.

Wood, R., Gilbert, G., Sharmina, M., Anderson, K. 2011. Shale gas: A provisional assessment of climate change and environmental impacts. Manchester: Tyndall Centre, University of Manchester.

Zulian, G., Paracchini, M., Maes, J., Liquete, C. 2013. ESTIMAP: Ecosystem services mapping at European scale. EUR 26474 EN. Luxembourg: Publications Office of the European Union.

25

CORINE Land Cover Outside of Europe

Gabriel Jaffrain

CONTENTS

25.1 Experience of the CORINE Land Cover Concept Outside of Europe

As of 2016 the European CoORdination of Information on the Environment (CORINE) Land Cover (CLC) project has 30 years of history, with nearly 40 participating countries. However, its database is the target of many detractors who believe this information is not accurate enough and whose scale and mapping thresholds must be improved.

Despite the availability and accessibility of a range of satellite data that is facilitated by different distribution services at an affordable price (sometimes even free of charge), many countries today do not have any land cover (LC) information of their own territory or this information is very difficult to access. Thus, many initiatives and projects are launched and financed to help countries to build capacity in the field of remote sensing technology and LC information. The concept of CLC is still a simple and effective tool that can be easily exported and transfer knowledge at low cost.

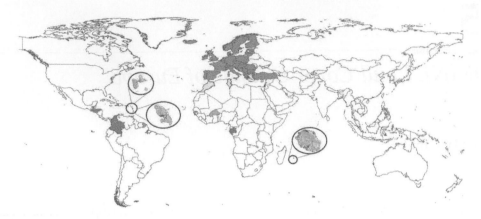

FIGURE 25.1
Illustration of CLC methodology applied outside of Europe.

The purpose of this chapter is not to demonstrate the usefulness of the European LC information, namely CLC. Thousands of CLC downloads (by means of the European Environment Agency Data Service) every single month and certainly tens of thousands since the first CLC1990 prove it. Dissemination of their national datasets by the member countries themselves should be added to this central service. Instead, the purpose of this chapter is to show the excellent adaptability of the CLC outside of Europe.

Indeed, the need of geographical information on the state of the environment cannot be restricted to what lies within the European boundaries, but concerns all countries facing and undergoing environmental stress due to human activities and other natural causes.

In parallel to the development of CLC in Europe, Institut Géographique National France International (IGN FI) was entrusted with several initiatives abroad (Figure 25.1) and has been actively involved in the development and transfer of CLC methodologies to different countries in Africa, Central America, and South America.

The cooperation between European Space Agency (ESA) and Food and Agriculture Organization of the United Nations (FAO) resulted in the report "CORINE Land Cover outside Europe. Nomenclature adaptation to other bio-geographical regions" (EEA/ETC/SIA, 2011). The report was written for the EEA and deals with CLC nomenclatures adapted to other biographical zones.

This chapter specifically relates a brief account of implementing a CLC project abroad and how to define an adapted CLC nomenclature according to the dominant non-European landscapes that is adjusted obviously in collaboration with national experts.

25.2 Key Factors for Success in Adapting a CLC Project Outside of Europe

The need for accurate and up-to-date spatial data is also a major concern for emerging countries. CLC-type projects have been developed in several countries around the world.

The main key factors for success are described in Figure 25.2 and include a serious involvement of foreign experts in acquiring an understanding of the local situation and a strong commitment of local experts in their fields (ecology, agronomy, geomorphology, etc.) in accordance with the local situation or needs.

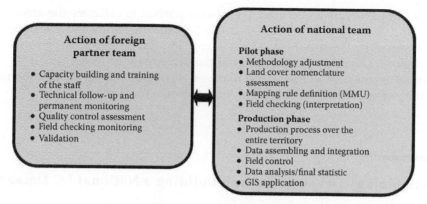

FIGURE 25.2
Cooperation and distribution of work between the French and the national team.

Implementation of a CLC project or other land monitoring project abroad and achievement of a successful transfer of the CLC concept (know-how) needs to follow several steps.

25.2.1 Global Understanding of the Context and State of the Art in the Country

First, the preliminary work consists of defining precisely the subject and the user or stakeholder requirement. The physical phenomena to be studied and the scale of the study (national, regional, or local) also need to be defined. On the basis of these requirements a methodology is defined and implemented. For instance, the resolution of spatial basic data used will not be the same if one studies a small watershed or the whole country.

Second, an inventory of the existing national LC/LU information and the collection of these data are needed. For instance, it is necessary to obtain information of dominant crops, natural area, and industrial plantation. The collection of LC/LU information includes the following steps:

- Bibliography research related to the country to gain an idea of the dominant landscape (natural area and dominant crops)
- Inventarization of previous projects in the field of LC and LU (dates, topics, scale, user)
- Collection of existing maps and research

25.2.2 Establishment of a National Team

A fundamental point is obviously the human resources available at the national level. One must make an estimate of the number of staff who will be involved in the project (skill/education level, specific training, etc.) along with their expertise and knowledge of their own territory. A selection of experts or specialists such as geographers, agronomists, and ecologists is made to carry out the project. A very close cooperation between the foreign experts and the local teams is obviously necessary.

25.2.3 Assessment of Logistics Needed

An assessment of the existing equipment, the availability of workspace (e.g., rooms), availability of computers (status, software and hardware), and information and communications technology infrastructure is required. It is also important to estimate the amount of field survey work, that is, the number of days in the field for the teams and the accessibility of field survey site. The logistics for the field survey (availability of cars, drivers, and other accommodations) should be provided by the client/country.

Each project is unique and the project manager during the entire production phase has to assess the potential risks that can delay or jeopardize the project.

25.3 Methodology Development and Building a National LC Database

25.3.1 Assessing the Methodology

National LC database production requires a feasibility study (called the pilot phase) before the real production process and its generalization to the entire territory starts. The feasibility study consists of the following phases:

1. Adjustment of the CLC production methodology to the national situation
2. Selection of the interpretation methodology: visual photointerpretation (computer-assisted photointerpretation) or automatic classification techniques, or a combination of the two methods
3. Definition of new LC classes, that is, LC classes that are not present in the European standard nomenclature feature, but characteristic of the dominant landscape of the country
4. Assessment during the pilot phase of the performance of the satellite image to identify the new LC classes such as industrial crops of bananas, sugar cane, or oil palm in Colombia or the natural formations of wooded savanna, shrub steppe found in the Sahel region in Burkina Faso.

25.3.2 Building a National LC Classification from the European CLC Nomenclature Basis

The European CLC nomenclature is a hierarchical nomenclature organized on three levels and containing five universal classes at the first level: 1 *Artificial surfaces*, 2 *Agricultural areas*, 3 *Forest and semi-natural areas*, 4 *Wetlands*, and 5 *Water bodies*. The entire surface of the Earth can be classified into one of these five categories. Landscape natural elements such as savannah, steppe, matorral, paramo, mangroves, and so forth or agricultural areas such as crops of banana, coffee, sugar cane, palm trees, and so forth can be included in one of the five specific universal classes of CLC. While adapting the CLC nomenclature to national standards, the idea is to maintain the first and second CLC level and add new classes only at the third level or introduce a fourth level.

The development of a national LC classification should be established a posteriori in collaboration with the national experts and must be tested in some pilot area and verified during field surveys. The nomenclature a posteriori must reflect the dominant national landscape. The classes defined a posteriori but not identified during the field survey or significant are deleted in the final nomenclature.

Section 25.4 describes some specific CLC implementations outside of Europe.

25.4 Some Specific Examples of CORINE Land Cover Implementation Outside of Europe

25.4.1 Tunisia (1990–1998)

From 1990 to 1998, IGN FI, together with the National Agency for Environmental Protection (ANPE) from the Ministry of Environment, produced an environmental database of 80,000 km² of the Tunisian coast. The first database is called MEDGEOBASE. Through this successful technical cooperation, the ANPE has acquired good knowledge on how to achieve an environmental database. In this context the ANPE has decided to (further) develop a monitoring tool that helps to improve the knowledge and management of the environment in Tunisia.

This operational decision support system for the management of natural resources and the environment (SAIDE) is based on basic data coming from the previous MEDGEOBASE project but this time it was extended to the whole country from Landsat TM coverage (IGN FI, ANPE—Life 95/TN/B2/TN/928/MED SAIDE August, 1998). It should become a tool for decision support in the field of LU planning and coastal protection for both the ANPE under the Ministry of Environment and Spatial Planning and for the new National Agency for Coastal Protection (Figure 25.3).

The LC nomenclature of Tunisia (see Figure 25.3) consists of a fourth level, and for natural areas (mainly the steppe formation) even a fifth level is introduced. The steppes are described at the dominant species level such as *Alfa* sp. steppe, *Armeria* sp. or *Thymus* sp. steppe, and other xerophyte steppe. To help the identification of this information we used

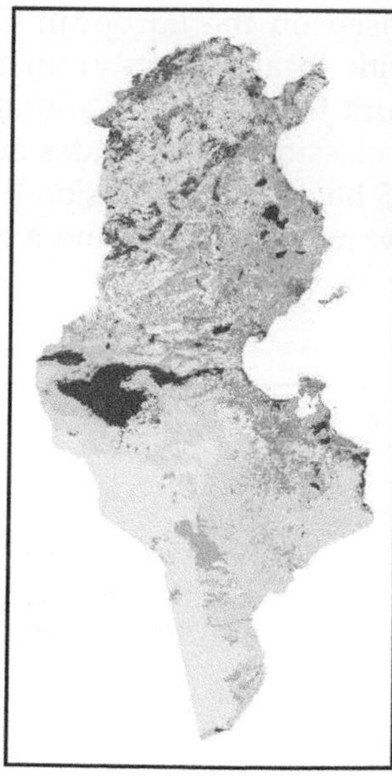

First level	Second level	Third level	Fourth level
2 Agricultural areas			
	22 Permanent crops		
		221 Vineyard	
		222 Fruit trees and berry plantation	
		223 Olive groves	
		224 Oasis	
	24 Heterogeneous agricultural areas		
		241 Annual crops associated with permanent crops	
			2411 Annual crops associated with olive groves
			2412 Annual crops associated with fruit trees
			2413 Annual crops associated with fruit trees

FIGURE 25.3
SAIDE project in Tunisia. The table in this figure is an example of an additional item for the agriculture field in the third and fourth CLC levels carried out in Tunisia (black: CLC European classes maintained; italics: Tunisian classes).

the vegetation map provided by the Tunisian forest Institution (based on field surveys). Aerial photographs were used also as ancillary data to help the photointerpreter. Many field surveys were organized to obtain a better knowledge of the dominant landscape. That is the reason that class 241 *Annual crops associated with permanent crops* (such as annual crops associated with olive groves—2411 and annual crops associated with other fruit trees—2412) could be detailed.

25.4.2 Central America (1999–2003)

In 1999, IGN FI conducted actions for the French Ministry of the Environment aimed at raising the awareness of CLC in the countries of the Caribbean and Central American isthmus. A feasibility study of the basin of Rio-Lempa was conducted to adapt the CLC nomenclature to the regional specifications and to launch environmental applications from this database. This project, called SHERPA (Suivi Hydrologique et Environnemental pour l'Amérique Centrale), was a great success owing to an excellent cooperation between the French institutes IGN FI, Institut Francais de l'Environment (IFEN), and Centre de Coopération International en Recherche Agronomique pour le Dévelopment (CIRAD) (the latter being in charge of applications for the second stage) and the local environmental agencies represented by the CACED (Central American Commission of the Environment and Development) as well as the local National Geographic Institutes. The information on land occupation was so important for the launch of a specific thematic application that an extension of the database to the entire Salvador territory was financed and completed. The illustration is provided in Figure 25.4. Some LC database applications have been published in a special Agricultural Atlas by the Ministerio de Medio Ambiente y recursos naturales (2005). An Environmental Atlas has been published by CIRAD-TERA, IGN FI (2003).

The land occupation database at a scale of 1:100,000 is based on the European CLC method. Its nomenclature was adapted in collaboration with local experts from San Salvador, Guatemala, and Honduras, while taking into account the landscape and agricultural diversity of the three countries (see Figure 25.4). The classification includes more than 40 items and the minimum mapping unit (MMU) is 25 ha. A new extension project for the entire Salvador territory was initiated with the new mapping rules and a new

First level	Second level	Third level	Fourth level
3 Forest and semi-natural areas	31 Forests	311 Broadleaves forest	
			3111 Deciduous forest
			3112 Evergreen forest
			3113 Semi evergreen forest
		312 Coniferous forest	
			3121 Dense coniferous forest
			3212 Open coniferous forest
		313 Mixed forest	
			3131 Dense mixed forest
			3232 Open mixed forest
		314 Mangroves	
		315 Forest plantation	

FIGURE 25.4
SHERPA project in Honduras, Guatemala, and San Salvador. The table in this figure is an example of an additional item for the forest in the third and fourth CLC levels carried out in San Salvador (black: CLC European classes maintained; italics: Salvador classes).

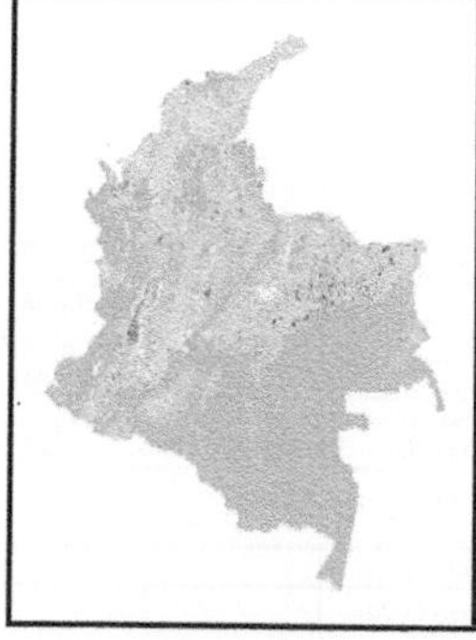

First level	Second level	Third level	Fourth level
3 Forest and semi-natural areas			
	32 Shrub and/or herbaceous vegetation associations		
		321 Herbaceous vegetation	
			3211 Natural grassland
			3212 Herbaceous savannah
			3213 Shrub/tree savannah
		322 Shrub and bushy vegetation/*matorrales*	
		323 Sclerophyllous vegetation	
		324 *"Paramo" formation*	
		325 *"Rupicole" vegetation*	

FIGURE 25.5

CLC in Colombia. The table in this figure is an example of additional item in the third and fourth CLC levels carried out in Colombia (black: CLC European classes maintained; italics: Colombian classes).

methodology including a nomenclature with more than 64 items and a MMU at 5 ha. The scale is compatible with 1:25,000.

25.4.3 Latin America: Colombia (2004–2007)

The implementation of CORINE Colombia began in June of 2004 over the entire hydrological basin of Rio Magdalena, covering a surface area of more than 274,000 km² (Figure 25.5). Based on a close collaboration between IGN FI's team and the Colombian agencies IGAC (Instituto Geografico Agustin Codazzi), IDEAM (Instituto De Hidrologia, Meteorologia y Medio Ambiente), and CORMAGDALENA (Corporacion Autonoma Regional del Rio Grande de La Magdalena), a link was established between the present quantities of carbon and the vegetation cover as well as the variability in time and space. The project took place in the context of the Kyoto Protocol. The methodology and nomenclature (IDEAM, IGAC, CORMAGDALENA, 2008) used is based on the European CLC hierarchical structure (see Figure 25.5).

The CORINE Colombia database, finished since the beginning of 2007, has become a reference for the establishment of new ecosystem databases. Its extension is forthcoming for the entire Colombian territory and negotiations are underway to establish a common system for Andean Community of Nations (ACN) countries in collaboration with all of the national geographical and environmental institutes.

25.4.4 West Africa: Burkina Faso (2003–2006 and 2012)

Since April of 2005, the National Program for Land Management (NPLM2) located in Burkina Faso has two land occupation databases Base de Donnée d'Occupation des Terres (BDOT) at its disposal that reflect the state of the territory in 1992 and 2002, respectively.

The project to produce those BDOT databases was funded by the Danish cooperation, launched and coordinated by the NPLM2 with support from the Ministry of Agriculture, Hydraulics and Halieutic Resources, the Ministry of the Environment and Quality of Life, and the PNGIM (Programme National de Gestion de l'Information sur le Milieu) network, and was entrusted to IGN FI. This BDOT is the European CLC concept adapted for the Sahelian conditions. The BDOT1992 and BDOT2002, and the change database, which is a product of the two BDOT databases, proved a fundamental source of information for the

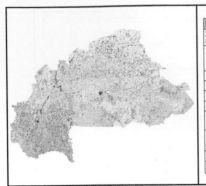

First level	Second level	Third level	Fourth level
3 Forest and semi-natural areas			
	32 Shrub and/or herbaceous vegetation associations		
		321 Herbaceous vegetation	
			3211 Natural grassland
			3212 Herbaceous savannah
			3213 Shrub/tree savannah
		322 Shrub and Bushy vegetation/*matorrales*	
		323 Sclerophyllous vegetation	
		324 "Paramo" formation	
		325 "Rupicole" vegetation	

FIGURE 25.6

LC database in Burkina Faso. The table in this figure is an example of European CLC adaptation in Burkina Faso (black: CLC European classes; italics: Burkina Faso classes).

implementation of biophysical accounts monitoring and diachronic analysis of the land occupation changes occurring on the steppes and savannahs during the period 1992–2002 (Jaffrain, 2007).

Figure 25.6 shows how the specific forest and savannah class characteristics of Burkina Faso territory have been integrated into the conventional European CLC nomenclature. Dense forest, open forest, gallery forest, and forest plantation have been included into the "Forests class" of the second level on European CLC nomenclature (see Figure 25.6). "Savannah" is included under the *Shrub and or herbaceous vegetation association* described at the third level of CLC nomenclature. Savannah formation is a grassland ecosystem characterized by the trees or shrubs being sufficiently widely spaced so that the canopy does not close. An herbaceous carpet is dominant in savannah formation. Three types of savannah have been identified and are included at the fourth level of the nomenclature (see Figure 25.6).

25.4.5 Overseas Departments (2007–2008)

After defining the nomenclature's technical characteristics for LC in French overseas departments, in partnership with the Service des Observations et des statistiques Ministère de l'Ecologie et du développement durable et de l'énergie (SOeS) and the French IGN (IGN FI, IGN, SOeS, 2010), the production of LC and LC change databases identified between 2000 and 2006/2007 was undertaken by the SIRS Company. The LC database at a scale of 1:100,000 was derived from the official CLC European method. The nomenclature and technical specificities were adapted for Guadeloupe, Martinique, La Reunion, and the coastal French Guiana and take into account the landscape diversity as well as the dominant agricultural practices of these new territories (Figure 25.7). To recall, a first study of CLC nomenclature adaptability had been already carried out by Braustein (1997) in the frame of the CARIGEOBASE project. Today the CLC database and its changes are available at: http://www.statistiques.developpement-durable.gouv.fr/.

Figure 25.7 illustrates new agricultural classes present in the overseas territories (Guyana, Guadeloupe, Martinique, and "La Reunion"). In this case, the additional classes sugar cane, banana plantation, and palm grove were included by a corresponding third level (italics) (see Figure 25.7).

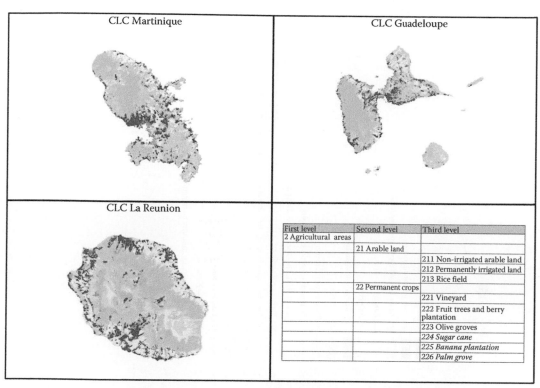

FIGURE 25.7
CLC overseas departments. The table in this figure is an example of an additional item in the third European CLC (black: CLC European classes; italics: overseas department classes).

25.4.6 Central Africa: Gabon (2010–2012)

As part of the "Emerging Gabon" initiated by President Ali Bongo as a strategy for enhancing human resource potential, natural and mine extraction of Gabon has been implemented. In this context, a National Geomatics Plan (NGP) was established to manage and coordinate the national spatial data infrastructure and the resulting key applications (land, planning, environmental risk, transport). The implementation of a multitemporal study (1990, 2000, and 2010) of LC and LC change in the Libreville estuary area was part of this NGP.

By crossing the LC information with a soil database and information dealing with legislation and licenses for areas such as mining, petrol extraction, or forestry it was possible to show the potential LU conflicts in this territory, and the soils potentially favorable to agriculture extension. This pilot project was presented in Libreville in December 2011 to the Gabonese Ministry of Budget and Agriculture and at the Environmental Systems Research Institute (ESRI) conference in Versailles in October 2012 (PNG, IGN FI, 2012).

A particular agricultural class has been added in the standard CLC nomenclature mentioned in the EEA/ETC/SIA (2012). "Class 245" represents a mosaic of agroforestry, food crops, and natural areas. They are areas principally occupied by agriculture within an agroforestry system associated with small food crops, interspersed with significant natural areas (Figure 25.8). These crops are generally subject to a rotation system every 10 years or more. The land use intensity is very high.

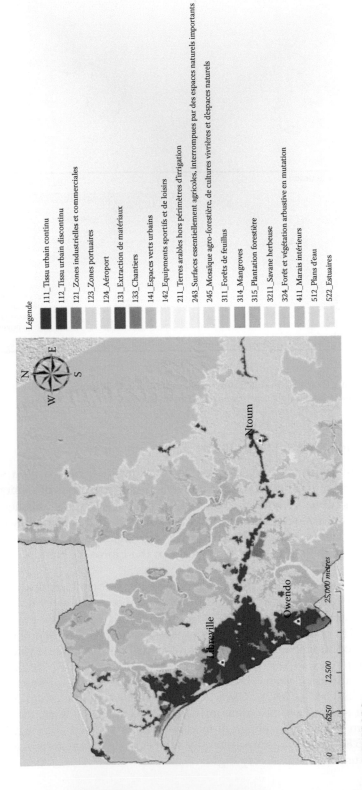

FIGURE 25.8
Project NPG: CLC applied in a part of Gabon—Region of Libreville.

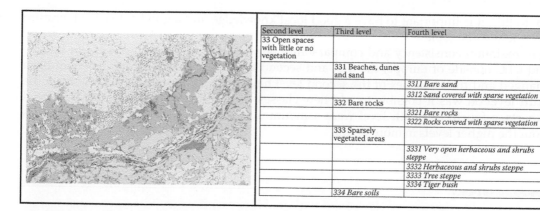

Second level	Third level	Fourth level
33 Open spaces with little or no vegetation		
	331 Beaches, dunes and sand	
		3311 Bare sand
		3312 Sand covered with sparse vegetation
	332 Bare rocks	
		3321 Bare rocks
		3322 Rocks covered with sparse vegetation
	333 Sparsely vegetated areas	
		3331 Very open herbaceous and shrubs steppe
		3332 Herbaceous and shrubs steppe
		3333 Tree steppe
		3334 Tiger bush
	334 Bare soils	

FIGURE 25.9

Land Cover layer database in Mali. Extract of Segou. The table in this figure is an example of European CLC adaptation in Mali (black: CLC European classes; italics: Mali classes).

25.4.7 Mali: Updating Topographic Maps at a Scale of 1:200,000 (2013–2015)

Most of western African countries made their topographic maps at a scale of 1:200,000 during the colonial period in the years 1950–1960. After 60 years, the need to update these data remains a priority for many governments in Africa. In this context a project to upgrade the topographic maps at a scale of 1:200,000 and the geodatabases in Mali was initiated and funded by the European Union.

IGN/IGN FI, in close collaboration with the Geographical Institute of Mali (IGM) (around 40 staff), produced an update of the topographical map and handled the production of the geodatabases. All (thematic) layers of the topographic map such as the hydrographic and roads network, contour lines, and the toponymy, has been updated from recent ortho-rectified SPOT 6 satellite images acquired in the year 2011/2012 having 1.5 m spatial resolution. A huge effort has been made to update the LC information. The LC classification was defined by the group of IGN, IGM, and IGN FI (2015). The nomenclature in Figure 25.9 is perfectly compatible with the basic principle of CLC and its hierarchical structure. New specific classes such as tiger bush, very open herbaceous and shrubs steppe, and rocks with sparse vegetation have been defined and identified through SPOT 6 satellite images. This new update allows, besides other things, identification of the phenomena of desertification in Sahel and monitoring of the strong human pressures on the natural environment and the protected forest areas.

25.5 Results and Conclusion

The various studies and projects conducted since 1990 in the aforementioned countries have proved the adaptability of the CLC concept to tropical and semi-tropical climates.

For each country a very good knowledge of the dominant landscape, the characteristics of the natural areas, the main industrial crops, the LU, and human practice of LC is a fundamental element in establishing the complementary classes of CLC nomenclature.

To do this, it is important to have a good local knowledge and a good extensive literature overview of the topic.

To maintain consistency and comparability between the application of CLC in Europe and CLC outside of Europe, the first and second levels of CLC in Europe have not been changed and the 14 classes of the second level have been fully preserved. Only from the third level on was it possible to add new classes. Depending on requests from national experts, a fourth and fifth level were sometimes established to describe certain classes from the higher levels in more detail.

References

Agency, Copenhagen, Denmark. Available at: http://projects.eionet.europa.eu/leac/library/cube /land_cover/presentation_leac_burkina_faso_ppt.

Braunstein, S. 1997. CARIGEOBASE Une base de donnée d'occupation du sol, outil régional au service de l'environnement et de la gestion de l'espace. Paris: Étude de Faisabilité.

CIRAD-TERA (Centre de Coopération International en Recherche Agronomique pour le Dévelopment–Département Territories, Environement et Acteurs, IGN FI (Institut Géographique National France International). 2003. Projet SHERPA, Applications Contribution au développement durable et à l'adaptation aux changements climatiques en Amérique Centrale. No. 20/04 November 2003.

EEA/ETC/SIA (European Environment Agency/(European Topic Centre on Spatial Information and Analysis). 2011. Nomenclature adaptation to other bio-geographical regions studies and project from 1990 to 2010, Task 2.6.1. Cooperation with ESA and FAO—CORINE Land Cover outside of Europe. Copenhagen: European Environment Agency.

EEA/ETC/SIA (European Environment Agency/(European Topic Centre on Spatial Information and Analysis). 2012. Proposal of land cover systems classification for ecosystem accounting, Final report—ETC/SIA. Copenhagen: European Environment Agency. Available at: http://forum .eionet.europa.eu/etc-sia-consortium/library/2012_subvention/261_land_data_sets/511 _cooperation/proposal-land-cover-systems-classification-ecosystem-accounting.

IDEAM (Instituto De Hidrologia, Meteorologia y Medio Ambiente), IGAC (Instituto Geografico Agustin Codazzi), CORMAGDALENA (Corporacion Autonoma Regional del Rio Grande de La Magdalena). 2008. Mapa de Cobertura de la Tiera Cuenca Magdalena-Cauca. Methodologia Corine land cover adaptada a la Colombia—Escala 1:100,000.

IGN FI (Institut Géographique National France International), ANPE (Agence Nationale de Protection de l'Environnement), 1998. Life 95/TN/B2/TN/928/MED SAIDE August, Final report.

IGN FI (Institut Géographique National France International), IGN, SOeS (Service des Observations et des statistiques Ministère de l'Ecologie et du développement durable et de l'énergie). 2010. Nomenclature Corine land cover pour les DOM rapport final. Available at: http://www .statistiques.developpement-durable.gouv.fr/donnees-ligne/li/1825.html.

IGN (Institut Géographique National), IGM (Geographical Institute of Mali), IGN FI (Institut Géographique National France International). 2015. Guide méthodologique Occupation du sol au Mali, carte 1/200.000.

Jaffrain, G. 2007. Diachronic analysis and land cover accounts in Burkina Faso, BDOT Presentation EIONET meeting, Copenhagen January 2006. European Environment.

Ministerio de Medio Ambiente y Recursos Naturales, 2005. Atlas de Agricultura y bosque, El Salvador.

PNG (Programme National de Gouvernance), IGN FI (Institut Géographique National France International). 2012. Projet pilote: Identification des terres agricoles et gestion durable du territoire—IGN FI, SIG 2012 Presentation ESRI conference in Versailles—4 October 2012.

Section V

CLC Perspective

26

Detailed CLC Data: Member States with CLC Level 4/Level 5 and (Semi-)Automated Solutions

Gerard Hazeu, György Büttner, Antonio Arozarena,
Nuria Valcárcel, Jan Feranec, and Geoff Smith

CONTENTS

26.1 Introduction

Alongside the European CoORdination of Information on the Environment (CORINE) Land Cover (CLC) data generation, activities for more detailed levels of land cover (LC) production take place in various European member states. All of these national LC data generation approaches vary in methodology, geometry, and thematic detail. The approaches range from land cover/land use (LU/LC) information derived directly from satellite data to LU/LC information derived largely from existing national data. The methodologies applied cover the range from computer-assisted photointerpretation (CAPI) to (semi-)automated generation (SAG) of LU/LC data out of a combination of data sources. The geometric detail/minimum mapping unit (MMU) varies from less than 1 ha to the 25 ha of CLC and the semantics are often more detailed for specific categories given the objective for which the national dataset is produced or the landscape specificities of the member states.

As well as the variation in national LU/LC data generation approaches, the relation between the national and European CLC approaches at the member states level also shows a great diversity within Europe, including single- and two-track approaches. In the single-track approach the national LU/LC data generation and European mapping are integrated. The national LU/LC data can serve as the basis for the European mapping (bottom-up) or the other way round (top-down). The dual track approach at the member states level, that is, national and European LU/LC data generation activities that are completely separated from each other, results in a categorization of LU/LC based on different nomenclatures (thematic resolution), on different geometrical standards (spatial resolution), and/or different reference years (temporal resolution). History, ownership within the member states, and different user needs are often the reasons for these variations between national and European LU/LC approaches at the member states level.

The objective of this chapter is to present examples of the following three cases:

1. European CLC as the basis for national enhanced CLC nomenclature (i.e., more detailed semantic description—CLC Level 4/5) (e.g., PHARE countries and Hungary)
2. (Semi-)automated solutions to generate the European CLC from (detailed) national data (e.g., Sistema de Información de Ocupación del Suelo en España [SIOSE]/ Land cover and land use information system of Spain, UK Countryside Survey)
3. Independent national LU/LC data and European CLC production (e.g., LGN [Landelijk Grondgebruiksbestand Nederland] and CLC—Netherlands)

Specific attention in the examples will be given to

1. Which national data are used for the CLC data generation activity (see also Chapter 4)
2. Which methodology is applied to generate the CLC data (see also Chapter 5)

Within the FP7 project HELM "Harmonised European Land Monitoring" an extensive inventory of commonalities, differences, and gaps in national and subnational land monitoring systems* took place. This chapter incorporates some of these results.

26.2 European CLC as the Basis for National CLC Level 4/5

26.2.1 Fourth Hierarchic Level of CLC Nomenclature for the PHARE Countries

The cognition and identification of LC in the context of the CLC project (Heymann et al., 1994; Bossard et al., 2000) gives a description of how the landscape appears at a scale of 1:100,000. More detailed information about LC on local and regional levels (i.e., in more

---

* HELM deliverable 3.1: Commonalities, differences and gaps in national and sub-national land monitoring systems (Ben-Asher et al., 2013).

spatial detail than a MMU of 25 ha) is often necessary in landscape research and national/ regional policy implementation. A finer scale (<1:100,000) initiated the development of a more detailed CLC nomenclature.

Preparation of a CLC nomenclature at the scale of 1:50,000 started in the Poland and Hungary Assistance for the Restructuring of the Economy (PHARE) Programme countries in 1995. The first draft of this nomenclature accepting the national particularities of LC in the Czech Republic, Hungary, Poland, and Slovakia has been complemented by experts from these countries and summarized at the Institute of Geography, SAS in Bratislava (Feranec et al., 1995). The report contains the methodology used for photointerpretation of satellite images at a scale of 1:50,000, a draft of CLC nomenclature for the given scale, definitions of LC classes, brief characteristics of options for linking up the CLC classes with the CORINE Biotopes classes, and a draft map with a representation of the fourth level of CLC classes. In 1995–1996, the application test of the proposed nomenclature in the Czech Republic, Hungary, Poland, and Slovakia (120 map sheets at a scale of 1:50,000) was carried out. After appraisal of the test and completion of the nomenclature adding the national particularities of LC in Bulgaria, Estonia, Latvia, Lithuania, Romania, and Slovenia, the fourth hierarchic level of CLC nomenclature with 105 classes acquired its final form (see Appendix 26A) (Feranec and Otahel, 1999). A map fragment M-34-133-C (Šurany, Slovak Republic) containing CLC classes of this detailed level is provided in Figure 26.1 (Feranec et al., 1994).

26.2.1.1 Delimitation Criteria Applied for Identification of the Fourth Level CLC Classes

The subdivision of CLC's third hierarchic class level (at a scale of 1:100,000) into the fourth class level (at a scale of 1:50,000) was carried out by the application of three delimiting criteria (Feranec and Otahel, 2004):

- A more detailed MMU of 4 ha better identifies areas of the new classes in the context of the proposed CLC nomenclature at a scale of 1:50,000. They will be relatively more homogeneous than the third hierarchic level classes, which often generalizes diverse features. For instance, areas of class 2111 arable land and class 2311* pasture with areas of 4 ha and more can be derived from the class 243 Land principally occupied by agriculture with significant areas of natural vegetation which, from its definition, contains areas of arable land, pastures, forests (at scale 1:100,000), and so forth.

- The use of morpho-structural and physiognomic attributes of LC objects differentiates the internal heterogeneity of CLC classes at a scale of 1:100,000. These attributes manifest themselves on satellite images via physiognomic characteristics (such as size and shape of buildings, arrangement and type of supplementary parts). For instance, 112 Discontinuous urban fabric requires the presence of gardens and settlement greenery and 313 Mixed forests has alternation of broadleaved and coniferous trees or groups of trees within the class, thus facilitating further itemization at a scale of 1:50,000.

- The incorporation of biogeographical (or regional) specificities (particularities) further differentiates the third level CLC classes. For instance, the nature of the soil can distinguish between broad-leaved forest on acid substrate and broadleaved forest on carbonate substrate.

---

* Explanation of the CLC Level 4/5 class codes can be found in Appendix 26A.

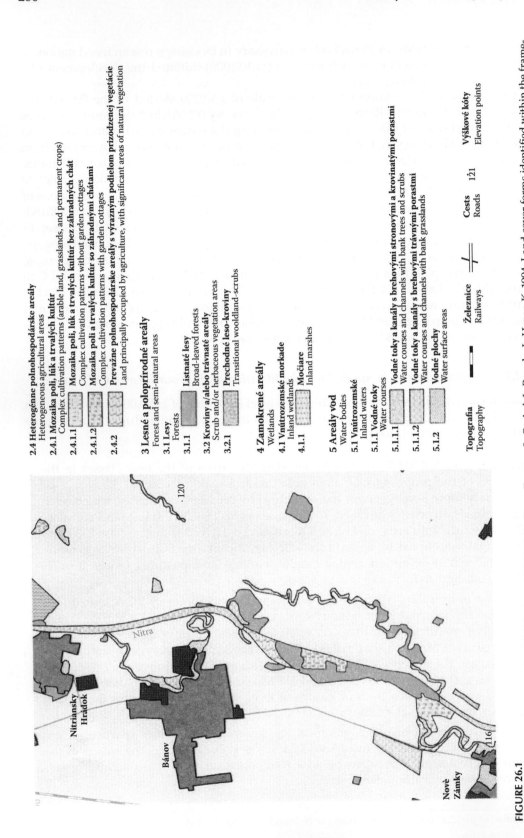

2.4 Heterogénne polnohospodárske areály
Heterogeneous agricultural areas

2.4.1 Mozaika polí, lúk a trvalých kultúr
Complex cultivation patterns (arable land, grasslands, and permanent crops)

2.4.1.1 **Mozaika polí, lúk a trvalých kultúr bez záhradných chát**
Complex cultivation patterns without garden cottages

2.4.1.2 **Mozaika polí a trvalých kultúr so záhradnými chátami**
Complex cultivation patterns with garden cottages

2.4.2 **Prevažne polnohospodárske areály s výrazným podielom prírodzenej vegetácie**
Land principally occupied by agriculture, with significant areas of natural vegetation

3 Lesné a poloprírodné areály
Forest and semi-natural areas

3.1 Lesy
Forests

3.1.1 **Listnaté lesy**
Broad-leaved forests

3.2 Kroviny a/alebo trávnaté areály
Scrub and/or herbaceous vegetation areas

3.2.1 **Prechodné leso-kroviny**
Transitional woooldland-scrubs

4 Zamokrené areály
Wetlands

4.1 Vnútrozemské morkade
Inland wetlands

4.1.1 **Močiare**
Inland marshes

5 Areály vôd
Water bodies

5.1 Vnútrozemské
Inland waters

5.1.1 Vodné toky
Water courses

5.1.1.1 **Vodné toky a kanály s brehovými stronovými a krovinatými porastmi**
Water courses and channels with bank trees and scrubs

5.1.1.2 **Vodné toky a kanály s brehovými trávnými porastmi**
Water courses and channels with bank grasslands

5.1.2 **Vodné plochy**
Water surface areas

Topografia
Topography

Železnice
Railways

Cesta 1̇21̇
Roads

Výškové kóty
Elevation points

FIGURE 26.1
Fragment of the map of LC at scale 1:50,000, sheet M-34-133-C. (From Feranec, J., Otahel, J., Pravda, J., Husar, K. 1994. Land cover forms identified within the framework of the project CORINE land cover. *Geografický časopis* [Geographical Journal] 46:35–48 [in Slovak].)

The basic draft of the CLC 1:50,000 nomenclature was prepared by J. Feranec and J. Otahel and during a meeting at Smolenice, Slovak Republic (January 1995) it was completed and approved.* This nomenclature was further elaborated into its final version with contributions of national CLC teams[†] (Feranec et al., 1998).

26.2.2 CLC50 in Hungary

The motivation to create a higher spatial resolution CLC has been to support Hungary's accession to the European Union (EU). In this context a strong need was recognized for detailed and up-to-date LC information. The standard European CLC1990 was considered already outdated (produced using satellite images taken in 1990–1992), in addition, it did not provide sufficient detail for national purposes. A resolution of the Hungarian government on the "Development of environmental information systems" (2339/1996(XII.6.)) ordered the "Setting up a CLC database at a scale of 1:50,000." In 1999, in the frames of the Acquis National Programme—the Ministry of Environment and the Ministry of Agriculture and Regional Development initiated the CLC mapping project at a scale of 1:50,000 (CLC50) (Büttner et al., 2001). The country had been mapped in this fashion by the end of 2003. The applied methodology and some of the results are presented in Sections 26.2.2.1 and 26.2.2.2 (Büttner et al., 2004).

26.2.2.1 Methodology of CLC50

The CLC50 can be considered as national enhancement of the standard European CLC product.

Thematic classes have been developed from the standard CLC Level 3 nomenclature, using recommendations of PHARE countries for 1:50,000 scale mapping (Feranec et al., 1995). This enhanced nomenclature (Level 4 and Level 5 classes) includes 79 categories, adapted for Hungarian conditions. Owing to the complexity of the nomenclature and the lack of updated and detailed environmental datasets, CLC50 has been derived by means of CAPI. Precisely ortho-rectified 20 m spatial resolution SPOT 4 satellite images taken in 1998–1999 and the visual photointerpretation on computer screen allowed for high positional accuracy of delineation of LC polygons. The 4 ha (1 ha for water) size of MMU provided enhanced geometric detail compared to the 25 ha MMU of the European CLC. Digitized topographic maps (1:50,000) and orthophoto prints were used by photointerpreters as primary ancillary data. Field checking has been part of the mapping process. Internal quality control has the main aim to harmonize the work of photointerpreter team (15 experts) to ensure similar understanding and application of the nomenclature and to provide a homogeneous database.

One of the innovations of CLC50 was the development of low-cost and easy to use CAPI software, as before 2000 there were no suitable tools available on the market. Consequently,

---

* With contributions from M. Baranowski, M. Bossard, G. Büttner, A. Ciołkosz, J. Kolar, and Ch. Steenmans.
[†] Bulgaria (D. Kontardiev), Czech Republic (D. Zdenkova), Estonia (K. Aaviksoo), Lithuania (R. Kaulakys), Latvia (H. Baranovs), Hungary (G. Büttner), Poland (M. Baranowski), Romania (V. Vajdea), Slovakia (J. Feranec and J. Otahel), and Slovenia (A. Kobler and B. Vrscaj).

CLC1990, the first standard European CLC inventory, was based on traditional photointer-pretation* (see Chapter 5). This procedure often caused several types of error:

- Geometrical errors given by an imprecise hardcopy image size, distortions of the hardcopy, improper alignment of overlay and image
- Geometrical errors of digitization
- Thematic mistakes introduced during the database coding phase
- Thematic errors because of the limitations of hard copy image

CAPI completely eliminates the first three of the aforementioned errors, and reduces the last one by optimal combination of capabilities of the human expert and that of the computer. The main benefits over the traditional approach include

- Magnification of the imagery using the zoom function
- Possibility of using multitemporal imagery
- More precise delineation of polygons
- Easy corrections (lines, codes)
- Automatic filtering out of invalid codes
- Automatic checking of polygon geometry (area, average width)
- Online nomenclature
- Possibility to use comments and remarks at the polygon level
- Data exchange via the Internet

The above functions were implemented under the ESRI ArcView software under the name InterView (Taracsák, 1999). The current version (called InterChange) is a stand-alone software tool that supports revision of CLC and mapping changes (Taracsák, 2012). It is used by more than half of the countries implementing CLC2012 as part of the Global Monitoring for Environment and Security (GMES) Copernicus Initial Operations (GIO) land program (Copernicus Land, 2014).

The other innovation of CLC50 was the external quality control. Experts from two national networks, the National Park Directorates and Plant Health and Soil Protection Service, were contracted to control the results of photointerpretation. These two networks represent a wealth of field knowledge and provide access to field data that were compared to the photointerpretation. About 50 external experts have contributed to CLC50 with their written, georeferenced remarks, which were used to correct the initial CLC50.

The main technical parameters of the Hungarian CLC50 are listed and compared to the standard European CLC in Table 26.1. Figure 26.2 is a flowchart of CLC50 data processing.

26.2.2.2 Results of CLC50

Naturally, the results of CLC50 include more details than the standard European CLC (Figure 26.3). The ratio of the total number of polygons in CLC50 to that of CLC1990 is 7.31.

---

* A transparent overlay was fixed on top of the satellite image hardcopy, and the photointerpreter drew polygons with a three-digit CLC code on it. After the interpretation was completed and supervised, the polygons were digitized, a topology was created, and the CLC codes were entered.

TABLE 26.1

Comparison of Main Parameters of European CLC and Hungarian CLC50

Parameter	CLC1990 Hungary	CLC50
Nomenclature	Standard EU Level 3	Extended Level 4/5
Methodology	Hardcopy PI	Softcopy PI
Satellite imagery	Landsat 5 TM (30 m pixel size)	SPOT 4 XS (20 m pixel size)
Area resolution (MMU)	25 ha for all categories	4 ha in general, 1 ha for lakes
Linear resolution, minimum feature width (MFW)	100 m	50 m
No. of classes	27 (out of 44)	79
No. of polygons	24.000	174.000
Positional accuracy	<100 m (RMS)	<20 m (RMS)
Thematic reliability	>85%	>90%
Supervision	Not documented: direct corrections on plastic overlays	Documented: remarks on polygon level (instructions for corrections)
External quality control	No	Yes
Final product	Topologically structured database in ArcInfo format	

Source: Büttner, G., Maucha, G., Bíró, M., Kosztra, B., Pataki, R. and O. Petrik, 2004. National land cover database at scale 1:50,000 in Hungary. *EARSeL eProceedings* 3(3), 323–330. http://www.eproceedings.org/. With permission from EARSeL.

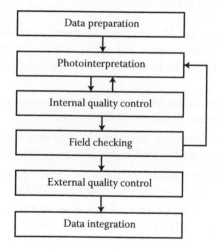

FIGURE 26.2
Flowchart of CLC50 project. (From Büttner, G., Maucha, G., Bíró, M., Kosztra, B., Pataki, R. and O. Petrik, 2004. National land cover database at scale 1:50,000 in Hungary. *EARSeL eProceedings* 3(3), 323–330. http://www.eproceedings.org/. With permission from EARSeL.)

The number of linear elements (rivers, channels, highways, railway sections) has increased significantly compared to CLC1990. Polygon numbers in some categories that are occupying relatively small size areas also increased significantly, such as CLC classes 121, 123, 132, 133, 141, 312, 324, and 411.* The large increase in the number of polygons for the arable land class (211) is explained by the introduction of two subclasses in CLC50 characterizing the dominant size of parcels. Concerning the comparison of class areas, the largest relative increase can be seen in the road and rail network class (122). The reason is that

---

* Explanation of CLC Level 3 class codes can be found in Chapter 3 (Table 3.1).

FIGURE 26.3
Comparison of the standard European CLC (left) and CLC50 (right) for Győr (NW Hungary). (From Büttner, G., Maucha, G., Bíró, M., Kosztra, B., Pataki, R. and O. Petrik, 2004. National land cover database at scale 1:50,000 in Hungary. *EARSeL eProceedings* 3(3), 323–330. http://www.eproceedings.org/. With permission from EARSeL.)

highways and several railway stations are included in the CLC50 database (and missing in the standard CLC) because of smaller geometrical limits.

Some of the important applications of CLC50 are

- CORINE Habitat Mapping: Natural/semi-natural classes of CLC50 are further detailed by botanists according to a habitat nomenclature, by using field data. The CORINE Habitat Map has been developed for most of the areas under nature conservation.

- Lake inventory to support the Ministry of Environment to prepare for the implementation of the EU Water Framework Directive.

- Important Bird Areas: Study of correlation between recorded bird occurrences and the landscape, represented by CLC50 (Báldi, 2003).

- Computation of agri-environmental indicators by comparing CLC1990 and CLC50 for three mezo-regions of Hungary (Büttner et al., 2002).

- CLC2000 project: The update of the first European CLC database in Hungary was achieved by using a generalized CLC50 (Büttner, 2003).*

26.3 Semiautomated Generation of CLC

An increasing number of countries have moved from conventional visual photointerpretation or CAPI toward semiautomatic approaches of CLC data production. They do so in particular to avoid high labour costs of photointerpretation, improve repeatability and increase

---

* It was a bottom-up solution, which is strongly supported by EEA. It means deriving the LC dataset for national purposes, which is to be converted for the EEA to support European needs. Unfortunately, CLC50 was not updated (until 2014) because of financial constraints, so it could not support the development of CLC2006 and CLC2012.

consistency with national datasets.* Methodological key elements of these semiautomated data production are discussed in Chapter 5. The semiautomated methods were initially applied in Finland, Iceland, Norway (Aune-Lundberg and Strand, 2010), Sweden, Switzerland, and the United Kingdom (Smith et al., 2007; Smith, 2014). In the case of CLC 2012 the methodology was also applied in Germany (Arnold, 2010; Hovenbitzer et al., 2014), Ireland, Austria, and Spain.

26.3.1 Sistema de Información de Ocupación del Suelo en España (SIOSE—LC and LU Information System of Spain)

26.3.1.1 Background

LC and LU information in Spain is produced and coordinated between different public administrations. The National Geographic Institute of Spain (Instituto Geográfico Nacional [IGN]) as the National Reference Centre on Land cover, Land use and Spatial Planning of the European Environment Information and Observation Network (NRC [EIONET]) coordinated the production of CLC (1990, 2000, and 2006 versions). CLC2000 was produced in cooperation between regional and national administrations using the approach described in Section 26.2, using the standard European CLC hierarchical model as the basis for a national CLC Level 4/5. CLC versions from 1990 to 2006 where created manually by visual photointerpretation over satellite images and orthophotos, supplemented by existing National thematic data such as National Crop Map and Forest Map, Cadastre, and so forth (Arozarena et al., 2004).

But even during the production of CLC2000, it became clear that there was a strong mismatch between the potential uses of CLC data and the real needs of national and subnational Spanish users. Therefore, it was necessary to establish a new information system to embrace Spanish requirements and build a structured flow of data, able to satisfy European, national, and subnational users' needs in a harmonized technological and organizational framework (bottom-up approach). With these aims, SIOSE was developed (Valcarcel et al., 2008). As well as in CLC2000, SIOSE production is made in cooperation between regional and national administrations, and its management, coordination, and quality control falls under the responsibility of IGN. The main objectives of SIOSE are

- Avoid data duplication and reduce costs of Geographic Information
- Provide a production and quality model in cooperation between national and regional administrations
- Satisfy Spanish National and Regional Administration requirements on LC/LU information
- Integrate Regional Administrations in the management, quality control, and productions of the LC and LU national database and information system
- Satisfy EEA's and EU requirements for future LC and LU information (e.g., CLC)
- Integrate LC and LU databases and information of the Spanish national institutions

Currently, three versions of SIOSE have been produced: SIOSE 2005, 2009, and 2011. These SIOSE versions make it possible to finally address the generation of CLC2012 via (semi-) automatic processes, by geometric and semantic generalization and geoprocessing,

---

* HELM deliverable 3.1: Commonalities, differences, and gaps in national and sub-national land monitoring systems.

applying the principle of the European Directive INSPIRE (Directive 2007/2/EC of the European Parliament and of the Council of March 14, 2007 establishing an Infrastructure for Spatial Information in the European Community) (see Chapter 27), capturing the data only once at the most efficient level, and use it repeatedly to generate LC/LU products required for both national and subnational users as European and global needs.

26.3.1.2 SIOSE Methodology

SIOSE is a geographic information system at an equivalent cartographic scale of 1:25,000 covering more than 504.000 km² with one unique layer of polygons, whose minimum area varies from 2 ha to 0.5 ha according to the LC class (SIOSE, 2009).

During the first SIOSE production (2005–2009) a great effort was made to integrate reference and thematic information (as an initial SIOSE "skeleton") such as IGN's reference topographic data (hydrography, roads, railway network), cadastre (urban limits, street axes), agricultural (National Crop Map, and the Spanish Land Parcel Information System [LPIS] database), and forest (National Forest Map) thematic reference data, plus other LC/LU databases in the Spanish regional administrations (if they exist). These reference data were used also for the SIOSE updates 2009 and 2011. Using this SIOSE "skeleton" as the initial geometric and thematic information layer, the 2005 photointerpretation was done by the regional teams using reference image satellite images (SPOT 5 2005 coverage, 2.5 m pixel) and national orthophotos (0.5 m pixel) as geometric and temporal reference. After the photointerpretation phase, a detailed quality control by the regional teams was followed by a final one by IGN, to ensure the adequate homogenization of data considering the entire Spanish territory. The SIOSE updates in 2009 and 2011 were based on semiautomatic photointerpretation, assisted by previous change detection processes using remote sensing techniques and conflation of reference data mentioned previously.

The methodology of class assignment is based on land description instead of classification. SIOSE polygons describe individually each piece of land by allocating a set of real landscape objects with percentages (Valcarcel et al., 2008; SIOSE, 2009; Delgado et al., 2012 as shown in Figure 26.4. This multipurpose philosophy builds on the object-oriented data model and makes possible diverse exploitations of the data depending on the user. This goal is achieved allocating one or more landscape objects to the same polygon, specifying

Continuous urban fabric
  85%—Buildings
  10%—Roads and other artificial surfaces
  5%—Green areas

Urban park
  95%—Green areas
  5%—Roads and other artificial surfaces

FIGURE 26.4
SIOSE data model allocation of more than one value of landscape object per polygon. Left and right images are two examples of different polygons. (From IGN, SIOSE data.)

the area percentage occupied by each object within the polygon. By working in this way, the semantic content inside a polygon is not limited by only one label or minimum size of geometry allowed. The SIOSE methodology makes it possible to store more information than traditional hierarchical classification systems or thematic maps at the same level of geometric detail. Furthermore, its information is also more versatile (Valcarcel et al., 2008). The SIOSE data model defines 45 classes of landscape objects (e.g., buildings, bridges, etc.), 40 LC classes, and more than 20 attributes for further description of LC characteristics (e.g., under construction, abandoned, etc.).

26.3.1.3 Generalization of CLC2012 Using SIOSE

One of the main objectives of SIOSE was to generate, as automatically as possible, CLC for the reference dates 2006 and 2012 out of the SIOSE data for references dates of 2005, 2009, and 2011. The allowed temporal ranges in the CLC2012 production is set as +1 year or −1 year from 2012. Therefore, the strategy is to generate CLC changes out of the SIOSE changes 2005–2011 according to CLC technical specifications.

The generation of CLC2012, performed by the IGN, is being done mainly by automatic geoprocessing, aided by two visual reviews of the resulting CLC2006 and CLC changes 2006–2012 (CLC$_{2006-2012}$) products to ensure data quality. Geoprocessing includes overlaps, geometric, semantic generalization, and aggregation processes, according to CLC guidelines. Figure 26.5 summarizes the steps taken.

This approach was possible because the SIOSE innovative object-oriented data model provides multipurpose outputs required by different types of users, in this case EEA and other European and global institutions. Generation of CLC is only one example, but further potential derived systems, nomenclatures, queries, or consultations coming from SIOSE data are also possible (e.g., crown cover density predominant typology of buildings, reclassification for topographic map representation, etc.). In this case, SIOSE provides a successful experience of bottom up production from national to European data, according to INSPIRE.

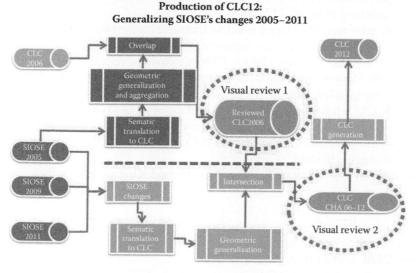

FIGURE 26.5
Production of CLC2012 from SIOSE. (From IGN.)

26.3.2 Countryside Survey UK (LCM2007 for CLC2006)

The United Kingdom has contributed to all CLC maps. The policy adopted until the current 2012 update has been to apply "bottom-up" techniques to generalize, transform, and edit the national LC product into the CLC specification. However, owing to changing technology and territorial responsibilities, no two UK productions of CLC have been the same, even though the policy is the same.

CLC1990 was produced for the United Kingdom in two administrative units. The CLC1990 for Great Britain (England, Scotland, and Wales) was derived by semiautomated generalization of the 1990 Land Cover Map of Great Britain (LCMGB) (Fuller et al., 1994), but not completed until 1998 (Brown et al., 1999, 2002). The CLC1990 for the Republic of Ireland and Northern Ireland was undertaken on an all-Ireland basis using the standard CLC methodology and was completed in 1993 using data from 1989 and 1990 (O'Sullivan, 1994). CLC2000 was produced on a full UK basis by semiautomated generalization of the Land Cover Map 2000 (LCM2000) (Fuller et al., 2005). CLC2006 was again produced on a full UK basis by semiautomated generalization (1Spatial, 2007) of the Land Cover Map 2007 (LCM2007) (Morton et al., 2011) for Great Britain and by conventional update of CLC2006 in Northern Ireland by visual photointerpretation.

The CLC1990, CLC2000, and CLC2006 production in the United Kingdom used broadly similar semiautomated generalization procedures; however, the input data differed and thus had impacts on the details of the processes required to generalize and thematically transform the results.

The CLC1990 used the raster-based LCMGB, and although the LCM2000 was vector based, it was converted to a raster for CLC2000 production to use adapted scripts from the CLC1990 production.

The CLC2006 production is described in more detail in the text that follows, as this is the most advanced version of the approach. The CLC and LCM specifications created a number of challenges for production of CLC based on national LCMs (Figure 26.6). At the spatial level, the UK LCMs have about 100 times the number of objects as the equivalent CLC product. Thematically, the 23 class UK nomenclature focuses on LC and habitats, whereas the CLC nomenclature (34 of the possible 44 classes were present in the United Kingdom) is a mixture of LC and LU classes.

In most cases, rules were developed to allow thematic conversion in an automated or semiautomated fashion. The generalization from the 0.5 ha MMU of LCM to the 25 ha of CLC is relatively straightforward for most classes. Others, however, required some form of manual intervention, particularly for the mixed classes, those classes that were based on LU and where CLC classes are impossible to map owing to their complex contextual definitions (e.g., airports).

All of the UK CLCs have used some form and amount of visual photointerpretation of satellite imagery as part of the process. Imagery was made available for CLC1990 from the LCMGB production, and for CLC2000 and CLC2006 the IMAGE2000 and IMAGE2006 datasets were extensively used.

The main steps in producing CLC2006 for the United Kingdom were therefore (1Spatial, 2007)

1. Extraction of the urban LU classes from CLC2000 (CLC Level 1 class 1**). This layer was updated with appropriate recent LU changes via visual interpretation against IMAGE2006 and topographic maps.

2. Generalization of the LCM2007 data into CLC classes as far as possible by thematic recoding and spatial merging.

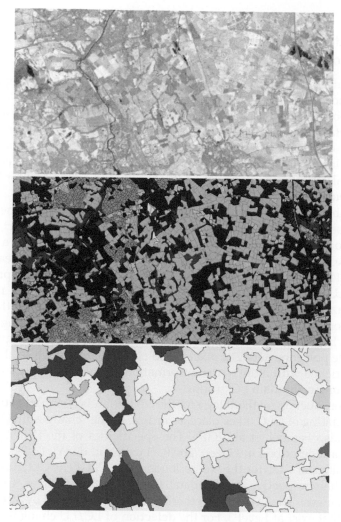

FIGURE 26.6
A comparison of input EO data (top), LCM2007 (middle, Morton et al., 2011) and CLC2006 (bottom).

3. Combination of generalized LCM2007 with the updated LU layer from step 1.

4. Checking with IMAGE2006 for geometric error, and so forth.

5. Detailed visual/interactive edit via photointerpretation, using IMAGE2006 and topographic maps to complete the conversion to the CLC spatial and thematic specification.

6. Edge match between tiles/working units and quality general checking of boundaries.

7. Final quality checking, editing, and delivery of CLC2006 stock product.

8. Automatic and interactive creation of CLC2006 changes database using CLC2000 and CLC2006 with IMAGE2000 and IMAGE2006 for photointerpretation.

9. Visual checking of changes and reconciling differences with the CLC2006 stock map.

26.4 Independent National and European LU/LC Data Generation (LGN and CLC—Netherlands)

In the Netherlands a number of different databases are responsible for mapping and monitoring LC. The most important ones are the topographical database (Top10NL), the LU database "Bestand BodemGebruik" (BBG), the national LU database (LGN), and the national CLC database (Hazeu, 2014). The LGN database combines information from the BBG and Top10NL databases and it integrates information from other additional sources in a second step. As it has a long history in mapping Dutch LC and its changes, a clear LC focus and integrates several other national data the LGN could be a good basis for bottom-up production of the Dutch CLC database.

The LGN database is a gridded database of 25×25 m cells reflecting the LC for 39 well-defined LC classes including urban areas, forest types, water, infrastructure, crop types, and several ecological classes. Since 1986 at regular intervals of 3–5 years a version of LGN was released presenting the LU/LC for the Netherlands for a specific year (Hazeu et al., 2011; Hazeu, 2014). Recently the seventh version of LGN (LGN7) was produced reflecting the Dutch LU/LC for the year 2012 (Hazeu et al., 2014). The LGN6 methodology as described in Hazeu (2014), and Hazeu et al. (2011) was used in a slightly adapted form for the production of LGN7 (Hazeu et al., 2014). The two main processing steps are (1) an object-oriented classification in which national databases are integrated and an LGN class is assigned to the polygons and (2) a pixel-based classification in which the homogeneous polygons are enriched with more spatial information. In the first step also the LC changes are detected among eight main LC groups.

The CLC2012 production methodology for the Netherlands is similar to the CAPI method in which the updating of CLC takes place according to the "changes first" as described in Chapter 5 (see also Feranec et al., 2007). The specificities of the Dutch CLC methodology and results are described in Hazeu (2003), Hazeu and de Wit (2004), and Hazeu et al. (2008). Four versions of CLC for the Netherlands exist: CLC1990, CLC2000, CLC2006, and CLC2012 with their respective change databases (CLC2000 changes, CLC2006 changes, and CLC2012 changes). The LC in the Netherlands is characterized by 30 out of the total of 44 CLC classes that exist at Level 3. For the detection of LC/LU changes satellite imagery of two time steps (t and $t + 1$) is the main input data source. The importance of additional data sources such as aerial photographs and national LC and LU databases in the updating of Dutch CLC is increasing with time. Although the national datasets do not form the (geometrical) basis of the Dutch CLC database they are used in the detection and interpretation of LU/LC changes.

Despite the existence of a broad range of LU/LC information and a long tradition in LU/LC mapping, the Dutch CLC activities are still independent from the production of national LU/LC databases (especially LGN). One of the reasons is that the long-term vision for updating of CLC is not yet stabilized to synchronize it with the national LC mapping (Hazeu, 2014). Furthermore, maintaining consistency with earlier versions, which makes monitoring of LU/LC changes possible, is an obstacle for integrating both CLC and (national) LGN activities. Next to the history of both databases, spatial detail and semantics are different, which makes harmonization difficult as discussed in, for example, Jansen et al. (2008) (see Figure 26.7). Next to technical issues, an important reason for having separate national and CLC is the economy of scale. The Netherlands

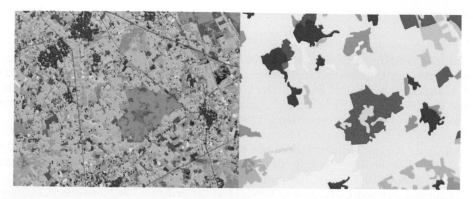

FIGURE 26.7
LGN7 and CLC2012 for an area in the province of Noord-Brabant showing the differences in spatial and thematic between both databases.

are small in surface area, which makes it economically not so cost effective to develop a bottom-up approach that fully integrates national LU/LC and the European activities (Dutch CLC database).

In Hazeu (2014) the semantics of both LGN and CLC are compared, showing that LGN has more thematic detail in agricultural and nature areas and less detail in the urban areas. The relations between LGN and CLC classes are of the 1:1, 1:m, and m:1 types. The following are some examples taken from Hazeu (2014):

- 1:1 relation: Salt marshes (LGN code 30) is 1:1 compatible with salt marshes (CLC code 421).
- 1:m relation: Grass in built-up areas (LGN code 23) can be mapped by construction sites, green urban areas, and sport and leisure areas (CLC codes 133, 141 respectively 142).
- m:1 relation: Maize (LGN code 2), sugar beet (LGN code 4), and cereals (LGN code 5) can be merged into the arable land (CLC code 211) or several LGN heathland classes are compatible with moors and heath lands (CLC code 322).

Other technical obstacles for a bottom-up approach are the incompatibility of the database formats (raster vs. vector), the different MMUs and application scales (25 × 25 m, 1:50,000 vs. 25 ha, 1:100,000), and the temporal synchronization that is often limited (no national equivalent of CLC2006) (Hazeu, 2014). Another major drawback for a bottom-up approach is the adaptation of the methodology, making comparison with previous databases difficult.

The national harmonization of different national databases into one national LU database (LGN) and the unique opportunity of having the same reference year for LGN7 and CLC2012 are positive conditions facilitating a bottom-up production of CLC. However, despite these positive developments the national LC/LU databases are still used only as an additional data source for updating CLC and not as input for a (semi-)automated production of CLC.

References

1Spatial, 2007. CORINE Land Cover Map Feasibility Study Final Project Report for Centre of Ecology and Hydrology. Unpublished 1Spatial report to Centre for Ecology and Hydrology and Defra.

Arnold, S., Busch, A., Grünreich, D., 2010. Das Projekt DLM-DE2009 Landbedeckung. Mitteilungen des Bundesamtes für Kartographie und Geodäsie 45. Arbeitsgruppe Automation in Kartographie, Photogrammetrie und GIS (AgA) Tagung 2009, 9–22.

Arozarena, A., Villa, G., del Bosque, I., Porcuna, A. 2004. CLC2000 in Spain. In: Workshop CORINE Land Cover 2000 in Germany and Europe and its use for environmental applications, January 20–21, 2004, Berlin, UBA Texte 04/04 (pp. 205–208). Available at: http://www.corine.dfd.dlr.de/media/download/ws-clc2000_arozarena_et-al.pdf.

Aune-Lundberg, L., Strand, G.-H. 2010. CORINE Land Cover 2006—The Norwegian CLC2006 project. Report from Norwegian Forest and Landscape Institute 11/2010.

Báldi, A. 2003. Land cover and breeding of bird species in the Important Bird Areas of Hungary (in Hungarian with and English and French summary). Budapest: Hungarian Ornithological and Naturea Conservation Society.

Ben-Asher, Z., Gilbert, H., Haubold, H., Smith, G., Strand, G.-H., 2013. HELM—Harmonised European Land Monitoring: Findings and Recommendations of the HELM Project. Tel-Aviv: The HELM Project.

Bossard, M., Feranec, J., Otahel, J. 2000. *CORINE Land Cover technical guide.* Addendum 2000, Technical Report 40. Copenhagen: European Environment Agency. Available at: http://www.eea.europa.eu/publications/tech40add.

Brown, N. J., Gerard, F. F., Fuller, R. M., Edmundson, D. S. 1999. Generalising the Land Cover Map of Great Britain to CORINE Land Cover by semi-automated means. Centre of Ecology and Hydrology report to DG XVI.

Brown, N. J., Gerard, F. F., Fuller, R. M. 2002. Mapping of land use classes within the CORINE Land Cover Map of Great Britain. *The Cartographic Journal* 39:5–14.

Büttner, G. 2003. CLC2000–Hungary—Phase-I. Final Report, FÖMI.

Büttner, G., Bíró, M., Maucha, G., Petrik, O. 2001. Land Cover mapping at scale 1:50.000 in Hungary: Lessons learnt from the European CORINE programme, 20th EARSeL Symposium, June 14–16, 2000, In *A Decade of Trans-European Remote Sensing Cooperation* (pp. 25–31). Dresden, Germany.

Büttner, G., Maucha, G., Kosztra, B. 2002. Towards agri-environmental indicators using land cover changes derived from CORINE Land Cover data. In *Building agro environmental indicators—Focussing on the European area frame survey LUCAS; JRC-EEA-EUROSTAT.* EUR Report 20521 EN.

Büttner, G., Maucha, G., Bíró, M., Kosztra, B., Pataki, R., Petrik, O. 2004. National land cover database at scale 1:50,000 in Hungary. *EARSeL eProceedings* 3(3):323–330. Available at: http://www.eproceedings.org/.

Copernicus Land, 2014. Available at: http://land.copernicus.eu/.

Delgado Hernández, J., Válcarcel Sanz, N., Fernández Villarino, X., Nuñez Maderal, E., Boluda Sánchez, A. 2012. Generalización de Nomenclaturas de Ocupación del Suelo desde el Sistema de Información de Ocupación del Suelo en España. Aplicaión al Mapa Topográfico Nacional 1:25.000. X Congreso TopCart 2012—I Congreso Iberoamericano de Geomática y C.C.de la Tierra COITTop 2012 (pp. 1–14). Madrid.

Feranec, J., Otahel, J. 1999. Mapping of land cover at scale 1:50,000: Draft of the nomenclature for the Phare countries. *Geografický časopis* (Geographical Journal) 51:19–44 (in Slovak).

Feranec, J., Otahel, J. 2004. The 4th level CORINE land cover nomenclature for the Phare countries. In G. Groom (Ed.), *Developments in strategic landscape monitoring for the Nordic countries* (pp. 54–63). Copenhagen: Nordic Council of Ministers.

Feranec, J., Otahel, J., Pravda, J., Husar, K. 1994. Land cover forms identified within the framework of the project CORINE land cover. *Geografický časopis* (Geographical Journal) 46:35–48 (in Slovak).

Feranec, J., Otahel, J., Pravda, J. 1995. Proposal for a methodology and nomenclatures scale 1:50,000. Final report. Contract No. 94-0893. Bratislava: Institute of Geography, Slovak Academy of Science (SAS).

Feranec, J., Otahel, J., Suri, M., Cebecauer, T. 1998. Final version of the 4th level CORINE land cover classes at scale 1:50,000. Technical Report. Bratislava: European Environment Agency Phare Topic Link on Land Cover and Institute of Geography, Slovak Academy of Science (SAS).

Feranec, J., Hazeu, G., Christensen, S., Jaffrain, G. (2007). Corine land cover change detection in Europe. Case Studies of the Netherlands and Slovakia. *Land Use Policy* 24:234–247.

Fuller, R. M., Groom, G. B., Jones, A. R. 1994. The Land Cover Map of Great Britain: An automated classification of Landsat Thematic Mapper data. *Photogrammetric Engineering and Remote Sensing* 60:553–562.

Fuller, R. M., Cox, R., Clarke, R. T., Rothery, P., Hill, R. A., Smith, G. M., Thomson, A. G., Brown., N. J., Howard, D. C., Stott, A. P. 2005. The UK Land Cover Map 2000: Planning, construction and calibration of a remotely sensed, user-oriented map of broad habitats. *International Journal of Applied Earth Observation and Geoinformation* 7:202–216.

Hazeu, G. W. 2003. CLC2000 Land Cover database of the Netherlands: Monitoring land cover changes between 1986 and 2000. Wageningen, Alterra, Green World Research. Alterra-rapport 775/CGI-rapport 03-006.

Hazeu, G. W. 2014. Operational land cover and land use mapping in the Netherlands. In I. Manakos and M. Braun (Eds.), *Land use and land cover mapping in Europe: Practices and trends* (pp. 282–296). Remote Sensing and Digital Image Processing, Dordrecht, the Netherlands: Springer.

Hazeu G. W., de Wit, A. J. W. 2004. CORINE Land Cover database of the Netherlands: Monitoring land cover changes between 1986 and 2000. Workshop on Remote Sensing of Land Use and Land Cover, May 28–29, 2004, Dubrovnik, Croatia. *EARSeL eProceedings* 3(3):382–387.

Hazeu, G. W., Dorland, G. J., Schuiling, C. 2008. CLC2006 Land cover database of the Netherlands. Land cover and Land Cover Changes 2000–2006. CLC2006–GMES-FTSP Land Monitoring. Deliverable for European Environment Agency, Copenhagen, Denmark.

Hazeu, G. W., Bregt, A. K., de Wit, A. J. W., Clevers, J. G. P. W. 2011. A Dutch multi-date land use data-base: Identification of real and methodological changes. *International Journal of Applied Earth Observation and Geoinformation* 13:682–689.

Hazeu, G. W., Schuiling, C., Dorland, G. J., Roerink, G. J., Naeff, H. S. D., Smidt, R. A. 2014. Landelijk Grondgebruiksbestand Nederland versie 7 (LGN7); Vervaardiging, nauwkeurigheid en gebruik. Wageningen, Alterra Wageningen UR (University and Research Centre), Alterra-rapport 2548.

Heymann, Y., Steenmans, Ch, Croisille, G., Bossard, M. 1994. *CORINE land cover. Technical guide.* Luxembourg: Office for Official Publications of European Communities.

Hovenbitzer, M., Emig, F., Wende, Ch., Arnold, S., Bock, M., Feigenspan, S. 2014. Digital land cover model for Germany—DLM-DE. In I. Manakos and M. Braun (Eds.), *Land use and land cover mapping in Europe: Practices and trends* (pp. 273–282). Remote Sensing and Digital Image Processing. Dordrecht, the Netherlands: Springer.

Jansen, L. J. M., Groom, G. B., Carrai, G. 2008. Land-cover harmonisation and semantic similarity: Some methodological issue. *Journal of Land-Use Science* 3(2–3):131–160.

Morton, D., Rowland, C., Wood, C., Meek, L., Marston, C., Smith, G., Wadsworth, R., Simpson, L. C. 2011. Final Report for LCM2007—The new UK Land Cover Map. CS Technical Report No. 11/07, Centre for Ecology and Hydrology, Natural Environment Research Council.

O'Sullivan, G. 1994. CORINE Land Cover Project (Ireland). Natural Resources Development Centre, Trinity College Dublin report to CEC, DG XI and DG XIV.

SIOSE. 2009. Sistema de Información de Ocupación del Suelo en España. Documento técnico. Available at: http://www.ign.es/siose/Documentacion/Guia_Tecnica_SIOSE/Doc_tecnico_SIOSE2005_v2.pdf.

Smith, G. 2014. Land use and land cover mapping in Europe: Examples from the UK. In I. Manakos and M. Braun (Eds.), *Land use and land cover mapping in Europe: Practices and trends* (pp. 273–282). Remote Sensing and Digital Image Processing. Dordrecht, the Netherlands: Springer.

Smith, G., Beare, M., Boyd, M., Downs, T., Gregory, M., Morton, D., Brown, N., Thomson, A. 2007. UK land cover map production through the generalisation of OS MasterMap. *The Cartographic Journal* 44(3):276–283.

Taracsák, G. 1999. InterView macro package for satellite image interpretation (V2.1). FÖMI internal report.

Taracsák, G. 2012. CLC2012 Support Package. http://clc2012.taracsak.hu/.

Valcarcel, N., Villa, G., Arozarena, A., Garcia-Asensio, L., Caballlero, M. E., Porcuna, A., Domenech, E., Peces, J. J. 2008. SIOSE, a successful test bench towards harmonization and integration of land cover/use information as environmental reference data. The International Archives of the Photogrammetry, Remote Sensing and Spatial Information Sciences. Vol. XXXVII. Part B8. Beijing 2008, pp. 1159–1164.

Appendix 26A: The Fourth Level CORINE Land Cover Nomenclature for the PHARE Programme Countries

This nomenclature was published in *Geograficky casopis*, 51, 1999, 1: 19–44. The Editorial Board of *Geograficky casopis* approves its publishing in the book *European Landscape Dynamics: CLC Data*.

1 Artificial surfaces

  1.1 Urban fabric

    1.1.1 Continuous urban fabric

1.1.1.1 Areas of urban centres

Areas of urban centres with public, administrative and commercial buildings, roads, parking lots and artificial surfaces (e.g., cemeteries without vegetation) cover more than 80% of the total surface. Urban greenery is exceptional.

1.1.1.2 Areas of ancient cores

Dense ancient cores (mainly residential buildings) with roads, parking lots, etc. Urban greenery is exceptional.

    1.1.2 Discontinuous urban fabric

1.1.2.1 Discontinuous built-up areas with multiflat houses prevailingly without gardens

Areas, substantial part of which are formed mainly by houses without more distinct representation of gardens. Also lawns and tree and shrub urban greenery, communications, parking lots, in lesser extent service buildings, cemeteries without vegetation, private family houses can be part of them. They are represented mainly by urban settlement with multistoreyed houses.

1.1.2.2 Discontinuous built-up areas with family houses with gardens

Areas, substantial part of which are formed mainly by family houses connected with gardens and lawns. There are fruit trees, vegetable, eventually agricultural crops. Other elements of the class are communications, various service buildings, parking lots, small squares reaching the area of approx. 20%–50% of the pattern area, cemeteries without vegetation. They are represented mainly by rural settlements and parts of urban settlements.

1.1.2.3 Discontinuous built-up areas with greenery

Areas mostly consisting of houses in forestland (dispersed houses in forest environment, e.g., Tapiola "Garden city").

1.2 Industrial, commercial and transport units

1.2.1 Industrial or commercial units

1.2.1.1 Industrial and commercial units

Areas of industrial enterprises, store houses, shops, agricultural farms (e.g., cattle-, pig-, poultry-, etc.), fair sites, exposition sites, power plants and unbuilt areas associated to industrial units, hospitals, university or school campus, etc.

1.2.1.2 Areas of special installations

Areas of technical infrastructure e.g. sewage plants, transformers, test fields of civil and military production industry, military base houses, etc.

1.2.2 Road and rail networks and associated land

1.2.2.1 Road network and associated land

Road network and associated areas. Lines of roads of minimum width 50 m with associated transport facilities (parking lots along the motorways, maintenance activities for roads, trenches, etc.).

1.2.2.2 Rail network and associated land

Rail network and associated areas. Lines of rail roads of minimum width 50 m with associated transport facilities (station buildings, maintenance activities for trains, trenches, etc.).

1.2.3 Port areas

1.2.3.1 Sea commercial, fishing and naval ports

Areas of commercial, fishery and naval ports with piers and associated infrastructure.

1.2.3.2 River and lake ports

Areas of ports situated on the lake shores or river banks with associated infrastructure of buildings and communications.

1.2.3.3 Shipyards

Areas formed by infrastructure of production and assembly halls with the associated water surface and communications.

1.2.3.4 Sport and recreation ports

Areas of sport and recreation ports with piers and associated infrastructure.

1.2.4 Airports

1.2.4.1 Airports with artificial surfaces of runways

Areas of airports with artificial surface of runways with associated grass areas and infrastructure of airport buildings.

1.2.4.2 Airports with grass surfaces of runways

Areas of airports with grass surface of runways with associated airport buildings.

1.3 Mine, dump and constructions sites

    1.3.1 Mineral extraction sites

1.3.1.1 Open cast mines

Areas of open cast coal mines, oil-shale mines, gravel, sand, and clay pits.

1.3.1.2 Quarries

Areas of quarries.

    1.3.2 Dump sites

1.3.2.1 Solid waste dump sites

Areas of public and industrial waste and dump sites of raw materials.

1.3.2.2 Liquid waste dumps

Areas of dump sites of liquid waste originating in mainly chemical industry.

    1.3.3 Construction sites

1.3.3.1 Construction sites

Areas under construction development for which earthworks and different stages of building constructions are typical.

1.4 Artificial, non agricultural vegetated areas

    1.4.1 Green urban areas

1.4.1.1 Parks

Areas of parks occurring within the settlements and formed mainly by lawns, tree and shrub greenery, strips of lanes and paths.

1.4.1.2 Cemeteries

Areas of cemeteries with vegetation.

    1.4.2 Sport and leisure facilities

1.4.2.1 Sport facilities

Areas of playgrounds within or outside the urban fabric, running tracks, race-courses, ski resorts, golf grounds, etc.

1.4.2.2 Leisure areas

Areas of leisure and recreation with cottages, spa buildings with parks, castle parks, zoo-gardens, forest-parks not surrounded by urban areas, historical open-air museums, skanzens, open-air theatres, etc.

2 Agricultural areas

    2.1 Arable land

        2.1.1 Non-irrigated arable land

2.1.1.1 Arable land prevailingly without dispersed (line and point) vegetation

Plots of arable land (where cereals, legumes, industrial crops, root crops and fodder crops, semi-permanent crops as strawberries, market gardens, kitchen gardens, flowers and trees nursery-gardens—non-forestry nurseries, are cultivated), with rare occurrence of scattered (line and point) greenery. This class includes also fallow lands (3–4 years abandoned). They can be seasonally irrigated.

2.1.1.2 Arable land with scattered (line and point) vegetation

Plots of arable land (where cereals, legumes, industrial crops, root crops and fodder crops, semi-permanent crops as strawberries, market gardens, kitchen gardens, flowers and trees nursery-gardens—non-forestry nurseries, are cultivated), with sporadic occurrence of scattered (line and point) greenery—less than 15%. This class includes also fallow lands (3–4 years abandoned). They can be seasonally irrigated.

2.1.1.3 Greenhouses

Areas of glass and plastic greenhouses.

2.1.2 Permanently irrigated land

2.1.2.1 Permanently irrigated land

The same definition than for the 3rd level.

2.1.3 Rice fields

2.1.3.1 Rice fields

Areas of rice fields.

2.2 Permanent crops

2.2.1 Vineyards

2.2.1.1 Vineyards

Areas of vineyards (single vineyard plots with area of 4 ha and more; if single vineyard plots smaller than 4 ha each, in total exceeding 60% of the area are mixed with e.g. fruit trees, arable land, meadows, priority will be given to 2211).

2.2.2 Fruit trees and berry plantations

2.2.2.1 Orchards

Areas of fruit orchards (apples, plums, pears, cherries, peaches, apricots, etc.) and ligneous crops (walnut, chestnut, hazel, almond, etc.).

2.2.2.2 Berry fruit plantations

Areas of plantations of berry fruits (black and red currants, raspberries, gooseberries, etc.).

2.2.2.3 Hop plantations

Areas of hop plantations.

2.2.2.4 Kiwi plantations

Plots planted with kiwi.

2.2.2.5 Oil-bearing rose plantations

Parcels planted with oil-bearing rose.

2.2.2.6 Wild willow plantations

Areas of wild willow plantations.

2.2.3 Olive groves

2.2.3.1 Olive groves

Areas of olive groves.

2.3 Pastures

    2.3.1 Pastures

2.3.1.1 Grassland (pastures and meadows) prevailingly without trees and shrubs

Areas of grassland prevailingly without trees and shrubs (less than 15%).

2.3.1.2 Grassland (pastures and meadows) with trees and shrubs

Areas of grassland with trees and shrubs (between 15%–40%).

2.4 Heterogeneous agricultural areas

    2.4.1 Annual crops associated with permanent crops

2.4.1.1 Annual crops associated with permanent crops

The same definition as for the 3rd level.

    2.4.2 Complex cultivation patterns

2.4.2.1 Complex cultivation patterns without scattered houses

Juxtaposition of small plots of diverse annual crops, pastures and/or permanent crops (fruit trees, vineyards and berry plantations) without scattered houses (settlements).

2.4.2.2 Complex cultivation patterns with scattered houses

Juxtaposition of small plots of diverse annual crops, pastures and/or permanent crops (fruit trees, vineyards and berry plantations) with scattered houses (settlements).

    2.4.3 Land principally occupied by agriculture, with significant areas of natural vegetation

2.4.3.1 Agricultural areas with significant share of natural vegetation, and with prevalence of arable land

Agriculturally cultivated areas with prevalence of arable land (over 50%) with a pronounced representation of natural vegetation, especially strips and patches of forest, grasslands and sporadic occurrence of water areas (artificial and natural).

2.4.3.2 Agricultural areas with significant share of natural vegetation, and with prevalence of grasslands

Agriculturally cultivated areas with prevalence of grasslands (over 50%) with representation of arable land, strips and patches of forest, grass communities and water areas (artificial and natural).

2.4.3.3 Agricultural areas with significant share of natural vegetation, and with prevalence of scattered vegetation

Agriculturally cultivated areas with prevalence of scattered greenery (woodland patches and bushes over 50%) with representation of arable land, grasslands and water areas.

2.4.3.4 Agricultural areas with significant share of ponds, and with presence of scattered vegetation

Agriculturally cultivated areas with prevalence of ponds (over 50%) with representation of arable land, grassland, strips and patches of forests.

2.4.3.5 Agricultural areas with significant share of permanent crops, and with presence of scattered vegetation

Agricultural areas with vineyards and orchards (to 50%) with representation of grasslands and strips of forests.

2.4.4 Agro-forestry areas

2.4.4.1 Agro-forestry areas

Areas of annual crops or grazing land under the wooded cover of forestry species.

3 Forest and semi-natural areas

3.1 Forests

3.1.1 Broad-leaved forests

3.1.1.1 Broad-leaved forests with continuous canopy, not on mire

Areas of broad-leaved woods forming continuos canopy (crown of trees overlap one another—continuous canopy is more than 80%).

3.1.1.2 Broad-leaved forests with continuous canopy on mire

Areas of broad-leaved woods forming continuos canopy (crown of trees overlap one another—continuous canopy is more than 80%).

3.1.1.3 Broad-leaved forests with discontinuous canopy, not on mire

Areas of broad-leaved woods forming discontinuous canopy (crown of trees do not owerlap one another—continuous canopy is less than 80%).

3.1.1.4 Broad-leaved forests with discontinuous canopy on mire

Areas of broad-leaved woods forming discontinuous canopy (crown of trees do not owerlap one another—continuous canopy is less than 80%).

3.1.1.5 Plantation of broad-leaved forests

Artificially planted areas of the same species of broad-leaved wood, e.g., poplar, etc. These plantations are cleared and replanted in regular intervals.

3.1.2 Coniferous forests

3.1.2.1 Coniferous forests with continuous canopy, not on mire

Areas of coniferous woods forming continuous canopy (crown of trees overlap one another—continuous canopy is more than 80%).

3.1.2.2 Coniferous forests with continuous canopy on mire

Areas of coniferous woods forming continuous canopy (crown of trees overlap one another—continuous canopy is more than 80%).

3.1.2.3 Coniferous forests with discontinuous canopy, not on mire

Areas of coniferous woods forming discontinuous canopy (crown of trees do not overlap one another—continuous canopy is less than 80%).

3.1.2.4 Coniferous forests with discontinuous canopy on mire

Areas of coniferous woods forming discontinuous canopy (crown of trees do not overlap one another—continuous canopy is less than 80%).

3.1.2.5 Plantation of coniferous forests

Artificially planted areas of the same species of coniferous woods, e.g., pine, larch, spruce, etc. These plantations are cleared and replanted in regular intervals.

3.1.3 Mixed forests

3.1.3.1 Mixed forests created by alternation of single trees with continuous canopy, not on mire

Areas of forest formed by alternation of the broad-leaved and coniferous woods (crown of trees overlap one another—continuous canopy is more than 80%).

3.1.3.2 Mixed forests created by alternation of single trees with continuous canopy on mire

Areas of forest formed by alternation of the broad-leaved and coniferous woods (crown of trees overlap one another—continuous canopy is more than 80%).

3.1.3.3 Mixed forest created by alternation of single trees with discontinuous canopy, not on mire

Areas of forest formed by alternation of the broad-leaved and coniferous woods (crown of trees do not overlap one another—continuous canopy is less than 80%).

3.1.3.4 Mixed forest created by alternation of single trees with discontinuous canopy on mire

Areas of forest formed by alternation of the broad-leaved and coniferous woods (crown of trees do not overlap one another—continuous canopy is less than 80%).

3.1.3.5 Mixed forests created by alternation of stands of trees with continuous canopy, not on mire

Areas of forest formed by alternation of the groups of broad-leaved and coniferous woods (crown of trees overlap one another—continuous canopy is more than 80%).

3.1.3.6 Mixed forests created by alternation of stands of trees with continuous canopy on mire

Areas of forest formed by alternation of the groups of broad-leaved and coniferous woods (crown of trees overlap one another—continuous canopy is more than 80%).

3.1.3.7 Mixed forests created by alternation of stands of trees with discontinuous canopy, not on mire

Areas of forest formed by alternation of the groups of broad-leaved and coniferous woods (crown of trees do not overlap one another—continuous canopy is less than 80%).

3.1.3.8 Mixed forests created by alternation of stands of trees with discontinuous canopy on mire

Areas of forest formed by alternation of the groups of broad-leaved and coniferous woods (crown of trees do not overlap one another—continuous canopy is less than 80%).

3.2 Scrub and/or herbaceous vegetation associations

3.2.1 Natural grasslands

3.2.1.1 Natural grassland prevailingly without trees and shrubs

Areas of natural grasslands without trees and shrubs (less than 15%). They are formed by grasslands of protected areas, alpine grasslands, military training area and abandoned low productivity grassland (e.g., karstic poljes meadows, etc.).

3.2.1.2 Natural grassland with trees and shrubs

Areas of natural grassland with trees and shrubs (between 15%–40%). They are formed by grasslands of protected areas, military training areas, alpine grasslands and abandoned low productivity grassland with trees and shrubs.

3.2.2 Moors and heathland

3.2.2.1 Heathlands and moorlands

Areas of heathlands, moorlands and transitory peat-bogs represented mainly by dense shrubs and herbaceous plants (Calluna vulgaris, Erica sp., Vaccinium sp., Genista sp., Rubus sp., Juniperus sp., etc.).

3.2.2.2 Dwarf pine

Areas of mountain dwarf pine (Pinus mugo ssp. mughus) or dwarf pine planted on dunes and mires (Pinus mugo ssp. uncinata).

3.2.3 Sclerophyllous vegetation

3.2.3.1 Sclerophyllous vegetation

Bushy sclerophyllous vegetation, including maquis and guarrigue.

3.2.4 Transitional woodland-scrub

3.2.4.1 Young stands after cutting (and/or clear cuts)

Areas of young stands planted by man after cutting, fire or natural disaster without vegetation.

3.2.4.2 Natural young stands

Areas of natural forest regeneration/recolonization.

3.2.4.3 Bushy woodlands

Areas formed by shrubs (Juniperus, Crataegus, Rosa, etc.) along with dispersed trees and grassland. Crowns of shrubs and trees do not form a continuous canopy.

3.2.4.4 Forest nurseries

Areas of forest nurseries (production areas of young forest trees).

3.2.4.5 Damaged forests

Areas of forests strongly affected by air pollution, biotical injurious agents or natural disasters.

3.3 Open spaces with little or no vegetation

3.3.1 Beaches, dunes, sands

3.3.1.1 Beaches

Areas of bank sand plains prevailingly without vegetation. They are adjacent to 5.2.3.

3.3.1.2 Dunes

Areas of dunes almost without vegetation or with sporadic occurrence of mainly thin grasses (occur in coastal zone as well as in inland).

3.3.1.3 River banks

Belts of river banks formed mainly by deposits of sands and gravel prevailingly without vegetation.

3.3.2 Bare rocks

3.3.2.1 Bare rocks

Areas of different rock outcrops, e.g., cliffs, screes, rock-face surfaces, etc.

3.3.2.2 Products of recent volcanism

Areas of recent lava streams and mud without vegetation

3.3.3 Sparsely vegetated areas

3.3.3.1 Sparse vegetation on sands

Areas of sand plains (dunes, glaciofluvial terraces) covered by sparse vegetation.

3.3.3.2 Sparse vegetation on rocks

Areas of xerothermic grasses and shrubs of karstic terrain, or areas of discontinuous alpine grasslands and partially dwarf mountain pine.

3.3.3.3 Sparse vegetation on salines

Areas of salines covered by sparse halophilic vegetation.

3.3.4 Burnt areas

3.3.4.1 Burnt areas

Areas (mainly forest and heathlands and moorlands) after recent fires.

3.3.5 Glaciers and perpetual snow

3.3.5.1 Glaciers and perpetual snow

Areas covered by glaciers or permanent snow.

4 Wetlands

4.1 Inland wetlands

4.1.1 Inland marshes

4.1.1.1 Fresh-water marshes with reeds

Areas of swamps with reed beds (more than 80%) and other water plants without peat deposition (peat layer is less than 30 cm thick) seasonally or permanently waterlogged with low mineral content.

4.1.1.2 Fresh-water marshes without reeds

Areas of swamps without reed beds (to 20%) and with other water plants without peat deposition (peat layer is less than 30 cm thick) seasonally or permanently waterlogged with low mineral content.

4.1.1.3 Saline (alkali) inland marshes with reeds

Areas of swamps with reed beds (more than 80%) and other water plants without peat deposition (peat layer is less than 30 cm thick) seasonally or permanently waterlogged (prevailingly arheic) with higher mineral content.

4.1.1.4 Saline (alkali) inland marshes without reeds

Areas of swamps without reed beds (to 20%) and with other water plants without peat deposition (peat layer is less than 30 cm thick) seasonally or permanently waterlogged (prevailingly arheic) with higher mineral content.

4.1.2 Peat bogs

4.1.2.1 Explored peat bogs

Areas of peat-bogs with extraction.

4.1.2.2 Natural peat bogs with scattered trees and shrubs, without pools

Characteristic for higher stage of bog development where conditions for pool formation are created (peat deposit more than 30 cm thick).

4.1.2.3 Natural peat bogs with pools communities

Areas of convex (raised) bog where zones of pools are perpendicular to water movement (peat deposit more than 30 cm thick).

4.1.2.4 Natural dwarf shrub bogs

Vegetation composed of shrubs (Calluna vulgaris, Ledum palustre, Empetrum nigrum, Andromeda polifolia, Oxycoccus sp.) and scattered pines and birches. Situated mainly in the margin zones of bogs (peat deposit more than 30 cm thick).

4.2 Maritime wetlands

4.2.1 Salt marshes

4.2.1.1 Salt marshes without reeds

Areas of coastal plains seasonally or permanently waterlogged covered by sparse halophilic vegetation.

4.2.1.2 Salt marshes with reeds

Areas flooded by sea water and colonized by Phragmites communis.

4.2.2 Salines

4.2.2.1 Salines

The same definition as for 3rd level.

4.2.3 Intertidal flats

4.2.3.1 Intertidal flats

The same definiton as for the 3rd level.

5 Water bodies

5.1 Inland waters

5.1.1 Water courses

5.1.1.1 Rivers

Natural water streams of minimum width of 50 m with meanders, usually without longer straight spells of banks which are often formed by deposits of gravel, sand or trees and shrubs.

5.1.1.2 Channels

Artificial water channels or regulated water streams of minimum width of 50 m, prevailingly straight.

5.1.2 Water bodies

5.1.2.1 Natural water bodies

Water areas of natural origin.

5.1.2.2 Artificial reservoirs

Water areas created by man with prevailing regular shape.

5.2 Marine waters

5.2.1 Coastal lagoons

5.2.1.1 Coastal lagoons

The same definition as for the 3rd level.

5.2.2 Estuaries

5.2.2.1 Estuaries

The same definition as for the 3rd level.

5.2.3 Sea and ocean

5.2.3.1 Sea and ocean

The same definition as for the 3rd level.

27

CLC in the Context of INSPIRE

Geir-Harald Strand and Stephan Arnold

CONTENTS

27.1 Background

This chapter explains the relationship between CoORdination of Information on the Environment (CORINE) Land Cover (CLC) and the Infrastructure for Spatial Information in the European Community (INSPIRE). The first section is a description of the data model developed for land cover (LC) data in INSPIRE, emphasizing how CLC data fits into the model. The section sets out to clarify the background of the INSPIRE data model, followed by a brief discussion of the thematic concept of "LC" in INSPIRE. The section continues with a systematic explanation of the constituent parts of the model, including the overall documentation of an LC survey, representation of geometry and temporality, LC observations, and the LC nomenclature. How CLC is handled with respect to each of these elements is explained. The section concludes with a summary of the data model applied to CLC. The second section pays particular attention to the question about LC classification and nomenclatures, explaining why INSPIRE does not set its own mandatory standard for LC classification in Europe. The section presents a short list of Pure Land Cover Components (PLCC) introduced in the INSPIRE data model and discusses how semantic translation can be used to facilitate data exchange when several classification systems are involved. The overall conclusion, drawn and explained in the third and final section of the chapter, is that CLC is fully INSPIRE-compliant.

27.2 LC in INSPIRE

In 2007 the European Parliament and the Council of the European Union decided to establish an Infrastructure for Spatial Information in the European Community (INSPIRE). The legal framework for this initiative is the European Directive 2007/2. The intention of the INSPIRE directive is to "establish a measure of coordination between the users and providers of the information so that information and knowledge from different sectors can be combined" (European Union [EU], 2007). Thirty-four thematic areas were listed in the Annexes of the INSPIRE directive. For each of these INSPIRE themes, data specifications have been developed as a technical basis for the implementation of the directive. Two of these thematic areas are LC and land use (LU). LC is concerned with the description of the physical cover of the Earth, while LU pertains to the actual or planned use of the land.

INSPIRE is not a mapping program and does not initiate any new mapping activities. INSPIRE is a framework for exchange and combination of data from existing and future mapping programs. As such, it is imperative that the INSPIRE data specifications are sufficiently flexible so as to allow exchange and integration of data from widely different sources and generated under different circumstances. An INSPIRE data specification must therefore be broad and inclusive rather than narrow and exclusive.

Keeping the data model broad and simple is not the same as simplifying the data themselves. A broad and inclusive model can accommodate data assembled from many different sources and with highly variable, complex, and detailed information content. Such comprehensiveness is important for a theme like LC that is involved in many application fields and used at different cartographic scales, a variability reflected in a multitude of LC specifications and classification systems.

CLC is the de facto standard for pan-European LC mapping and monitoring used by the European Environment Agency (EEA) (Bossard et al., 2000; Feranec et al., 2010). CLC provides snapshots of the LC situation in Europe in 1990, 2000, 2006, and 2012 (Büttner et al., 2010). An INSPIRE data specification for LC unable to accommodate the incorporation of CLC into the European spatial data infrastructure would be incomplete and fairly meaningless.

CLC is, on the other hand, far from being the only LC monitoring system in Europe. Eurostat operates its own sampling based monitoring system Land Use and Cover Area frame Survey (LUCAS) (Eurostat, 2003; Gallego and Delincé, 2010) and several European countries have implemented LC mapping and monitoring systems at the national and subnational levels (Ben-Asher et al., 2013; Martinez, 2013). Some, but far from all, of these national systems are built on—and compatible with—the CLC specification.

It is widely acknowledged that CLC has some weak points. The system has been criticized for its bias toward Mediterranean LC classes; the muddle of LC and LU concepts; classes representing mixtures of different LC; thematic incompleteness; selective handling of temporal phenomena; and lack of options for parameterization and attribution. It is therefore likely that the CLC monitoring system will be improved in the future, for example, by adaption to meet present standards and requirements or fulfill new needs. An INSPIRE data specification must therefore support the current CLC but at the same time be careful not to become an obstacle for future improvements of the system.

27.2.1 Development of INSPIRE Specifications

The purpose of the INSPIRE directive is to create a common framework for sharing environmental data across sectors and across national boundaries and thus to establish a common geospatial data infrastructure for Europe. As a preparatory step for the development of the implementing rules for the directive, a set of Thematic Working Groups were established, each responsible for the development of a data model for one or more themes. One of these groups was the Thematic Working Group for Land Cover (TWG-LC). TWG-LC started its work in May 2010 and the final version of the data specification was completed in February 2013 (TWG-LC, 2013). In parallel, another TWG was set up to draft the data specification for LU.

The development process was driven by interactivity between an abstract data model; the applied experience of the wider LC monitoring community; and practical tests carried out by voluntary laboratories across Europe. The final INSPIRE data specification for LC is therefore a compromise between idealism and pragmatism. The specification is on one hand attempting to create a generic, "ideal" data model based on a solid theoretical approach to LC modeling. On the other hand, the data model must be practical and operational in everyday situations in which data producers and users, occasionally with limited expertise, are handling information using software with reputable, but still imperfect functionality. Being realistic, these circumstances are expected, and one principle of INSPIRE is to establish a geospatial data infrastructure based on existing data and available technology without overburdening the affected institutions. In an ideal manner, equilibrium between sufficient complexity and feasible simplicity of the data specifications had to be targeted.

27.2.2 What Is Land Cover?

"You know what land cover is but does anyone else? ..." is the rhetorical title of an article by Comber et al. (2005a) in which the authors recognize that LC is perceived differently by different disciplines. They are therefore concerned about the impact of user ignorance when LC data obtained for one particular purpose are used in another context with completely different scale and/or objectives. In a related article Comber et al. (2005b) described LC as the result of an abstraction process linked to the social and political context of the different agencies involved. These considerations are particularly relevant within the context of INSPIRE, where data from many different sources are collected by different providers for a number of applications and shared in a broad common spatial data infrastructure. Consequently, it is important to acknowledge the context-specific character of the LC classification and emphasize the importance of adequate metadata.

Reflecting on these considerations, TWG-LC applied the following definition:

LC is an abstract description of the physical characteristics of the surface of the Earth.

TWG-LC thereby recognized LC classification as an abstraction, dependent on the observer's perspective and intention. There are many different systems for LC characterization and classification, reflecting a multitude of purposes and objectives. Each characterization method or classification system can be judicious within its own well-defined framework. It is therefore of no purpose to discuss the merit of various measurement techniques and classification systems in search for a single system to solve them all.

LC classification systems may also—for pragmatic reasons and justified by the purpose of the LC mapping exercise and the choice of cartographic scale—allow heterogeneous mixtures of vegetation; combinations of artificial, natural, and semi-natural elements; and even consider the LU or other external factors as a defining element of some LC classes. It is also inherently clear from the definition that LC does not necessarily represent any identifiable objects. The geometry of the LC observations as well as the semantics of LC descriptions must be chosen and justified against the background of a particular purpose and a corresponding cartographic scale. In this context, the objective of an LC data model for INSPIRE is not to impose a certain LC mapping system on the European land monitoring community, but to develop a model that provides a structure wherein this multitude of LC data can be represented, shared, and exchanged in a meaningful way.

The difference between LC and LU is further elaborated upon in Chapter 3 of this book. The INSPIRE directive as well as the INSPIRE data specification for LC do recognize this difference, noting that "LC is different from LU [...] which is dedicated to the description of the use of the earth's surface" (TWG-LC 2013: 12). CLC does, by name, present itself as an LC dataset. The dataset does, however, also include information on LU and these LU aspects are not clearly separated from the LC aspects. Besides, the nomenclature is not complete in its representation of the landscape, neither from the LC nor from the LU perspective. It is, owing to this partial lack of information, not feasible to split CLC into two separate datasets to be handled individually as one LC and another LU dataset in INSPIRE. CLC must be handled either as an LC dataset or as an LU dataset, accepting that either way some aspects of LC respectively LU are missing in the whole picture.

The INSPIRE data specification does acknowledge this difficulty, noting that "LC and LU are [...] related to each other and often combined in practical applications. Data combining LU and LC information often emphasize LU aspects in intensively used areas (e.g., built-up or industrial areas, artificial land) and LC aspects in extensively used areas (e.g., natural vegetation, forest areas)" (TWG-LC 2013: 12). Overall, the INSPIRE data specification for LC therefore emphasize the fact that LC is an abstraction and pursues a pragmatic practice allowing CLC and similar hybrid datasets to be included in the European infrastructure as LC data.

27.2.3 LC Survey Initiative

The long-term reference frame for the creation of LC data is an LC survey initiative, that is, the administrative context where LC data collection and management is carried out by a mandated organization. Examples of LC survey initiatives are the CLC program (Bossard et al., 2000) implemented by the EEA and the LUCAS area frame survey (Eurostat, 2003) implemented by Eurostat. Many national and subnational governmental authorities also conduct one or more LC survey initiatives serving national and regional needs for land information and land monitoring (Arnold, 2009; Hazeu et al., 2011; Blanes and Green, 2012; Strand 2013).

From a meta-conceptual point of view, an LC survey initiative embodies a linkage among the real world, users, documentation, and data. The *user* is the institution, agency, organization, or people requesting information about the LC. *Documentation* is the collection of technical explanatory descriptions of the data collection methods; feature type definitions; and rules for measurements, classification, and generalization. The documentation is a kind of metadata, usually available as text documents and of limited interest to the general public but indispensable for the consistent and professional use of the data. *Data* within the LC survey initiative are the encoded information and are organized into entities described in the data model as *LC coverages*. A coverage is defined by the International Organization for Standardization (ISO) as a feature that acts as a function to return values from its range for any direct position within its spatial, temporal, or spatiotemporal domain (ISO 2005c, 2009).

In the context of INSPIRE, CLC is an *LC Survey Initiative*, while the individual CLC datasets created by the initiative (CLC1990, CLC2000, CLC2006, and CLC2012) are *LC Coverages*.

The CLC *documentation* consisting of a nomenclature specification, technical guidelines, and metadata are provided online by EEA as PDF files. This documentation is sufficient to meet the requirements according to the INSPIRE data specification, although documentation in a machine-readable format would be a preferable improvement.

27.2.4 Geometry and Temporality

A CLC dataset is a collection of classified polygons. The corresponding INSPIRE conceptualization is that an LC Coverage is a collection of LC observations with a fixed (static) geometry consisting of points, polygons, or tessellation units (regular grid cells, triangles, hexagons, etc.). A CLC dataset is therefore, in INSPIRE terminology, simply an LC Coverage where the geometry representing the LC observations consists of *polygons* (Jones, 1997).

Two important constraints in the INSPIRE data model are that polygons are not allowed to overlap and that the individual LC Coverage is a static geometry that does not change. If a geometry is changed (because a polygon is split or two polygons are merged), the result is a new version of the dataset and therefore a new LC Coverage. These two constraints also apply to CLC. CLC does not allow overlapping polygons and each instance of CLC mapping results in a separate dataset (CLC2000, CLC2006, CLC2012, etc.). CLC is therefore fully INSPIRE compliant with respect to geometry.

The individual geometrical features in an LC Coverage are called LC Objects. An LC Object is the geometry of a single LC observation. During the CLC mapping process, an LC observation leads to the delineation and classification of a polygon. This polygon is the georeferenced location of the observation. The CLC polygons thus correspond to the more general INSPIRE concept of an LC Object.

At this point, INSPIRE introduces a mechanism for handling temporality. A time stamp, for example, a date, can be assigned to the LC Object. The INSPIRE data model does not specify the exactitude of the time stamp. It could be registered down to the exact moment when the observation is made (an example is the acquisition time of a satellite image used as the source of information), or it could simply be a reference year (as in CLC).

Observation dates are of vital importance in other monitoring programs. The data are used to create observation timelines when a fixed set of spatial units are measured several times. An example is Eurostat's LUCAS survey, in which permanent sampling points on the ground are visited every few years. The result is several observations at the same location. In the INSPIRE LC data specification, this time stamp is called the "Situation."

The Situation represents the state of a land unit at a certain point in time (avoiding the term "observation date" because it could have many different interpretations). An LC Object can have several Situations if the LC has been observed several times. CLC is in this context a specific case, where each polygon only is observed once and polygons often change from one CLC survey to another. The introduction of the Situation in the model does, however, make the data model more general and provides an opportunity for future development of CLC as well as for data from other mapping initiatives.

The time stamp defining the "Situation" for CLC will be the reference year. This repetitive information (where every polygon in a CLC2012 dataset has the time stamp "2012") is admittedly redundant. CLC is, however, also in this respect INSPIRE compliant and the INSPIRE data model provides an opportunity to introduce information about the image acquisition time (down to the date, hour, minute, and second if required) into the dataset in the future.

The relevance of the Situation item for CLC is clearer with respect to the CLC Changes (CLCCs) dataset. This dataset was first produced during the CLC2000 survey (Feranec et al., 2007). The method was consolidated in the CLC2006 survey when analysts used satellite images from 2000 and 2006 and delineated areas where the LC had changed. For these polygons, they would register two attributes: LC in the year 2000 and LC in the year 2006. The resulting dataset is also an INSPIRE compliant LC Coverage, although limited to areas where changes have been observed. Each polygon in this dataset is an LC Object with information about two Situations. The first Situation is the reference year 2000, while the second Situation is the reference year 2006. The INSPIRE item Situation may therefore by redundant with respect to CLC, but it is necessary to make several other datasets—like CLCCs—INSPIRE compliant.

27.2.5 The LC Observations

Classification in CLC means to assign a coded label from the nomenclature to each polygon, implying that this class is the only (or dominant) class observed in the polygon. This is compliant with the INSPIRE specification, which requires that one or more LC classes should be recorded for each Situation documented for an LC Object. Allowing more than one LC class to be present in each LC Object is necessary to support survey systems where mosaics are permitted. Mosaics are used in mapping systems to allow two or more LC classes to coexist in each LC Object. Mosaics is not an issue in CLC, but is frequently used in national mapping systems (Bryn, 2006; Strand, 2013). The mosaic can simply be a list of the LC classes recorded in the polygon, or a coverage value (percentage of coverage) can be attached to each class in the mosaic.

The INSPIRE data model allows each Situation in an LC Object to be described with a list of LC classes, each optionally qualified with a percentage coverage value. CLC is a special case where a single LC class is recorded for each polygon. CLC is thereby INSPIRE compliant in this respect as well.

27.2.6 The LC Nomenclature

The INSPIRE data specification does not prescribe or recommend any particular LC nomenclature for use in INSPIRE. During the development of the data model, the Thematic Working Group asked the European LC mapping community for their preference: Should INSPIRE require the use of a single, standardized European LC nomenclature or should the model be open-ended and allow data classified according to any nomenclature? The answers divided the community into two distinct groups of approximately equal size: 50% in favor of setting a single mandatory nomenclature as the European standard, 50% in favor of an open-ended solution. There was, however, no consensus among those in favor of a single nomenclature about the choice of nomenclature to be used as a standard. A few respondents argued in favor of the CLC nomenclature, but an equal number made an explicit remark that the CLC nomenclature should in no case be selected. A wide range of other existing nomenclatures and classification systems were proposed as the INSPIRE standard.

The conclusion drawn by TWG-LC was that all of the respondents actually argued in favor of a data specification that would allow them to continue using an LC nomenclature of their own choice. Half of the respondents also acknowledged that others should be entitled to use a nomenclature by their own preference. The remaining 50% wanted to impose their own preferred nomenclature as a European standard, but there was no agreement among this group as to the choice of a standard nomenclature.

The INSPIRE data model simply requires LC to be represented using a classification system organized as a legend, that is, a list of codes with accompanying explanations. The CLC nomenclature is a legend and fits well into the INSPIRE model. The INSPIRE data model for LC allows the data provider to use any nomenclature and classification system, as long as it is well documented. The documentation could be anything from a text document to a computer readable list. The nomenclature used in CLC is well documented and CLC is therefore, also in this respect, INSPIRE compliant. It is important that due to the open-ended approach to nomenclatures, INSPIRE does not block future changes or development of the CLC system. Furthermore, INSPIRE allows countries that have developed more detailed national nomenclatures based on CLC to make these data available to the wider European community next to the standard CLC data within the framework of INSPIRE.

27.2.7 The Core INSPIRE LC Model

The INSPIRE core model for LC is represented schematically in Figure 27.1. A LC dataset is always associated with a code list in which the nomenclature is documented. The code list consists of legal class values and their associated class names. The LC dataset is composed of a number of spatial LC units, each represented as a point, a polygon, or a raster cell. The LC of one of these units can be observed repeatedly, each time recorded as one or more LC values from the code list and optionally with a covered percentage associated with each code.

The CLC version of the INSPIRE core model is shown in Figure 27.2. A CLC dataset is associated with the CLC nomenclature, consisting of legal class values and their associated class names. The CLC dataset is composed of a number of CLC polygons in which the LC situation is observed in a particular reference year and recorded as a CLC class value from the code list. It is clearly seen by comparing Figures 27.1 and 27.2 that CLC fits nicely into the INSPIRE core model for LC.

The difference between the current CLC format and CLC in INSPIRE is simply that information about the "Reference year" (e.g., 1990/2000/2006) is added to every polygon. This may appear as redundant compared to attaching the recorded reference year only once to the dataset, but constitute no obstacle to integrating CLC in INSPIRE. It also eases the work with the data on polygon level. Furthermore, the attribute can, if it is seen as practical and useful, be used to record the acquisition date of the imagery (instead of the reference year) or ancillary data used to delineate the polygon.

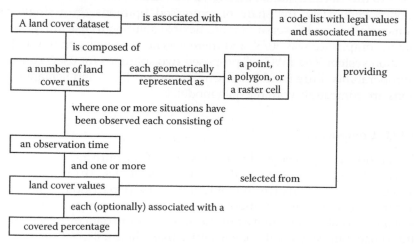

FIGURE 27.1
The INSPIRE core model for LC. Adapted from TWG-LC (2013).

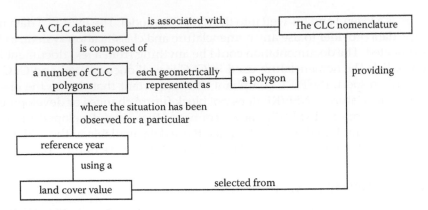

FIGURE 27.2
CLC version of the INSPIRE core model for LC.

27.3 Classification Systems

The choice of a classification system is a recurrent issue in LC monitoring. There is a multitude of different ways to describe LC. This is attributable partly to the wide range of aspects of the environment embraced by LC, but also to the many different uses of LC data (di Gregorio and O'Brien, 2012). There is only one "real world" but many different portrayals of this world depending on the aims, methodology, and terminology of the observer. It is therefore a misguided approach to enforce a single classification system as the common classification system for Europe. Consequently, there is no particular prescribed or recommended LC nomenclature for use in INSPIRE.

The approach taken by the INSPIRE data specification is instead to allow several different LC nomenclatures to coexist in the context of INSPIRE. The owners of the various code lists are encouraged to document their code lists by using the ISO standard 19144-2 Land Cover Meta Language (ISO, 2009) or by using a feature catalogue (ISO, 2005a,b,c) and provide access to this documentation through a web link.

As mentioned previously, the share of expressed opinions on a single mandatory INSPIRE nomenclature for LC versus a more flexible solution to use any given system to provide data in compliance was 50/50 and there was no consensus on the choice of a unified classification system. The TWG-LC proposed two solutions for users concerned about LC data interoperability: Pure Land Cover Components (PLCC) and Semantic translation. Both solutions are compatible with the data model.

27.3.1 Pure LC Components

The simple solution is represented by a collection of 18 pure LC components (PLCC; see Table 27.1) included in an annex to the data specifications document. The PLCCs are assumed to represent an exhaustive and mutually exclusive partition of the domain "LC." It shall be clearly stated that these components are not in any way mandatory in INSPIRE, but represent an alternative encoding available as a service to data providers who for some reason want to use a more general nomenclature than their native system. The future of the PLCC depends on how useful they are found to be by the community. Any existing

TABLE 27.1

Pure Land Cover Classes (PLCC)

PLCC Code	Component Name
001	Artificial constructions
002	Consolidated bare surface
003	Unconsolidated bare surface
004	Arable land
005	Permanent woody and shrubby crops
006	Coniferous forest trees
007	Broad-leaved forest trees
008	Shrubs
009	Herbaceous plants
010	Lichens and mosses
011	Wetlands and marshes
012	Organic deposits (Peatland)
013	Chemical deposits
014	Intertidal flats
015	Fresh water course
016	Fresh water bodies
017	Salt or brackish water
018	Permanent snow and ice

Source: TWG-LC (Thematic Working Group for Land Cover). 2013. D2.8.II.2 INSPIRE Data Specification on Land cover—Draft Technical Guidelines v3.0rc3. Available at: http://inspire.jrc.ec .europa.eu/documents/Data_Specifications/INSPIRE_Data Specification_LC_v3.0rc3.pdf.

Note: The 18 PLCC described in the INSPIRE data specification can be used by data providers to translate their data to a common classification system. The system is not a mandatory part of INSPIRE.

nomenclature can be converted to PLCC, but always with some—and sometimes with major—loss of information. Still, if compatibility is a core issue, loss of information on a higher level of semantic aggregation may be an acceptable price to pay.

CLC classes can in some cases be translated directly into PLCC codes. Examples are CLC class 311 *Broad-leaved forest*, which correspond with PLCC code 007 and CLC class 312 *Coniferous forest*, which correspond with PLCC code 006. There is, however, no PLCC code that correspond with CLC class 313, *Mixed forest*. The solution is the mechanism by which INSPIRE allows several class values to be attached to each LC Unit. A polygon with CLC class 313 can be recoded to PLCC by applying both PLCC codes 006 and 007. Following this approach, it is possible to recode a CLC dataset into a PLCC dataset and release the PLCC dataset in INSPIRE along with or instead of the original CLC dataset.

27.3.2 Semantic Translation

The second solution proposed by TWG-LC is to use semantics to provide a platform for interoperability (Bishr, 1998). Semantics is the study of meaning and focuses on the relation between signifiers (e.g., words or class names) and their denotation (i.e., what they represent). Semantic interoperability is achieved through translation based on measuring the similarity between individual classes in two or more classification systems (Rodriguez and Egenhofer, 2003). This task can be rather challenging, because most classification systems to some extent rely on subjective assessment; classes with identical names may have different specifications; and classes with different names may carry very similar meanings.

Semantic similarity between LC classes can be defined as a feature matching process, which requires that the classes are described in terms of components that can be compared (Tversky, 1977). The degree of semantic similarity between two classes from different classification systems is based on the number or amount of features they share, and may also employ a relative importance factor. An example based on similarity assessment, comparing classes from the United States National Vegetation Classification Standard (NVCS) and the USGS Anderson Classification System, is found in Feng and Flewelling (2003).

This approach was taken one step further, incorporating explicit representation of semantic uncertainty, by Ahlqvist (2005) in an article presenting a "CLC map of North America" based on semantic translation from the LC database of North America (using NVCS) into the CLC nomenclature. The vehicle for translation was the FAO Land Cover Classification System (LCCS) (di Gregorio and Jansen, 2000), found useful for the purpose because of its formal structure and reliance on parameterized classes (Jansen and di Gregorio, 2002). This translation exercise was backed up by an article discussing the parameterized approach to classification and interoperability in which a number of modifications to LCCS were proposed (Ahlqvist, 2008). The purpose of the proposed modifications was to facilitate the use of LCCS as a tool for semantic translations between other classification systems.

The use of LCCS as a documentation and translation system was also foreseen by the editing committee of ISO standard 19144-2. The background for this standard was a proposal from FAO to set LCCS as an international standard for LC classification (Herold et al., 2006). The proposal was voted down by the ISO member countries and an editing committee for the standard was established. This committee decided to promote the development of a Land Cover Meta Language (LCML) based on LCCS, instead of an LC classification system (ISO, 2012). The rationale was that no single classification system can fulfill the full range of information requirements represented by past, present, and future LC monitoring programs, but a standard for description, documentation, and possibly also transformation between these systems would be welcome. Such an approach was also thought to be more in line with the intentions behind ISO standardization.

The ISO LCML can be used in the context of INSPIRE for documentation of proprietary LC classification systems, and probably also as a basis for semantic translation between LC classification systems used by data providers in INSPIRE. Attempts to do so with CLC as the test bed have, however, revealed a number of challenges. These challenges are partly attributed to the presence of LU information in the CLC nomenclature, partly to fuzzy class definitions in CLC and partly to shortcomings in LCML itself. There is thus a need for further development of this concept, either in terms of an enlarged "European" version of LCML (which is explicitly allowed and encouraged by the standard itself) or by an alternative system.

One attempt to fill this gap is initiated by the EAGLE group (EIONET Action Group on Land monitoring in Europe)—an open assembly of land monitoring experts participating in the EIONET (European Environment Information and Observation Network) network on LC organized by the EEA (http://land.copernicus.eu/eagle). The group is drafting an object-oriented data model with several objectives: (1) to elaborate a future-oriented conceptual solution supporting the European information capacity for land monitoring built on existing or future national data sources; (2) to replace classification with a descriptive approach based on land characteristics; and (3) to facilitate semantic translation between different land monitoring initiatives (Arnold et al., 2013). The technical aim of the EAGLE group is to provide the conceptual basis for a data model to support a European Land Monitoring System in which CLC is included in its present as well as possible future manifestations.

27.4 Conclusions

CLC is, as we have shown in this chapter, fully INSPIRE compliant. Delivery of CLC as INSPIRE dataset simply requires that the reference year is added as an attribute to each polygon and that the data are converted from the proprietary ESRI® formats often used outside INSPIRE, into the technical format prescribed for INSPIRE in general. The open specification for LC data provided by INSPIRE will allow CLC to coexist with other land monitoring systems, including future modifications of the CLC monitoring system itself. INSPIRE does not offer specific tools for thematic integration of LC data (or spatial data in general) utilizing different nomenclatures, but the infrastructure does provide a framework where users have access to the information needed to implement such tools.

The INSPIRE data specification for LC does not specify any particular standard LC nomenclature for Europe as mandatory. Interoperability and comparability must therefore be attained by semantic translation. The simple solution is to recode data to the PLCCs proposed by TWG-LC. According to the principle of simplicity, a short list of broad components was thought to be an advantage and useful at the pan-European level. The disadvantages are loss of information content and the absence of precise definitions and instructions for the translation. A more elaborate solution is to employ a more detailed formal semantic translation. The advantages of this latter approach are the documentation and repeatability of the method and the capacity to retain more details from the nomenclature. The disadvantage is the methodological complexity.

The INSPIRE is now in an implementation phase lasts until 2019. Member states will have to implement the INSPIRE directive and make the relevant data, probably including CLC, available in the infrastructure during this phase. The special circumstance for CLC is that the data are standardized throughout Europe and created in a coordinated action organized by EEA. A valid question is therefore to ask if CLC also should be made available as an INSPIRE dataset through a centralized implementation organized by EEA. Member states could still make national subsets available through national INSPIRE portals on a voluntary basis. Such a centralized service would alleviate the burden of the INSPIRE implementation from member states and improve the ease of access for users, especially those looking across national borders.

References

Ahlqvist, O. 2005. Using uncertain conceptual spaces to translate between land cover categories. *International Journal of Geographic Information Science* 19:831–857.

Ahlqvist, O. 2008. In search of classification that supports the dynamics of science: The FAO Land Cover Classification System and proposed modifications. *Environment and Planning B: Planning and Design* 35:169–186.

Arnold, S. 2009. Integration of remote sensing data in national and European spatial data infrastructures—Derivation of CORINE Land Cover data from the DLM-DE. *Photogrammetrie—Fernerkundung—Geoinformation* 2:129–141.

Arnold, S., Kosztra, B., Banko, G., Smith, G., Hazeu, G., Bock, M., Sanz, N. V. 2013. The EAGLE concept—A vision of a future European Land Monitoring Framework. In R. Lasaponara, N. Masini, and M. Biscione (Eds.), *Towards Horizon 2020* (pp. 551–568). Paris: EARSeL.

Ben-Asher, Z., Gilbert, H., Haubold, H., Smith, G., Strand, G.-H. 2013. *HELM—Harmonised European Land Monitoring: Findings and recommendations of the HELM Project*. Tel-Aviv: The HELM Project.

Bishr, Y. 1998. Overcoming the semantic and other barriers to GIS interoperability. *International Journal of Geographic Information Science* 12:299–314.

Blanes, N., Green, T. 2012. Panorama of European Land Monitoring, HELM Deliverable 1.1. Available at: http://www.fp7helm.eu.

Bossard, M., Feranec, J., Otahel, J. 2000. *CORINE land cover technical guide*. Addendum 2000. Technical Report No 40. Copenhagen: European Environment Agency.

Bryn, A. 2006. Vegetation mapping in Norway and a scenario for vegetation changes in a Norwegian mountain district. *Geographica Polonica* 79:23–37.

Büttner, G., Kosztra, B., Sousa, A., Steenmans, C. 2010. CLC2006: Mapping land cover of Europe under GMES. In I. Manakos and C. Kalaitzidis (Eds.), *Imagin [e.g.] Europe: Proceedings of the 29th Symposium of the European Association of Remote Sensing Laboratories* (pp. 26–34). Chania: IOS Press.

Comber, A., Fisher, P., Wadsworth, R. 2005a. You know what land cover is but does anyone else?... An investigation into semantic and ontological confusion. *International Journal of Remote Sensing* 26:223–228.

Comber, A., Fisher, P., Wadsworth, R. 2005b. What is land cover? *Environment and Planning B: Planning and Design* 32:199–209.

European Union. 2007. Directive 2007/2/EC of the European Parliament and of the Council of 14 March 2007 establishing an Infrastructure for Spatial Information in the European Community (INSPIRE), Official Journal of the European Union L 108, 25.04.2007.

Eurostat. 2003. The Lucas survey: European statisticians monitor territory. Luxembourg: Office for Official Publications of the European Communities.

Feng, C. C., Flewelling, D. M. 2003. Assessment of semantic similarity between land use/land cover classification systems. *Computers, Environment and Urban Systems* 28:229–246.

Feranec, J., Hazeu, G., Christensen, S., Jaffrain, G. 2007. CORINE land cover change detection in Europe (case studies of the Netherlands and Slovakia). *Land Use Policy* 24:234–247.

Feranec, J., Jaffrain, G., Soukup, T., Hazeu, G. 2010. Determining changes and flows in European landscapes 1990–2000 using CORINE land cover data. *Applied Geography* 30:19–35.

Gallego, F. J., Delincé, J. 2010. The European land use and cover area-frame statistical survey. In R. Benedetti, M. Bee, G. Espa and F. Piersimoni (Eds.), *Agricultural survey methods* (pp. 151–168). Chichester: John Wiley & Sons.

di Gregorio, A., Jansen, L. J. M. 2000. *Land cover classification system: LCCS: Classification concepts and user manual*. Rome: Food and Agriculture Organization of the United Nations.

di Gregorio, A., O'Brien, D. 2012. Overview of land-cover classifications and their interoperability. In P. Ch. Giri (Ed.), *Remote sensing of land use and land cover: Principles and applications* (pp. 37–48). Boca Raton, FL: CRC Press.

Hazeu, G. W., Bregt, A. K., de Wit, A. J. W., Clevers, J. G. P. W. 2011. A Dutch multi-date land use database: Identification of real and methodological changes. *International Journal of Applied Earth Observation and Geoinformation* 13:682–689.

Herold, M., Latham, J. S., di Gregorio, A., Schmullius, C. C. 2006. Evolving standards in land cover characterization. *Journal of Land Use Science* 1(2–4):157–168.

ISO (International Organization for Standardization). 2005a. Geographic information—Rules for application schema, ISO 19109:2005. Geneva: International Organization for Standardization.

ISO (International Organization for Standardization). 2005b. Geographic information—Methodology for feature cataloguing, ISO 19110:2005. Geneva: International Organization for Standardization.

ISO (International Organization for Standardization). 2005c. Geographic information—Schema for coverage geometry and functions, ISO 19123:2005. Geneva: International Organization for Standardization.

ISO (International Organization for Standardization). 2009. Geographic information—Classification systems. Part 1: Classification system structure, ISO 19144–1:2009. Geneva: International Organization for Standardization.

ISO (International Organization for Standardization). 2012. Geographic information—Classification systems. Part 2: Land Cover Meta Language (LCML), ISO 19144–2:2012. Geneva: International Organization for Standardization.

Jansen, L. J. M., di Gregorio, A. 2002. Parametric land cover and land-use classifications as tools for environmental change detection. *Agricultural Ecosystems and Environment* 91:89–100.

Jones, B. Ch. 1997. *Geographical information systems and computer cartography*. Essex: Longman.

Martinez, L. I. 2013. Improving the use of GPS, GIS and RS for setting up a master sampling frame. Rome: Food and Agriculture Organization of the United Nations.

Rodriguez, M. A., Egenhofer, M. J. 2003. Determining semantic similarity among entity classes from different ontologies. *IEEE Transactions on Knowledge and Data Engineering* 15:442–456.

Strand, G-H. 2013. The Norwegian area frame survey of land cover and outfield land resources. *Norwegian Journal of Geography* 67:24–35.

Tversky, A. 1977. Features of similarity. *Psychological Review* 84:327–352.

TWG-LC (Thematic Working Group for Land Cover). 2013. D2.8.II.2 INSPIRE Data Specification on Land cover—Draft Technical Guidelines v3.0rc3. Available at: http://inspire.jrc.ec.europa.eu /documents/Data_Specifications/INSPIRE_DataSpecification_LC_v3.0rc3.pdf.

28

Future of Land Monitoring in Europe

Stefan Kleeschulte, Tomas Soukup, Gerard Hazeu,
Geoff Smith, Stephan Arnold, and Barbara Kosztra

CONTENTS

28.1 Background

The history of European land monitoring is a story of permanent evolution of approaches and concepts, resulting in different solutions for individual needs and specifications. In this context the CoORdination of Information on the Environment (CORINE) Land Cover (CLC) specification provided the first set of European-wide accepted de facto standards, that is, a nomenclature of land cover (LC) classes, a geometric specification, an approach for land cover change (LCC) mapping, and a conceptual data model (Heymann et al., 1994).

In this sense, CLC represents a unique European-wide, if not global, consensus on land monitoring, supported by 39 countries. Nevertheless, a revision and modernisation of this European level land monitoring concept, its origin dating back to the early 1980s, is urgently needed.

With the development of new national and regional land monitoring approaches, particularly over the last decade (Büttner et al., 2004; CBS, 2008; Valcarcel et al., 2008; Arnold, 2009; Aune-Lundberg and Strand, 2010; CEH, 2011; Hazeu et al., 2011; Banko et al., 2013), the arrival of new Earth Observation (EO) sensor systems (e.g., the Sentinel fleet, Landsat 8, etc.) and advancement of techniques in the analysis of remote sensing data, a new basis for European land monitoring information has become a necessity. Instead of the production of ad hoc LC maps with irregular update intervals, varying organizations, and short-term funding, as for CLC1990, CLC2000, and still for CLC2006, the establishment of a sustainable, reliable, and continuous land monitoring process with a secured long-term funding mechanism, based on existing information and a common model in line with Infrastructure for Spatial Information in the European Community (INSPIRE), needs to be put in place. The integration of CLC2012 under the Global Monitoring for Environment and Security (GMES) Initial Operations was a first step in that direction.

The precondition for such a common approach is the development of a harmonized land monitoring concept that could replace the current practices and also include a new core product that would be as compatible as possible with previous CLC.

The first steps in this direction have already been taken from different starting points: (1) the implementation of the Copernicus program and data policy (European Commission [EC], 2010), (2) the development of scalable national data models, and (3) the development of a tool for analytic decomposition of class definitions and semantic translation between recent or future LC nomenclatures (i.e., the EIONET Action Group on Land monitoring in Europe [EAGLE] data model).

28.2 Copernicus Developments

Copernicus, previously known as Global Monitoring for Environment and Security (GMES), is the European Program for the establishment of a European capacity for EO. The Copernicus program comprises a space component (i.e., satellites and associated ground segment with missions observing land, atmospheric and oceanographic parameters), an in situ data component (i.e., ground-based and airborne data gathering networks), and a services component that combines these to provide information essential for monitoring a range of themes including terrestrial environments (http://gioland.eionet.europa.eu/).

The principal objective of Copernicus (http://www.copernicus.eu/pages-principales /overview/copernicus-in-brief/) is to provide high-quality geo-information services, which are built on accurate data in the field of the environment and security and which support policymakers and other stakeholders with information required to take informed decisions and to define strategies related to different policy questions. It addresses a wide range of policies such as environment, agriculture, regional development, spatial planning, transportation, and energy at European Union (EU) level, as well as European commitments to international conventions. The Copernicus Land Service consists of three main components: global, continental (pan-European), and local (http://land.copernicus.eu/).

28.2.1 The Copernicus Land Service

The Land Monitoring service has begun its Initial Operations phase following the entry into force of Regulation (EU) No 911/2010 of September 22, 2010 of the European Parliament and of the Council on the European Earth monitoring program (GMES) and its initial operations (GIO), 2011 to 2013.

In line with Article 4(5) of Regulation (EU) No 911/2010 and a delegation agreement signed on May 25, 2011 between the EU and the European Environment Agency (EEA), the EEA is entrusted with the coordination of the pan-European continental and local land monitoring components (EC, 2010). This includes activities related to the organization and production of CLC2012 in the EEA-39 countries as well as the procurement, coordination, and quality control of the High-Resolution Layers (HRLs) and the local component products focused on riparian zones (see Chapter 9).

The HRLs comprise a series of EO-based products related to "thematic" LC characteristics addressing (1) the degree of imperviousness, (2) tree crown cover density and forest type, (3) permanent grasslands, (4) temporary wet areas, and (5) permanent water bodies.

In addition, the Urban Atlas (see Chapter 19) includes high spatial resolution LC maps of some 305 EU cities and agglomerations (referred to as larger urban zones [LUZs]) in version 1 (based on 2009 image data) and an updated version of some 697 cities based on 2012 image data.

While the CLC inventories are produced in a decentralized approach by the countries with EEA's organization and technical coordination, the HRLs, Urban Atlas, and other local components are being produced in a centralized manner with public procurements via framework and service contracts. All of these implementations are fully INSPIRE compliant.

After the end of its Initial Operations phase (2011–2013) the Copernicus program entered into its full operational phase (2014–2020) on 27 of April 2014, encompassing a total budget of 4.3 billion euros, including some 33 million euro for implementation of current and future land monitoring services (e.g., green linear features or NATURA 2000 hotspot mapping).

The complementarity of the component products of the Copernicus Land Service forms a multiscale information resource to support a range of complex landscape analyses, such as biodiversity or ecosystem assessments. In this way large-scale landscape features such as riparian zones and river basins can be compared to more generalised landscape structure provided by CLC, but then enhanced with the addition of the HRLs and the local component riparian zone and Urban Atlas products (Figure 28.1).

A broad range of European land monitoring activities can benefit from the Copernicus Land Services, as they provide

- Detailed information available at more frequent intervals that can/will serve European needs (policy assessments, EU spatial analysis)
- Harmonized and centralized data procurement covering Europe without reflecting national production constraints and specificities
- Monitoring of LC/land use (LU) changes at specific intervals for policy-relevant thematic classes

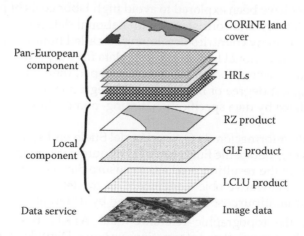

FIGURE 28.1
Multiscale information resource. GLF, green linear features; LCLU, land cover/land use product; RZ, riparian zone delineation, e.g., Urban Atlas.

28.2.2 Copernicus Space Component

From the very first CLC1990 to today, the satellite input data has undergone significant evolution, from 80 m spatial resolution of Landsat MSS data in the 1980s to 30 m spatial resolution Landsat TM and finally to the 20 m spatial resolution of the Satellite Pour l'Observation de la Terre (SPOT, Satellite for observation of Earth) and Indian Remote Sensing (IRS) data used today.

While the satellite images for CLC1990 were still acquired on a country by country basis, from the year 2000 onwards the European Space Agency has been providing a full European coverage (IMAGE2000) and from 2006 even for two distinct seasonal acquisition periods for the given reference year (±1 year). In total, three satellite image campaigns have been carried out with the aforementioned conditions (IMAGE2006, IMAGE2009, and IMAGE2012; for details see Chapter 4).

Similar to the technical progress, the European satellite data policy has undergone even more significant changes. From very specific project-related user licences to multiuse licences for public administrations involved in CLC production has already been a major improvement. With the Sentinel satellites, the European Delegation Act on Copernicus data and information policy (EC, 2013) will provide free, full and open access to users of environmental data from the Copernicus program, including data from the Sentinel satellites.

28.3 National Developments

In terms of land monitoring, European countries show a considerable diversity from countries (see also Chapter 26), where CLC is the sole or major land monitoring activity and where other LC/LU data either do not exist or are insufficient, to those with a well-developed national mapping system (i.e., Norway, Finland, Sweden, the Netherlands). In these countries the parallel production of CLC and national LC maps was considered inefficient (Ben Asher et al., 2013), and better synergies of national and European land monitoring activities have been explored to avoid high labor costs of photointerpretation and improve repeatability and consistency with national datasets.

This has resulted in many countries developing so-called bottom-up approaches, using more and more existing national land monitoring data for the production of CLC. Some of these countries (Spain, Germany, Austria) have developed their national LC datasets with a view toward a highest degree of flexibility and integration, ensuring that a bottom-up production is facilitated by data specification/timing/nomenclature:

- The Sistema de Información de Ocupación del Suelo en España (SIOSE) program in Spain aims to generate the European CLC by automatic geoprocessing, aided by visual reviews of the resulting products to ensure data quality. This was possible because of the innovative object-oriented data model used at the national level that allows the multipurpose outputs required by different types of users.
- In Germany, the topographic reference data ATKIS Basis-DLM (Amtliches Topografisch-Kartografisches Informationssystem—Digitales Basis Landschaftsmodell) is used as the basis for the national Digital Land Cover Model (DLM-DE), which is used subsequently to derive the contribution to the European CLC. The

topographic reference data are continuously updated by the individual federal states' land surveying authorities in a rolling fashion, so the most recent data are brought up to date for the specific CLC reference year with additionally acquired satellite imagery.

- The national Land Information System Austria (LISA) was recently adopted as a standard for nation-wide roll-out by the Austrian Federal Office of Metrology and Surveying (BEV). The overall aim of LISA is to provide current and detailed geospatial information of the status and development of LC and LU in Austria, derived from very high resolution EO and in situ data. The LISA object-oriented data model is designed to allow direct geometric and semantic transformation and generalization toward the CLC nomenclature, fully conformant to INSPIRE.

Others (e.g., Norway, Finland, Iceland, United Kingdom) have developed their own national land monitoring schemes independent of CLC, but are (often operationally) exploring the ways to derive CLC based on these datasets:

- Norway applied generalization and database merging techniques of existing sources, image classification, and visual control by satellite images to create a CLC base layer that was compatible with national information.

- Finland combined automated interpretation of satellite images and data integration with existing digital map data. Map data provided the information describing LU and soils. Satellite images were used for the estimation of continuous variables describing vegetation type and coverage, as well as for updating map data. Continuous LC variables were transformed into discrete CLC classes using thresholds of these variables according to class descriptions in CLC nomenclature.

- The United Kingdom has an environment monitoring program called Countryside Survey that includes a national LC map and is updated on an 8- to 10-year cycle. The UK LC is not aligned to CLC specifications but can be transformed and generalized by automated and manual processing.

Even though similar approaches exist in a number European countries, it should be noted that most European countries (30 out of 39 EEA member states) do not have their own national LC mapping program and that in those countries CLC with its "top-down" approach is often the only source of regularly updated LC information (FP7 HELM*: Commonalities, differences and gaps in national and subnational land monitoring systems [http://www.fp7helm.eu/fileadmin/site/fp7helm/HELM_3_1_Commonalities_and _Differences.pdf.]). Trends, however, show that the number of countries following the bottom-up approach is likely to increase further in the future.

---

* The overall goal of the HELM project, composed of and building on the experience of national and regional public organisations responsible for land monitoring as well as some private companies, was to explore ways how to make European land monitoring more productive by increasing the alignment of national and subnational level land monitoring endeavours and by enabling their integration to a coherent European LU and LC data system (http://www.fp7helm.eu/ms/fp7helm/mp2/helm_goals/).

28.4 A New LC Data Model

The EAGLE working group (EIONET [Environmental Information and Observation Network] Action Group on Land monitoring in Europe) is a network of LC/LU experts from different countries formed in response to the need for the elaboration of solutions for a better integration and harmonization of national mapping activities with European land monitoring initiatives (i.e., CLC). The main objective and key achievement of the group to date has been the initial elaboration of a future-oriented conceptual model for the land monitoring, applying an object-oriented data modeling approach.

Referred to as the EAGLE concept (Arnold et al., 2013), it applies a strict separation of LC (i.e., real-world physical landscape elements such as buildings, trees, water) and LU, the "territory characterized according to its current and future planned functional dimension or socio-economic purpose (e.g., residential, industrial, commercial, agricultural, forestry, recreational)" in the INSPIRE Directive (EC, 2007).

The main outputs of the EAGLE working group include

- *EAGLE matrix*: A tool for semantic comparison between the class definitions of different classification systems by decomposing them to LC components, LU attributes, and further landscape characteristics, in the form of an Excel table.
- *EAGLE data model*: A UML (Unified Modeling Language) model representation of the conceptual data model, visualized in the form of a graphical UML chart. It follows the ISO standard 19109 (Geographic information—Rules for application schema) similar to that applied for INSPIRE (INSPIRE, 2013a,b,c).

The two outputs (matrix and data model) contain the same information and are based on the same considerations and model elements.

The EAGLE concept (matrix and data model) is understood as a model to describe landscape units by decomposing them into their elementary "building blocks" and can be used

- For analytical decomposition of nomenclature/class definitions
- For semantic translation between different classification systems
- As a data model for new mapping initiatives

Going beyond the pure semantic and landscape modeling benefits, the EAGLE concept has a great potential as a conceptual framework for the integration of national and European land monitoring initiatives. The concept envisions the EAGLE matrix and data model as a centerpiece of a harmonized European land monitoring framework (Figure 28.2) acting as data collection, harmonization, and redistribution interface between national and European levels.

As a first application step, the EAGLE concept has been used to enhance CLC nomenclature guidelines, in this way contributing to a more consistent understanding of CLC classes and thus allowing more harmonized production by means of both traditional visual photointerpretation and semi-automated bottom-up methods. In a similar manner, semantic analysis can be a basis of improving the specifications of the HRLs to fulfill criteria of completeness and exclusiveness (see Section 3.4 in Chapter 3).

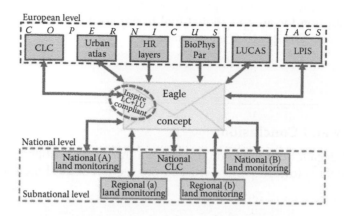

FIGURE 28.2
Vision of an integration scheme for a European land monitoring framework with the EAGLE model as the central vehicle for data integration and exchange of land information. (From Arnold, S., Kosztra, B., Banko, G., Smith, G., Hazeu, G., Bock, M., Valcarcel Sanz, N. 2013. The EAGLE concept—A vision of a future European Land Monitoring Framework. In R. Lasaponara, L. Masini, and M. Biscione [Eds.], *Towards Horizon 2020: Earth observation and social perspectives. 33th EARSeL Symposium Proceedings* [pp. 551–568]. Matera, Italy: European Association of Remote Sensing Laboratories and Consiglio Nazionale delle Richerche.)

28.5 The Way Forward

Considering the diversity of national approaches and programs, a proposed future land monitoring approach should be able to integrate information flows from countries with very detailed national data, meanwhile allowing others to fulfill European data requirements by contribution with a CLC-like dataset. Transition from traditional CLC toward a proposed future land monitoring approach should be gradual. With a short- to mid-term perspective, CLC has to be modified in such a way that more differentiated information about landscape can be stored. This can be achieved for example by adding higher thematic detail (Level 4/5 classes), or by opening the way for feature attribution ("population" of CLC polygons with parameters of LC and LU). In the long term, CLC is foreseen to be replaced by a more consistent systematic approach that is capable of overcoming the CLC concept's shortcomings, while maintaining backward compatibility with former CLC time series.

In such a system a core dataset (minimum requirement) should be defined that is feasible to be completed by countries with no national land monitoring activity/data (in the mid-term this can be CLC). The system should strive to provide the information content of CLC required by users instead of focusing on how to reproduce CLC dataset itself.

The prerequisite for a better integration of national activities with European land monitoring goals in the area of land monitoring is the widely held conviction that close collaboration is beneficial for all parties involved (Ben Asher et al., 2013). The envisaged land monitoring system must be based on common interests of all involved parties so as to enable them to jointly take advantage of mutual benefits. Therefore, the common strategy does not necessarily call for a "one and only" single opinion in every applicable aspect, but rather, it implies that all partners respect, accept, and value the diversity of views among the different (sub-)national approaches. Here the EAGLE concept offers a first and important contribution for an operational tool to enhance semantic translations of and between national and European LC classifications.

28.6 Summary and Conclusions

During the past decades the evolution of land monitoring in Europe has undergone significant developments toward a more efficient production of land monitoring data characterized by

- A more open data policy and a better integration of national mapping programs to better serve national and regional needs
- More automation—aiming at taking advantage of recent achievements in EO data provisioning and processing
- A flexible, multiscale, multipurpose approach moving from a rigid nomenclature to a flexible, object-oriented data modeling approach

Looking to the satellite input data side, the EO data policy in the context of CLC production has evolved to more open access to EO data, at least for public administrations and their reuse of the EO data for other applications under their mandate. With the launch of the Sentinel satellites the European EO data policy will finally complete this step toward a free, full, and open data policy for Sentinel data products under the Copernicus program (https://sentinel.esa.int/web/sentinel/sentinel-data-access).

On the land monitoring side, similarly important developments are being undertaken, from an independent, parallel (top-down) production of CLC in the early years to an increasingly integrated production based on existing national LC information in several European countries. This bottom-up approach uses synergies with already existing data to derive European land monitoring information in a more efficient manner that is more comparable to national LC change statistics than was possible with the traditional CLC approach in the past. The production of the HRLs to support CLC mapping is a promising approach, as it aims to capture additional, more detailed and complementary information about landscape features by measuring more parametric characteristics rather than limiting the output to thematic classes.

The main aim of the future of European land monitoring should therefore be to integrate all of the above—to better utilize current and future EO data potential, to integrate and serve national activities, and to set up more flexible future-oriented conceptual model for land monitoring.

Currently multiple activities are in line with this overall aim, but the main future challenge is to streamline, coordinate, and organize them in a way that will fulfill all of the aforementioned expectations but still keep the current main asset—the Europe-wide consensus on a harmonized European coverage. This is the same strong vision as for the original CLC program, but as in the past it needs the same strong governance: clear objectives, practical organizational support, realistic stepwise implementation plan, and long term budget line.

This is obviously a not easy and rather challenging task, but one of the promising aspects of 30 years of CLC heritage is that besides the institutional support required there exists a wide, strong, and enthusiastic land monitoring expert community behind the program pushing and supporting the CLC evolution forward.

Therefore to deliver an improved land monitoring in Europe in the future the following steps are needed:

- Creation of a standard model that is capable of storing more detailed information based national datasets (INSPIRE, EAGLE) compared to current systems
- A combined approach of bottom-up processes supplemented by top-down provision of EO data, preprocessed products, and layers of potential changes
- Optimized processing taking full advantage of synergies of centralized and de-centralized approaches and their respective benefits
- Access to in situ data as well as capacity building

In practice it means that an update of the HRLs is currently implemented for the reference year 2015 and in 2018 another HRL and CLC update will take place. Depending on the maturity of the EAGLE data model, more and more countries with a national land monitoring program will start deriving CLC from existing national data sources by then, while countries without a national land monitoring approach will follow the traditional CLC approach on a parallel track.

References

Arnold, S. 2009. Digital landscape model DLM-DE—Deriving land cover information by integration of topographic reference data with remote sensing data. ISPRS Workshop Hannover 2009 Proceedings XXXVIII-1-4-7_W5. Available at: http://www.isprs.org/proceedings/XXXVIII/1_4_7-W5/paper/Arnold-167.pdf.

Arnold, S., Kosztra, B., Banko, G., Smith, G., Hazeu, G., Bock, M., Valcarcel Sanz, N. 2013. The EAGLE concept—A vision of a future European Land Monitoring Framework. In R. Lasaponara, L. Masini, and M. Biscione (Eds.), *Towards Horizon 2020: Earth observation and social perspectives. 33th EARSeL Symposium Proceedings* (pp. 551–568). Matera, Italy: European Association of Remote Sensing Laboratories and Consiglio Nazionale delle Richerche.

Aune-Lundberg, L., Strand, G.-H. 2010. CORINE Land Cover 2006—The Norwegian CLC2006 project. Report from Norwegian Forest and Landscape Institute, 11. As: Norsk Institut for Skog og Landskap.

Banko, G., Franzen, M., Ressl, C., Riedl, M., Mansberger, R., Grillmayer, R. 2013. Bodenbedeckung und Landnutzung in Österreich—Umsetzung des Projektes LISA zur Schaffung einer nationalen Geodateninfrastruktur für Landmonitoring. In J. Strobl et al. (Eds.), *Angewandte Geoinformatik 2013:Beiträge zum 25. AGIT Symposium*, Salzburg (pp. 14–19). Berlin: Wichman.

Ben-Asher, Z., Gilbert, H., Haubold, H., Smith, G., Strand, G.-H. 2013. HELM—Harmonised European Land Monitoring: Findings and Recommendations of the HELM Project. Tel-Aviv: The HELM Project. Available at: http://www.FP7HELM.eu/results.

Büttner, G., Maucha, G., Bíró, M., Kosztra, B., Pataki, R., Petrik, O. 2004. National land cover database at scale 1:50,000 in Hungary. *EARSeL eProceedings* 3(3), 323–330. Available at: http://www.eproceedings.org/.

CBS NL. 2008. Centraal Bureau voor de Statistiek—Statistics Netherlands. Bodemgebruik Nederland BBG2008. Available at: http://www.cbs.nl/nl-NL/menu/themas/dossiers/nederland-regionaal/cijfers/cartografische-toegang/geoviewer.htm.

CEH (Centre for Ecology and Hydrology). 2011. Land Cover Map 2007 Dataset documentation, version 1.0.

EC (European Commission). 2007. Directive 2007/2/EC of the European Parliament and of the Council establishing an Infrastructure for Spatial Information in the European Community (INSPIRE).

EC (European Commission). 2010. Regulation (EU) No. 911/2010 of the European Parliament and of the Council on the European Earth monitoring programme (GMES) and its initial operations (2011 to 2013).

EC (European Commission). 2013. Commission delegated Regulation (EU) No. 1159/2013 supplementing Regulation (EU) No. 911/2010 of the European Parliament and of the Council on the European Earth monitoring programme (GMES) by establishing registration and licensing conditions for GMES users and defining criteria for restricting access to GMES dedicated data and GMES service information.

Hazeu, G. W., Bregt, A. K., de Wit, A. J. W., Clevers, J. G. P. W. 2011. A Dutch multi-date land use database: Identification of real and methodological changes. *International Journal of Applied Earth Observation and Geoinformation* 13:682–689.

Heymann, Y., Steenmans, Ch., Croissille, G., Bossard, M. 1994. *Corine Land Cover*. Technical guide. Luxembourg: Office for Official Publications of the European Communities.

INSPIRE (Infrastructure for Spatial Information in the European Community). 2013a. D2.8.II.2 Data specification on land cover—Draft technical guidelines version 3 (Identifier D2.8.II.2_v3.0rc3).

INSPIRE (Infrastructure for Spatial Information in the European Community). 2013b. D2.8.III.2 Data specification on buildings—Draft technical guidelines version 3 (Identifier D2.8.III.2_v3.0rc3).

INSPIRE (Infrastructure for Spatial Information in the European Community). 2013c. D2.8.III.4 Data specification on land use—Draft technical guidelines version 3 (Identifier D2.8.III.4_v3.0rc3).

Valcarcel, N., Villa G., Arozarena A., Garcia-Asensio, L., Caballlero, M. E., Porcuna, A., Domenech, E., Peces, J. J. 2008. SIOSE—A successful test bench towards harmonization and integration of land cover/use information as environmental reference data. *The International Archives of the Photogrammetry, Remote Sensing and Spatial Information Sciences* 38(Part B8):1159–1164.

29

Conclusions

Jan Feranec, Tomas Soukup, Gerard Hazeu, and Gabriel Jaffrain

CoORdination of Information on the Environment (CORINE) Land Cover (CLC) represents a unique Europe-wide consensus in land monitoring, unique even from global perspective. With the finalization of CLC2012 it should be emphasized that for those interested in the European landscape an important source of environmental information is available showing the dynamics for the period of almost 25 years. In our information society this is an asset comparable with the other strategic resources available. The greatest benefit of CLC is that because of harmonized content and compatibility in time it allows for the perception of continent-wide spatial and temporal patterns in land cover/land use (LC/LU) changes and thus it provides insight into the heterogeneous European landscape structure and related processes.

Individual parts of this monograph explain main elements of the CLC projects, which record landscape attributes by means of satellite data. Methodology for the generation of CLC1990, CLC2000, CLC2006, and CLC2012 data layers; assessment of their precision; and dissemination policy, as well as generation of three layers of changes $CLC_{1990-2000}$, $CLC_{2000-2006}$, and $CLC_{2006-2012}$, are also carefully documented. These LC data provide information about LC classes (their area and occurrence) in a particular time horizon of the European landscape, but they also suggest the trend of changes in the tracked period. The significance of the obtained knowledge is expected even to rise during future monitoring activities. A substantial part of the monograph is that dedicated to examples of the CLC data applications—various environmental assessments in which the CLC data play an important role (e.g., landscape fragmentation, ecosystem mapping, national accounting, etc.).

The final chapters deal with the perspectives of LC observation in Europe. Future European land monitoring will have to deal most probably with new approaches for updating and upgrading of the European LC information that need to be in line with the new national and regional approaches (e.g., EIONET Action Group on Land monitoring in Europe [EAGLE] concept), with the INSPIRE (Infrastructure for Spatial Information in the European Community) initiative, but also with the data generated by the new satellites (Sentinel, Landsat 8, and others).

If the monograph succeeds in convincing users about the assets of the European CLC program and information generated to contribute to thorough knowledge of the European landscape and monitoring of its dynamics then it achieved its aim. Hopefully the efforts of the editors to summarize the activities connected with the implementation of CLC projects and application of CLC data to solutions of environmental problems resulted in a readable overview to be used in future studies related to CLC.

As indicated in the book, there is a solid 30 years of CLC heritage to build on, but there are also challenges ahead to integrate on board new emerging approaches, concepts, and technological achievements. One of the promising and most important aspects is that beside the institutional support there is also a strong, enthusiastic and vital community of land monitoring experts behind the program pushing and supporting the CLC evolution forward. We hope that this publication makes readers certain that this is the real guarantee of the future of the CLC program.

Index

Note: Page numbers ending in "f" refer to figures. Page numbers ending in "t" refer to tables.

Printed and bound by CPI Group (UK) Ltd, Croydon, CR0 4YY

01/11/2024

01782601-0007